BREAST EPITHELIAL ANTIGENS

Molecular Biology to Clinical Applications

BREAST EPITHELIAL ANTIGENS

Molecular Biology to Clinical Applications

Edited by

Roberto L. Ceriani

John Muir Cancer and Aging Research Institute
Walnut Creek, California

SPRINGER SCIENCE+BUSINESS MEDIA, LLC

Library of Congress Cataloging in Publication Data

International Workshop on Monoclonal Antibodies and Breast Cancer (4th: 1990: San Francisco, Calif.)
  Breast epithelial antigens: molecular biology to clinical applications / edited by Roberto L. Ceriani.
      p.   cm.
  "Proceedings of the Fourth International Workshop on Monoclonal Antibodies and Breast Cancer, held November 5–6, 1990, in San Francisco, California."—T.p. verso.
  Includes bibliographical references and index.
  ISBN 978-1-4613-6665-2   ISBN 978-1-4615-3740-3 (eBook)
  DOI 10.1007/978-1-4615-3740-3
  1. Breast—Cancer—Immunodiagnosis—Congresses. 2. Breast—Cancer—Immunotherapy—Congresses. 3. Tumor antigens—Congresses. I. Ceriani, Roberto L. II. Title.
  [DNLM: 1. Antibodies, Monoclonal—diagnostic use—congresses. 2. Breast Neoplasms—diagnosis—congresses. 3. Breast Neoplasms—therapy—congresses. WP 870 I614b 1990]
  RC280.B8I58   1990
  616.99′4490756—dc20
  DNLM/DLC   91-24095
  for Library of Congress   CIP

ISBN 978-1-4613-6665-2

© 1991 Springer Science+Business Media New York
Originally published by Plenum Press in 1991
Softcover reprint of the hardcover 1st edition 1991

4TH INTERNATIONAL WORKSHOP ON MONOCLONAL ANTIBODIES AND BREAST CANCER

San Francisco, California
November 5-6, 1990

Organized by the John Muir Cancer & Aging Research Institute, with the
cooperation of the International Association for Breast Cancer Research

WORKSHOP CHAIRPERSON

   Dr. Roberto L. Ceriani
      John Muir Cancer & Aging Research Institute

ORGANIZING COMMITTEE

      Chairperson: Dr. Roberto L. Ceriani

      Members: Dr. Jerry A. Peterson
                    Ms. Carolyn Klinepeter

ACKNOWLEDGEMENTS

The Organizing Committee for the 4th International Workshop on Monoclonal Antibodies and Breast Cancer, together with the John Muir Cancer & Aging Research Institute gratefully acknowledge the support of the following in making the Workshop possible:

SPONSORS

COULTER IMMUNOLOGY, HIALEAH, FLORIDA
ABBOTT LABORATORIES
CYTOGEN CORPORATION
TRITON BIOSCIENCES
Burroughs Wellcome Company
Bio-Rad Laboratories
Toys 2 Go (Discovery Toys) - Judy Corley
In Memory of Fred Vinson
American Speedy Printing - Walnut Creek
John Muir Medical Center
Heublein Fine Wine Group consisting of:
  Beaulieu Vineyards
  Inglenook - Napa Valley
  Christian Brothers - Greystone Cellars
  Quail Ridge Cellars
  Gustave - Niebaum Collections
  Rutherford Estate Cellars

PREFACE

The wealth of research results in the area of breast cancer diagnosis and therapy with monoclonal antibodies presented in previous workshops is now complemented with a new rendition of the proceedings of the 4th Workshop on Monoclonal Antibodies and Breast Cancer, held in San Francisco on November 5-6, 1990. Basic science findings reported in previous workshops have now percolated to the clinical level and have become immunoassays and imaging and therapy reagents, as the program shows us. Thus, the latest discoveries in immunology, biochemistry and molecular biology of breast epithelial antigens and their corresponding antibodies have produced newer diagnostic tests and therapeutic approaches that are altering and improving the way we attack breast cancer.

The recent spectacular and rapid advancements in the molecular biology of several of the breast epithelial antigens are presented in this volume. The way in which the final assembly of different components of the breast antigens is achieved and their functions are now within our grasp as a result of new understanding of molecular structure of these breast antigens. In addition, newer immunoassays aiming at the earliest detection of the disease are also described that integrate with promising attempts at imaging and radioimmunotherapy to set the stage for new oncological possibilities in breast cancer treatment. All these areas of intense involvement of scientists with diverse specialties are presented in this volume, which proves the need for multidisciplinary approaches to increase our chances for success in this field of medical research.

The publication of these collected papers represents the cutting edge of an area of rapid scientific and clinical development. They are grouped into 4 chapters that include the molecular biology of breast epithelial antigens, their cell biology and immunology, the use of anti-breast epithelial antibodies in serum tests, and imaging and therapeutic application of these novel drugs in breast cancer.

R.L. Ceriani

CONTENTS

SESSION III

SESSION IV

MOLECULAR ANALYSIS OF H23 EPITHELIAL TUMOR ANTIGEN – DIFFERENTIALLY SPLICED FULL LENGTH cDNAs AND GENE

D. H. Wreschner, I. Tsarfaty, M. Hareuveni, J. Zaretsky, N. I. Smorodinsky, M. Weiss[1], S. Zrihan, M. Burstein, J. Horev, P. Kotkes, R. Lathe[2], C.A. Hart[3], K. McCarthy[3], C. Williams[4], A. Dion[4] and I. Keydar

Dept. Microbiology/Cell Biology, Tel Aviv University, Ramat Aviv, Israel 69978; [1]Dept. Medicine, Tel Hashomer, Israel; [2]AFRC, University of Edinburgh, King's Bldgs, Edinburgh, U.K.; [3]Medical Microbiology, University of Liverpool, Prescott St., Liverpool, U.K.; [4]Center for Molecular Medicine and Immunology, Newark, New Jersey, U.S.A.

ABSTRACT

The isolation and characterization of the complementary DNAs (cDNAs) and gene which code for an epithelial tumor antigen (H23-ETA), aberrantly expressed in human breast tumor tissue, are described here. A diversity of H23-ETA protein forms is generated by a series of alternative splicing events that occur in regions located upstream and downstream to a central tandem 20 amino acid (aa) repeat array (TRA) that is rich in proline, serine and threonine residues. The upstream region shows that differential usage of alternative splice acceptor sites generates two protein forms containing putative signal peptides of varying hydrophobicities located at the NH_2 terminus. The region downstream to the tandem repeat array indicates that one mRNA transcript is collinear with the gene and defines a 160 aa open reading frame (secreted or sec form). A second cDNA correlates with a mRNA that is generated by a series of splicing events and codes for 149 aa downstream to the TRA, identical with the aa sequence of the unspliced cDNA, after which it diverges and continues for an additional 179 aa. This sequence (transmembrane or tm form) contains a highly hydrophobic transmembrane domain of 28 aa followed by a hydrophilic "transfer-stop signal" (Arg Arg Lys) and a cytoplasmic domain of 72 aa. The various protein forms (alternative signal sequences, secreted and transmembrane) are likely routed to different cytoplasmic, cell membrane and extracellular compartments. Reverse PCR indicates that the relative ratios of the alternatively spliced forms vary in different epithelial tissues. To identify the individual protein species, monoclonal antibodies (mAb) are being generated against synthetic peptides unique to each form. The H23-ETA gene was also isolated and sequenced, demonstrating a putative promoter region that includes a 'TATA' box, Spl binding elements and an upstream putative hormone responsive element. Commensurate with these findings, H23-ETA expression was increased following hormonal treatment of BT549 breast tumor cells. These molecular studies have unravelled novel H23-ETA protein and gene structures, and facilitate future investigations that will focus on H23-ETA function and interaction with other cellular proteins.

Breast Epithelial Antigens, Edited by R.L. Ceriani
Plenum Press, New York, 1991

INTRODUCTION

A number of monoclonal antibodies (mAbs) have been developed that recognize an antigen of epithelial cell origin that is expressed primarily in breast tumor tissue and, to lesser extents, in other malignant epithelial tissues (1-11). This antigen is likely a constituent of normal epithelial cells which, in malignant breast tissue, undergoes quantitative and/or qualitative changes. The H23 mAb detects the breast cancer associated antigen (H23-ETA) in both cytoplasm of human breast cancer cells and body fluids of breast cancer patients (11,12). Since H23-ETA serum levels have clinical significance and correlate with the severity of disease (12), an analysis of H23-ETA at the molecular level was undertaken.

The cDNA coding for the epitope recognized by H23 is composed of tandem 60 base pair (bp) repeating units that code for a 20-amino-acid (aa) repeat motif rich in proline, serine, threonine and alanine (13,14). Other mAbs (DF3, HMFG1 and HMFG2, MAM6) that are also directed against a breast cancer antigen, were used to isolate cDNAs that are composed of almost identical 60-bp tandem repeats (15-19). Southern blot analyses indicated that the H23-ETA gene is highly polymorphic and 70-80% of individuals are heterozygous at this locus (13,14,20). The different allelic sizes result from variation in the number of repeating 60-bp units present within the genomic repeat array and the polymorphism detected in the H23-ETA protein products correlates with the various allelic sizes (14).

In addition to its presence in serum, the cellular localization of the breast tumor associated antigen has been variously designated as apical, membranous, intracytoplasmic or focal (1-11). It appears therefore that varying forms of H23-ETA may localize to different cellular and extracellular compartments. In order to understand the postulated different H23-ETA protein forms and to clarify its overexpression in breast cancer, both full length cDNAs (13,21) and the H23-ETA gene were isolated and characterized (21-24). These molecular studies, presented here, have unravelled novel protein and gene structures, that may elucidate H23-ETA function in tumor progression, as well as the regulatory mechanisms responsible for its overexpression in breast cancer tissue.

RESULTS AND DISCUSSION

H23-ETA amino acid sequence – NH_2 terminal to tandem repeats

A number of cDNAs correlating with the region 5'upstream to the central tandem 20 aa (60 bp) repeat array (TRA) as well as with the 3' downstream region were isolated, sequenced and characterized (Figs. 1, 2 and 3). The NH_2 terminal aa sequence deduced from the pSe 2 cDNA (Figs. 1, 2 and 3) open reading frame (ORF), demonstrates the initiating methionine followed by a highly hydrophobic 13-aa peptide that includes 5 tandem leucine residues. Because of its size, hydrophobicity and proximity to the amino terminus it seems likely that this domain represents a signal sequence. H23-ETA secretion or its insertion into the plasma cell or/and cytoplasmic membranes may be mediated by this classical signal sequence. Interestingly, the ORF determined by a second 5' cDNA (pSe 4, Fig. 1) indicates diversity in the signal sequence domain (Fig. 2A and B). Alternative usage of two splice acceptor sites generates the variability between the pSe-2 and pSe-4 cDNA sequences. The splice event generating the pSe-2 cDNA sequence determines a Thr-Val-Val peptide. The ORF of the pSe-4 cDNA sequence, however, interrupts this tripeptide and determines an in-frame insertion of nine aa (Fig. 2A and B). As the residues introduced by this alternative splice change the hydrophobicity of the signal peptide region, cellular routing and targeting of H23-ETA may also be affected.

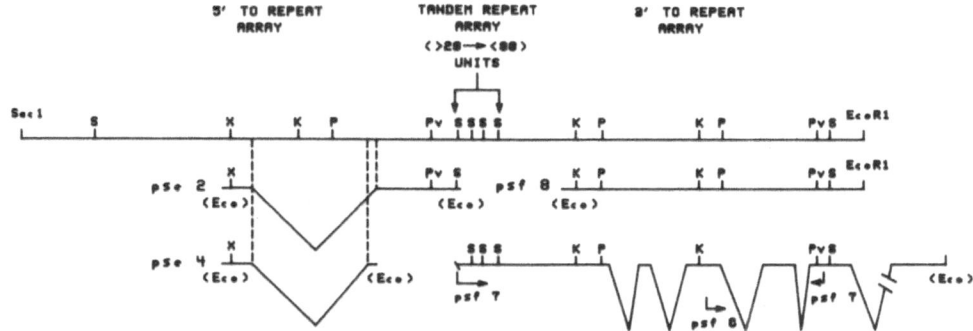

Fig. 1 Restriction map of genomic fragment coding for H23-ETA and localization of isolated cDNA inserts. The top line represents the restriction map of a 5.3-kb SacI-EcoRI genomic fragment part of which codes for H23 ETA. The various cDNA inserts obtained are shown below the genomic restriction map. For convenience the tandem array is depicted as three repeat units when, in fact, it ranges over 20-80 repeats. The restriction enzymes KpnI, PstI, PvuII, SmaI and XmnI are represented by K, P, PV, S and X respectively, and the synthetic EcoRI sites are shown in parentheses. The genomic fragments used to reprobe the cDNA library are the SmaI-PstI and PstI-PstI, PstI-EcoRI fragments located 5' and 3' to the repeat array, and are designated pse and psf, psg respectively. Pse 2 and 4 and psf6, 7 and 8 are the cDNA inserts isolated.

H23-ETA amino acid sequence - tandem repeats

The ORF determined by the in-frame initiation methionine extends pSe 2 into the 60-bp repeat unit which codes for a highly conserved proline-rich 20-aa repeat motif that also contains 3 threonine, 2 serine, 2 glycine and 2-5 alanine residues (Fig. 2 and 3).

The epitope in H23-ETA recognized by H23-ETA mAb is situated within this repeat motif. Synthetic peptides correlating with different parts of the 20 aa repeat motif, determined by the ORF, were synthesized and analyzed for immune reactivity with H23 mAb. The Pro-Asp-Thr-Arg sequence was found to be essential for recognition by H23 mAb, although flanking residues on both the NH_2 and COOH termini are also required for maximal immune reactivity. It.is notable that this tetrapeptide represents the hydrophilic segment and possibly, therefore, the most antigenic part of the 20 aa repeat motif.

H23-ETA amino acid sequence - COOH terminal to tandem repeats

The region downstream to the TRA (Fig. 1, psF8, psF7 and psF6) indicates that one mRNA transcript (cDNA psF8) is unspliced and collinear with the gene and defines a 160 aa ORF (secreted or sec Form). This H23-ETA protein form contains a short 12 amino acid hydrophobic region bounded by potential N-glycosylation sites at positions 289 and 315 (Fig. 3B). The small size of this hydrophobic region is insufficient to support transmembrane localization, but it may function as a signal for glycolipid mediated membrane anchorage. The aa sequence of the sec form, on the carboxyl side of the TRA, is serine rich and contains four potential N-linked glycosylation sites (NxS/T).

3

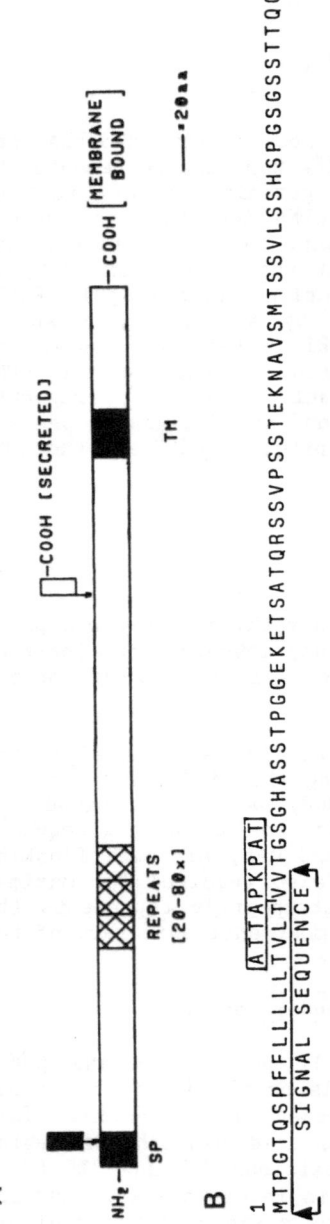

A

NH₂ — SP

REPEATS [20-80x]

TM

COOH [SECRETED]

COOH [MEMBRANE BOUND]

~20aa

B

1
MTPGTQSPFFLLLLLTVLTVVTGSGHASSTPGGEKETSATQRSSVPSSTEKNAVSMTSSVLSSHSPGSGSSTTQGQDVTL
SIGNAL SEQUENCE [ATTAPKPAT]

81
APATEPASGSAATWGQDVTSVPVTRPALGSTTPPAHDVTSAPDNKPA|PGSTAPPAHGVTSAPDTRPP|PGSTAPPAHGVTS
REPEAT UNIT 20-80x

161
APDTRPPP|GSTAPAAHGVTSAPDTRPAP|GSTAPPAHGVTSAPDNRPALASTAPPVHNVTSASGSASGSASTLVHNGTSAR

241
ATTTPASKSTPFSIPSHHSDTPTTLASHSTKTDASSTHHSTVPPLTSSNHSTSPQLSTGVSFFLSFHISNLQFNSSLED

321
PSTDYYQELQRDISEMFLQIYKQGGFLGLSNIKFRPGSVVVQLTLAFREGTINVHDVETQFNQYKTEAASRYNLTISDVS
VSIGLSFPHLP C-terminus (SECRETED)

401
VSDVPFPFSAQSGAGVPGWGIALLVLVCVLVALAIVYLIALAVCQCRRKNYGQLDIFPARDTYHPMSEYPTYHTHGRYVP
TRANSMEMBRANE

481
PSSTDRSPYEKVSAGNGGSSLSYTNPAVAATSANL C-terminus (MEMBRANE-BOUND)

Fig. 2 Scheme demonstrating the structure and sequence of various ETA forms. A) Different domains are presented schematically: the hydrophobic signal peptide (SP) and transmembrane domain (TM) are in black and the highly conserved 20 amino acid repeat units are crosshatched. The small boxes above the membrane bound form represent the protein variants resulting from differential splicing. B) The complete ETA amino acid sequences.

A second cDNA form (psF7 and psF6, Fig. 1) codes for 149 aa downstream to the TRA that are identical with the aa sequence of the unspliced cDNA, after which it diverges and continues for an additional 179 aa. This sequence (transmembrane or tm form) contains a highly hydrophobic transmembrane domain of 28 aa followed by a cytoplasmic domain of 72 aa (Figs. 2, 3 and 5). Viable intact breast tumor cells can be immunofluorescently stained with H23 mAb, indicating that the tandem 20 aa repeats are located in the extracellular domain of the H23-ETA tm form. The presence of 5 potential N-linked glycosylation sites (AsnxSer/Thr) in this region lends credence to this hypothesis.

That the H23-ETA tm form may be involved in signal transmission and serve as a receptor for an as yet unidentified ligand, is supported by the fact that the cytoplasmic domain comprises a long 72 aa tail. Were the H23-ETA tm form solely a structural transmembrane protein, a significantly smaller cytoplasmic domain would be sufficient to anchor the protein securely in the cell membrane. It therefore seems plausible that the cytoplasmic domain may interact with cellular proteins and elicit changes in cell behaviour by transmitting signals from the cell exterior. Whether ligand binding to the extracellular domain or direct interaction of the ETA-tm form with extracellular matrix, substratum or adjacent cells induces such signals, remains to be determined.

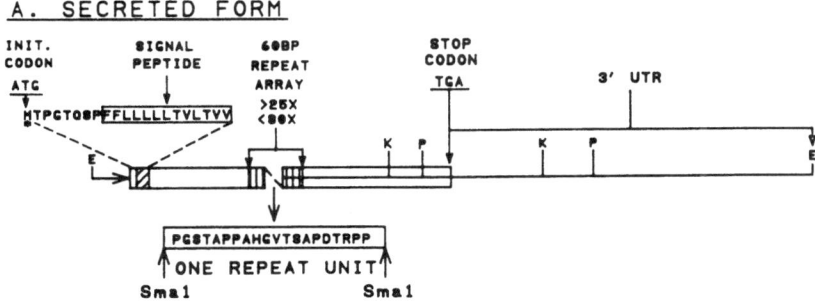

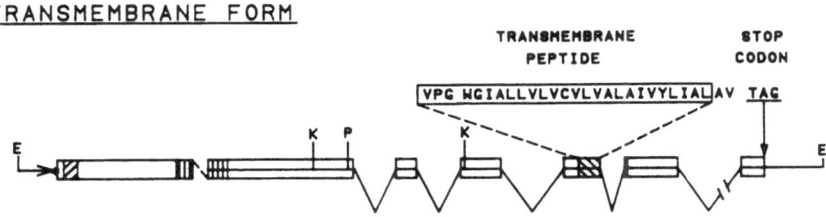

Fig. 3 Scheme of differential splicing that may generate secreted and membrane-bound forms of the epithelial tumor antigen. The initiation codon (ATG) and stop codons (TGA and TAG, secreted and transmembrane forms respectively) are indicated. The restriction sites for KpnI and PstI are designated by K and P. The one-letter code for amino acids is used and the region bracketed by 3'UTR represents the 3' untranslated region.

Multiple H23-ETA forms - multiple localization sites

As H23-ETA can be detected in the serum of patients, it is obviously secreted into the peripheral circulation. Furthermore, immunohistochemical analyses of breast tumor sections stained with H23 mAbs demonstrated primarily intracytoplasmic staining and an immunofluorescence study indicated that H23-ETA is also membrane bound (11). H23-ETA localization to different sites is presented in Fig. 4. Immunohistochemical staining of T47D breast tumor cells and a breast tumor tissue paraffin section (Figs. 4A and B) demonstrates both membrane and cytoplasmic staining. On the other hand, staining of a benign breast fibroadenoma paraffin section (C) shows extracellular glycocalyx staining at the ductal apical surfaces.

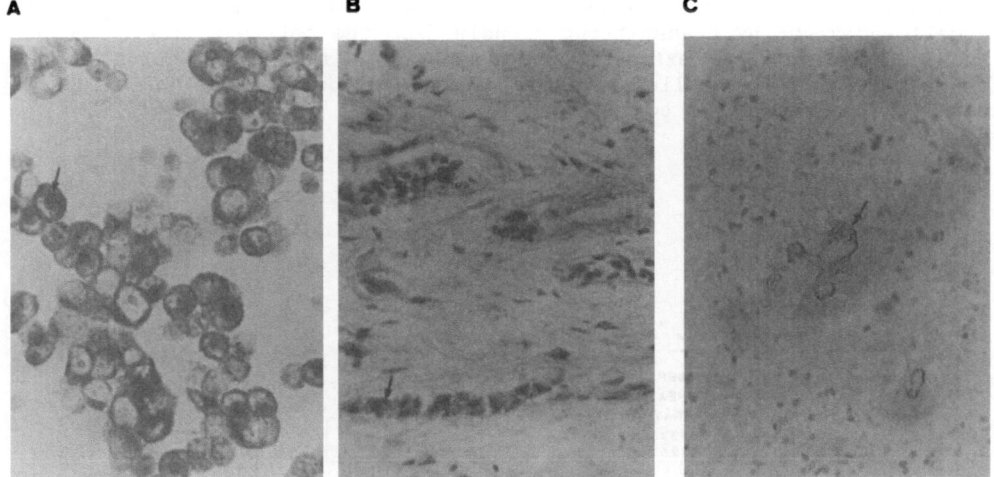

Fig. 4   Immunohistochemical staining of breast tissue and tumor cells. T47D breast tumor cells (A) and paraffin sections of an infiltrating ductal breast carcinoma (B) and a benign breast fibroadenoma (C) were indirectly immunoperoxidase stained with H23 mAb. Note the cytoplasmic and membrane staining in (A), predominantly cytoplasmic staining in (B) and glycocalyx staining at the ductal apical surfaces in (C).

The aa sequences of the different H23-ETA forms, as presented here, provides us with a molecular rationale for multiple localization sites (Figs. 3 and 5). The presence of a signal sequence at the NH_2 terminus and sec sequence downstream to the repeat array may determine protein secretion. On the other hand, presence of the NH_2 signal sequence and transmembrane sequence downstream to the repeat array could target the protein for cell membrane localization. The effect of nonapeptide insertion into the signal sequence on post-translational processing and cellular routing are, at present, not known. It may well be, however, that insertion of the nine residue peptide disrupts signal sequence cleavage, rendering the uncleaved hydrophobic NH_2 terminus, a lipophilic endoplasmic reticulum membrane anchor, thereby explaining H23-ETA cytoplasmic localization.

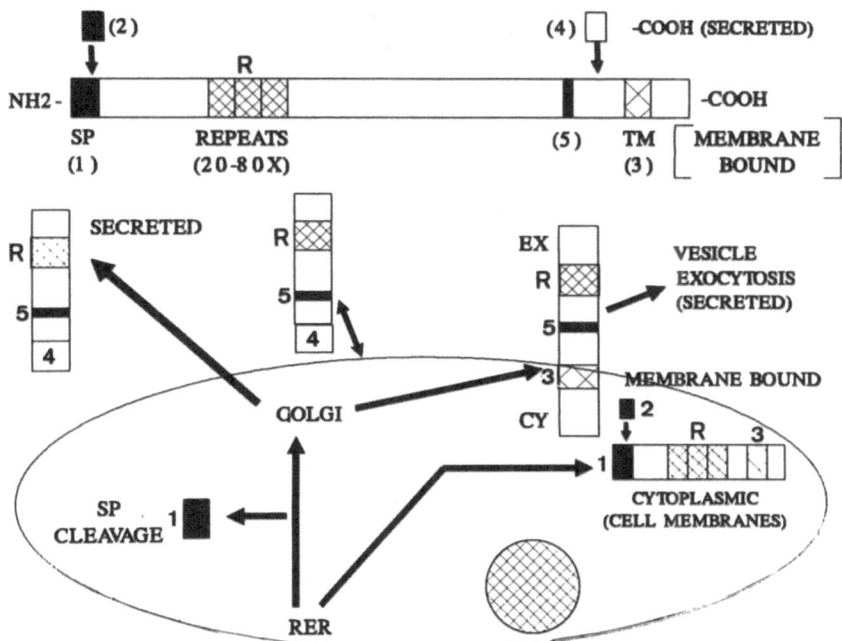

Fig. 5 Scheme demonstrating possible localization sites of the various
H23-ETA forms. The H23-ETA protein is illustrated at the top of
the figure with NH_2 and COOH termini at the left and right borders
respectively. The signal peptide (SP) is indicated by the left
black box (1) near the NH_2 terminus and the 9 aa insertion in the
signal sequence, determined by a 5' alternative splice, is
designated by the black box (2). The 3 densely hatched boxes (R)
in the middle represent tandem 20 aa repeats which may occur in
H23-ETA from 20 to 80 times. The large 28 aa highly hydrophobic
domain (TM) present in the H23-ETA transmembrane form is
represented by the sparsely cross-hatched box (3) close to the COOH
terminus. The 11 aa sequence, unique to the putative secreted
H23-ETA form and representing its COOH terminus (COOH secreted), is
illustrated by the open box (4). The shorter 12 aa hydrophobic
region present both in the transmembrane and secreted H23-ETA forms
is designated by the slender black box (5). Attachment of the
secreted form to the cell membrane may be mediated by this short
hydrophobic region (5). The extracellular and cytoplasmic domains
of the transmembrane form are indicated by Ex and Cy respectively.
The diagram is schematic and is not drawn to scale.

Relative expression of alternatively spliced H23-ETA forms varies in different epithelial tissues

It is of obvious interest to know whether all H23-ETA forms are
equally expressed in epithelial tissues or, alternatively, whether
different epithelial tissues preferentially express certain H23-ETA forms.
This was investigated by using reverse PCR technology in which the mRNA is
reverse transcribed into cDNA which is then amplified with a chosen pair of
downstream and upstream primers (Fig. 6). The sizes of the PCR products

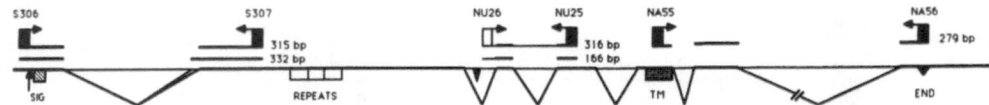

Fig. 6 Relative expression of alternatively spliced H23–ETA forms varies
in different epithelial tissues. The indicated pairs of upstream
and downstream oligonucleotide primers were used to PCR cDNA
obtained by reverse transcription of mRNA isolated from a number of
different epithelial tissues. The PCR products (indicated in bp)
were visualized by electrophoresis through 8% polyacrylamide gels.
The top line of the scheme designates the various oligonucleotide
codes – below this the black and white boxes show the positions of
oligonucleotides located in exons and introns respectively. The
horizontal arrows define the cDNA segment that underwent PCR
amplification whereas the heavy and light horizontal lines
represent exon and intron sequences respectively. The initiation
and termination codons are indicated by the vertical upward facing
arrow; signal peptide and transmembrane region are designated SIG
and TM respectively.

reflect the alternatively spliced forms of mRNA (see Fig. 6). The analysis
performed with RNA from a variety of different epithelial tissues confirmed
that the relative ratios of expression of the alternatively spliced forms
vary in different epithelial tissues.

For example, the pair of primers designated S306 and S307 will lead to 315
bp and/or 332 bp reverse PCR products depending on the alternative splice
taken within the signal sequence region. An analysis with this primer pair
demonstrated that whereas some epithelial tissues express both
alternatively spliced forms, other tissues exclusively expressed one or the
other species. It is clear, therefore, that certain tissues may
preferentially express specific H23–ETA forms.

Evidence supporting the existence of a secreted H23–ETA form

As previously noted a cDNA, psF 8 (Fig. 1), located in the region
downstream to the TRA was isolated, and nucleotide sequencing indicated
that the mRNA transcript from which it is derived is unspliced and
collinear with the gene (Fig. 1). Retention of the intron immediately
downstream to the TRA introduces an in-frame termination codon that will
result in a protein product that, as lacking the transmembrane domain,
could be secreted from the cell (Figs. 2 and 3).

Precedent exists for intron inclusion in mature mRNA in one tissue,
whereas the same intron is removed by splicing in other tissues, resulting
in a mRNA with a longer open reading frame that produces a larger protein
product (26–28) (for review see 25). For example, P element transposition
in Drosophila depends on germ-line pre-mRNA splicing of the P element third
intron, producing a germline mRNA that is translated into an active 87Kd
transposase protein. In somatic tissues, however, this same P element
third intron is not spliced out resulting in a mature mRNA that contains an
in-frame termination codon within the unspliced intron and is translated
into a truncated 66 Kd repressor protein that functions as a negative
regulator of transposition (27,28).

It is of course possible that the isolated psF 8 cDNA described here
represents a pre mRNA transcript present within the population of

8

polyadenylated RNAs used to prepare the cDNA library, thereby explaining gene collinearity and intron presence. Alternatively, it may represent a bona fide mature mRNA transcript coding for a secreted H23-ETA form. Support for this latter possibility is derived from several lines of evidence.

Firstly, Northern blot analysis of RNA extracted from both breast tumor cell lines and primary breast tumor tissue demonstrated hybridization to sequences within the introns of the transmembrane mRNA (29).

Secondly, reverse PCR was performed on RNA extracted from a variety of epithelial tissues using an upstream oligonucleotide NU26, that is located within the first intron 3' to the TRA and a downstream oligonucleotide Nu25 (Fig. 6). Two PCR products were observed - one sized at 316 bp that is collinear with the gene and a second smaller 166 bp product that corresponds to a mRNA that is spliced in the second intron downstream to the TRA (Fig. 6). As the upstream NU26 primer is located in the first intron downstream to the TRA, both the 316 bp and 116 bp reverse PCR products obviously contain this intron downstream to the TRA (Fig.6)

Thirdly, mAbs have been recently generated against the amino acid sequence SIGLSFPMLP that represents the C-terminal 10 aa, unique to the putative secreted H23-ETA form (Fig. 2). These sec specific mAbs, when analyzed in an ELISA assay, not only recognized the synthetic peptide with which the mice were immunized, but also bound H23-ETA secreted into culture medium by T47D breast tumor cells. Furthermore, the binding of these sec specific mAbs to H23-ETA was competed out by preincubation of the mAbs with the synthetic sec specific peptide.

These data support the hypothesis that a secreted H23-ETA form exists, and work on its further characterization is presently in progress.

Isolation and characterization of the H23-ETA gene - Identification of a hormonal responsive element

To understand mechanisms that regulate expression of the H23-ETA gene and to elucidate its overexpression in breast tumor tissue, a genomic library prepared with DNA from the MCF7 breast tumor cell line was used to isolate and subsequently characterize the H23-ETA gene (22). The H23-ETA EcoRI-EcoRI 7.5 Kb genomic fragment isolated was designated I7.5. It contained a tandem repeat array of 2.3 Kb and unique sequences both 5' upstream and 3' downstream to the TRA. A SacI-EcoRI 5.3 Kb fragment of I7.5 was sequenced and comparison with cDNA nt sequences identified different domains in the H23-ETA gene. As in the cDNA sequence, the genomic sequence contains the same consensus initiation sequences, putative signal sequence divided by an intron, TRA and unique sequences 3' to the TRA.

Comparison of the cDNA and gene sequences revealed one intron located 5' to the TRA that divides the Thr and Val residues towards the end of the signal peptide. The 499 bp intron contains a highly purine rich region (78% AG) over the first 419 bp, followed by a pyrimidime rich sequence (75% CT) for the remaining 80 bp. A nucleotide homology search revealed a putative enhancer sequence situated within the purine rich region of the intron that showed 86% identity with a 28 bp murine cellular enhancer of retroviral gene expression.

The genomic sequence upstream from the 5' terminal cDNA sequences revealed a putative promoter region consisting of TATA sequences (Hogness box) flanked on the 5' and 3' sides by G+C rich regions that include several Sp1 binding sites.

Interestingly, a partially palindromic sequence is located 300 bp upstream to the TATA box. The 5' tetranucleotide, AGGA, and the 3' pentanucleotide, GACCT, within this sequence are identical to the hormone responsive consensus element and it is postulated that this sequence may also be a hormonal responsive element within the H23-ETA gene.

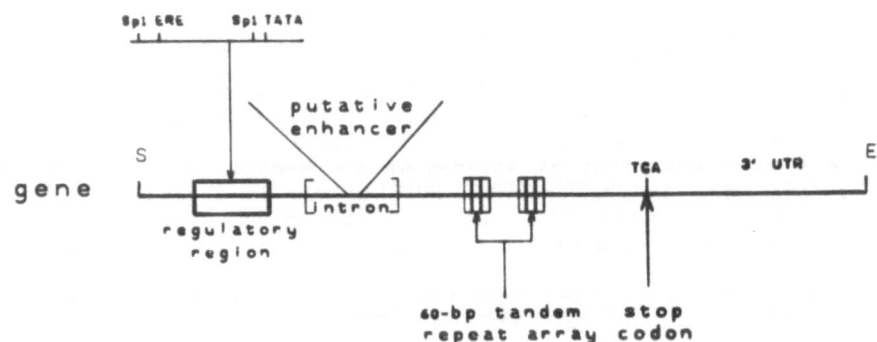

Fig. 7 Primary structure of the H23-ETA gene. SacI and EcoRI sites are designated by S and E respectively. Transcription initiates downstream to the boxed regulatory region. Putative regulatory sequences in the regulatory region refer to Sp1 binding sites (Sp1), hormone responsive element (ERE) and the TATA site. The intron and stop codon refer to the secreted H23-ETA cDNA form.

Commensurate with these findings, BT549 human breast tumor cells have been shown to exhibit enhanced expression and secretion of H23-ETA when cultivated in a medium supplemented with hydrocortisone and insulin (24).

Conclusions and Perspectives

The molecular analysis of H23-ETA at the cDNA and genomic levels has unravelled the complete amino acid sequences of different forms of this clinically important tumor antigen and revealed potential regulatory sequences that will enable us to investigate cis and trans acting factors regulating H23-ETA gene expression.

Analysis of the tumor antigen by the powerful techniques of recombinant DNA technology allows us now to address issues relating to a) H23-ETA function, b) H23-ETA interaction with both extracellular ligands and intracellular proteins, c) the development of H23-ETA form specific mAbs, possibly more effective in the early diagnosis of breast cancer and d) mechanisms regulating the normal and aberrant expression of H23-ETA gene in human breast tissue.

Acknowledgments

Parts of this work were supported by grants to DHW from The Israel Cancer Research Fund, and Israel Cancer Association.

References

1. Johnson, V.G., Schlom, J., Paterson, A.J., Bennett, J., Magnani, J.L. and Colcher, D. Analysis of a human tumor associated glycoprotein (TAG-72) identified by monoclonal antibody B72.3. Cancer Res. 46: 850 (1986).

2. Burchell, J., Gendler, S., Taylor-Papadimitriou, J., Girling, A., Lewis, A., Millis, R. & Lamport, D. Development and characterization of breast cancer reactive monoclonal antibodies directed to the core protein of the human milk mucin. Cancer Res. 47: 5476 (1987).

3. Kufe, D.W., Nadler, L., Sargent, L., Shapiro, P., Hand, P., Austin, F., Colcher, D. and Schlom, J. Biological behavior of human breast carcinoma-associated antigens expressed during cellular proliferation. Cancer Res. 43: 851 (1983).

4. Tjandra, J.J. and McKenzie, I.F.C. Murine monoclonal antibodies in breast cancer: an overview. Br. J. Surg., 75: 1067 (1988).

5. Bramwell, M.E., Bhavanandan, V.P., Wiseman, G. and Harris, H. Structure and function of the Ca antigen. Br. J. Cancer, 48: 177 (1983).

6. Ceriani, R.L., Peterson, J., Lee, J.Y., Moncada, R. and Blank, F.W. Characterization of cell surface antigens of human mammary epithelial cells with monoclonal antibodies prepared against human milk fat globule Somatic Cell Genet., 9: 415 (1983).

7. Hilkens, J., Buijs, F., Hilgers, J., Hageman, P., Calafat, J., Sonnenberg, A. & van der Valk, M. Monoclonal antibodies against human milk-fat globule membranes detecting differentiation antigens of the mammary gland and its tumors. Cancer, 34: 197 (1984).

8. Schechter, R.L., Major, P.P., Kovac, P.E., Ishida, M., Kovalik, E.C., Dion, A.S., Langleben, A., Boileau, G., Boos, G., Panasci, L. and Margolese, R. Double antibody radioimmunoassay for monitoring metastatic breast cancer. Br. J. Cancer, 58: 362 (1988).

9. Price, M.R., Edwards, S., Owainati, A., Bullock, J.E., Ferry, B., Robins, R.A. and Baldwin, R.W. Multiple epitopes on human breast-carcinoma-associated antigen. Int. J. Cancer, 36: 567 (1985).

10. Ceriani, R.L., Thompson, K.E., Peterson, J.A. and Abrahams, S. Surface differentiation antigens of human mammary epithelial cells carried on the human milk fat globule. Proc. Natl. Acad. Aci. USA, 74: 582 (1977).

11. Keydar, I., Chou, C.S., Hareuveni, M., Tsarfaty, I., Sahar, E., Seltzer, G., Chaitchik, S. & Hizi, A. Production and characterization of monoclonal antibodies identifying breast tumor associated antigens. Proc. Natl. Acad. Sci. USA, 86: 1362 (1989).

12. Tsarfaty, I., Chaitchik, S., Hareuveni, M., Horev, J., Hizi, A., Wreschner, D.H. & Keydar, I. H23 monoclonal antibodies recognize a breast cancer tumor associated antigen: Clinical and molecular studies. In Ceriani R.L. (ed.) Breast Cancer Immunodiagnosis and Immunotherapy, pp. 161-169 New York: Plenum, 1988.

13. Wreschner, D.H., Tsarfaty, I., Hareuveni, M., Zaretsky, J., Smorodinsky, N., Weiss, M., Horev, J., Kotkes, P., Zrihan, S., Jeltsch, J.M., Green, S., Lathe, R. & Keydar, I. Isolation and characterization of full length cDNA coding for the H23 breast tumor associated antigen. In: Rich, M.A., Hager, J.C. and Keydar, I. (eds.) Breast Cancer: Progress in Biology, Clinical Management and Prevention, pp. 41-59. Boston: Kluwer Academic Publishers, 1989.

14. Hareuveni, M., Tsarfaty, I., Zaretsky, J., Kotkes, P., Horev, J., Zrihan, S., Weiss, M., Green, S., Lathe, R., Keydar, I. & Wreschner, D.H. A transcribed gene, containing a variable number of tandem repeats, codes for a human epithelial tumor antigen - cDNA cloning, expression of the transfected gene and over-expression in breast cancer tissue. Eur. J. Biochem., 189: 475 (1990).

15. Gendler, S.J., Burchell, J.M., Duhig, T., Lamport, D., White, R., Parker, M. and Taylor-Papadimitriou, J. Cloning of partial cDNA encoding differentiation and tumor-associated mucin glycoproteins expressed by human mammary epithelium. Proc. Natl. Acad. Sci. USA, 84: 6060 (1987).

16. Gendler, S.J., Lancaster, C.A., Taylor-Papadimitriou, J., Duhig, T., Peat, N., Burchell, J., Pemberton, L., Lalani, E.-N. and Wilson, D. Molecular cloning and expression of the human tumour-associated polymorphic epithelial mucin, PEM. J. Biol. Chem. 265: 15286 (1990).

17. Siddiqui, J., Abe, M., Hayes, D., Shani, E., Yunis, E. & Kufe, D. (1988) Isolation and sequencing of a cDNA coding for the human DF3 breast carcinoma associated antigen. Proc. Natl. Acad. Sci. USA. 85, 2320 (1988).

18. Abe, M., Siddiqui, J. and Kufe, D.W. Sequence analysis of the 5' region of the human DF3 breast carcinoma-associated antigen gene. Biochem. Biophys. Res. Comm. 165: 644 (1989).

19. Ligtenberg, M.J.L., Vos, H.L., Gennissen, A.M.C. and Hilkens, J. Episialin, a carcinoma associated mucin, is generated by a polymorphic gene encoding splice variants with alternative amino termini. J. Biol. Chem. 265: 5573 (1990).

20. Swallow, D.M., Gendler, S., Griffiths, B., Corney, G., Taylor-Papadimitriou, J. and Bramwell, E. The human tumor-associated epithelial mucins are coded by an expressed hypervariable gene locus PUM. Nature, 328: 82 (1987).

21. Wreschner, D.H., Hareuveni, M., Tsarfaty, I., Smorodinsky, N., Horev, J., Zaretsky, J., Kotkes, P., Weiss, M., Lathe, R., Dion, A.S., and Keydar, I. Human epithelial tumor antigen cDNA sequences - Differential splicing may generate multiple protein forms. Eur. J. Biochem. 189: 463 (1990).

22. Tsarfaty, I., Hareuveni, M., Horev, J., Zaretsky, J., Weiss, M., Jeltsch, J.M., Garnier, J.M., Lathe, R. Keydar, I. and Wreschner, D.H. Isolation and characterization of an expressed hypervariable gene and cDNA coding for a breast cancer associated antigen. Gene. 93: 313 (1990).

23. Zaretsky, J.Z, Weiss, M., Tsarfaty, I., Hareuveni, M., Wreschner, D.H. and Keydar, I. Expression of genes coding for pS2, c-erbB2, estrogen receptor and the H23 breast tumor associated antigen. A comparative analysis in breast cancer. FEBS, 265: 46 (1990).

24. Williams, C.J., Wreschner, D.H., Tanaka, A., Tsarfaty, I., Keydar, I. and Dion, A.A. 1990. Multiple protein forms of the breast tumor-associated epithelial membrane antigen (EMA) are generated by differential splicing and induced by hormonal stimulation. Biochem. Biophys. Res. Commun. 170: 1331 (1990).

25. Latchman, D.S. Cell-type-specific splicing factors and the regulation of alternative RNA splicing. The New Biologist, 2: 297 (1990).

26. Laski, F.A., Rio, D.C. and Rubin, G.M. Tissue specificity of Drosophila P element transposition is regulated at the level of mRNA splicing. Cell 44: 7 (1986).

27. Siebel, C.W. and Rio, D.C. Regulated splicing of the Drosophila P transposable element third intron in vitro: somatic repression. Science, 248: 1200 (1990).

28. Craig, N.L. P element transposition, Cell, 62: 299 (1990).

29. Hareuveni, M., Gautier, C., Kieny, M.P., Wreschner, D.H., Chambon, P. and Lathe, R. Vaccination against tumor cells expressing breast cancer epithelial tumor antigen. Proc. Natl. Acad. Sci. USA. In Press (1990).

CHARACTERIZATION AND EVOLUTION OF AN EXPRESSED HYPERVARIABLE
GENE FOR A TUMOR-ASSOCIATED MUCIN, MUC-1

Sandra J. Gendler, Andrew P. Spicer, Lucy Pemberton, Carole A. Lancaster, Trevor Duhig, Nigel Peat, Joyce Taylor-Papadimitriou and Joy Burchell

Imperial Cancer Research Fund, P O Box 123, Lincoln's Inn Fields London WC2A 3PX, U.K.

INTRODUCTION

Mucins, present on highly polarized, secretory epithelial cells, have gained prominence in recent years as many monoclonal antibodies selected for their reactivity on differentiated or normal tissues react with epitopes present on these molecules. Although difficult to analyze biochemically because of their large size and large amount of O-linked carbohydrate, recent cloning studies from a number of different laboratories have produced structural information of the core protein. In most cases cDNA clones were obtained from λgt11 expression libraries following the development of antibodies to the stripped core protein. Thus far, three human mucin partial or full-length cDNA clones have been characterized as well as the porcine submaxillary mucin and a Xenopus integumentary mucin. In each case a domain of the core protein was found to consist of tandem repeats of a defined length. Although no homology exists between the tandem repeats of the various mucin genes at either the DNA or protein level, the repeated sequences in each case code for molecules which could be highly O-glycosylated. One characteristic feature of mucins is the presence of between 50 and 90% carbohydrate which is linked to serines or threonines via an O-glycosidic linkage to N-acetylgalactosamine. A second characteristic is the presence of prolines which along with glycosylation help to provide the extended core protein structure characteristic of mucins. All of the predicted proteins coded for by the mucin clones contain these features.

MUCINS

The first mucin to be cloned was the mammary gland or milk mucin called PEM or MUC 1 (Gendler et al., 1987; 1988;1990; Siddiqui et al., 1988; Abe and Kufe, 1989; Ligtenberg et al., 1990; Wreschner et al., 1990;) and this is the only human mucin for which a full length cDNA exists. Partial cDNAs for two intestinal mucins have been reported (Gum et al., 1988; 1989) as well for the porcine submaxillary mucin (Timpte et al., 1988) and the Xenopus integumentary mucin (Probst et al., 1990).

Tandem repeats appear to be a characteristic of mucins, being found in all 5 mucins described so far. The repeat units of the different mucins show no similarity to each other in either sequence or number of amino acids in the repeat(Fig. 1), although in each case serines and/or threonines make up a high percentage of the amino acids, giving the molecules the potential to be highly glycosylated. The variability observed in the size of the mucin molecules suggests that length is not crucial to mucin function, but rather that the core protein exists in an extended form as a scaffold for O-linked carbohydrate. The carbohydrate side chains may differ depending on the tissue studied or on the change to malignancy.

The sequences of the tandem repeats are shown in fig. 1; the designation of MUC 1, 2 and 3 is as proposed by Gum (1990) with the numbers reflecting the order in which the human genes were characterized. The porcine submaxillary and Xenopus integumentary mucins are

Breast Epithelial Antigens, Edited by R.L. Ceriani
Plenum Press, New York, 1991

MUC 1	GSTAPPAHGVTSAPDTRPAP (20 AA)
MUC 2	PTTTPITTTTTVTPTPTPTGTQT (23 AA)
MUC 3	HSTPSFTSSITTTETTS (17 AA)
pMUC 4	GAGPGTTASSVGVTETARPSVAGSGTTGTVSGASGSTGSSSG
	SPGATGASIGQPETSRISVAGSSGAPAVSSGASQAAGTS (81 AA)
xMUC 5	GESTPAPSETT (11 AA)

Fig. 1.   Sequences comprising the tandem repeat units of the five mucin genes characterized. Sequences contain a high proportion of serines and/or threonines to which the linkage sugar N-acetylgalactosamine may be attached and prolines which provide an extended core protein to allow for up to 90% of the molecule to consist of carbohydrate.

designated pMUC 4 and xMUC 5, respectively. The number of amino acids in the repeat is given in the brackets following the sequence.

MUC 1 CORE PROTEIN

The encoded core protein of the mammary gland mucin consists of 3 distinct regions: the N-terminus with a putative signal peptide and degenerate tandem repeats; the major portion of the protein which is the tandem repeat region; the C-terminus consisting of degenerate repeats and unique sequence containing a transmembrane region and a cytoplasmic tail. Potential O-glycosylation sites (serines and threonines), the signpost of a mucin, make up more than one-fourth of the amino acids. The tandem repeat carries the epitopes recognized by the monoclonal antibodies HMFG-1, HMFG-2 and SM-3 as well as by a number of other well-characterized antibodies (see Burchell and Taylor-Papadimitriou, 1989, for review). Surprisingly enough, when clones were obtained for the pancreatic mucin core protein, the sequence was found to be identical, although it is glycosylated differently in the pancreas and presents a different profile of epitopes (Lan et al., 1990; Khorrami et al, 1989). The pancreatic mucin aggregates to much greater size than the mammary gland mucin, presumably due to the increased length and complexity of the carbohydrate side chains and possible interactions of the carbohydrate structures. Even on the mucin synthesized in a single tissue, ie. the mammary gland, carbohydrate structures of great diversity have been reported, varying from structures consisting of 8 to 14 oligosaccharides in the neutral milk mucin (Hanisch et al., 1989) to a tetrasaccharide unit present on the DF3 antigen in normal milk and a disaccharide on a breast cancer cell line BT20, all of which contain varying amounts of sialic acid (Hull et al., 1989). These glycosylation differences have been indirectly detected previously with the development of a monoclonal antibody SM3 which showed a high degree of tumour specificity (Burchell et al., 1987; 1989). The exposure of core protein epitopes in the cancer-associated mucin which were masked in the normally processed mucin suggested that there was a basic alteration in the glycosylation occurring in cancer cells.

That the mucin is indeed a transmembrane molecule can be demonstrated using polyclonal antiserum (CT-1) made to the last 17 amino acids of the cytoplasmic tail. In immunofluorescent analysis of cultured cells, the antiserum CT-1 reacts only with permeabilized cells confirming that the carboxy terminal region is a cytoplasmic domain (Fig. 2).

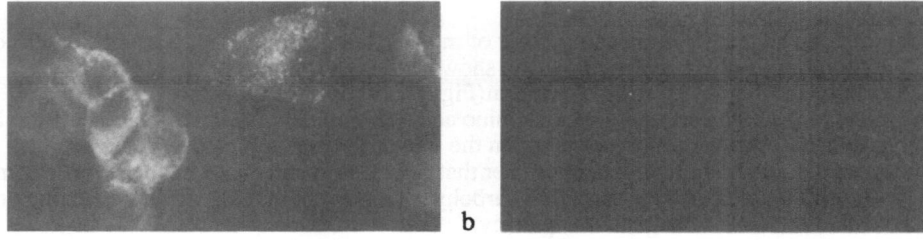

a   b

Fig. 2.   Reactivity of polyclonal antiserum CT-1 with methanol: acetone fixed a) and live b) MCF-7 cells using indirect immunofluorescence staining. X400.

Although a secreted form of the mucin, presumed to be produced by alternative splicing, has been described by Williams et al. (1990), we have observed no secretion of the mucin in our breast cancer cell lines using radioimmunoassays to test the culture supernatants (E.-N. Lalani and M. Boschell, unpublished results). Moreover, this alternative form of cDNA was not detected by Lan and colleagues (1990) in the HPAF cell line which releases large amounts of mucin in the supernatant. Large quantities of mucin can be obtained from human milk; whether this mucin is secreted or results from the dissolution of the milk fat globule membrane (which is known to be highly unstable) has yet to be shown definitively. Experiments are in progress to express the secreted form as designated by Wreschner to ascertain whether this form can exit from the cells.

MUC 1 GENE STRUCTURE

The human MUC 1 or PEM gene is unusually compact, spanning approximately 4 to 7 kb of genomic DNA (the size being variable according to the size of the VNTR unit in exon 2) and contains 7 exons (Lancaster et al., 1990) (Fig. 3). The N-terminal region contains a hydrophobic signal sequence. Within intron I an alternative splice site to that found by ourselves has been observed for the intronI/exon 2 boundary in the partial 5' sequence of the DF3 antigen, the H23 antigen and episialin (all three being the same molecule, although named differently; Abe and Kufe, 1989; Wreschner et al., 1990; Ligtenberg et al., 1990;). The use of this splice site results in exon 2 having an additional 27 bp which alter the codon where the sequence is inserted (at the penultimate amino acid in the potential signal sequence), but does not affect the reading frame of the translated product. The biological significance of these splice variants is not yet known. It is conceivable that differences in the signal sequence could affect the transport to the plasma membrane and/or secretion of the molecule.

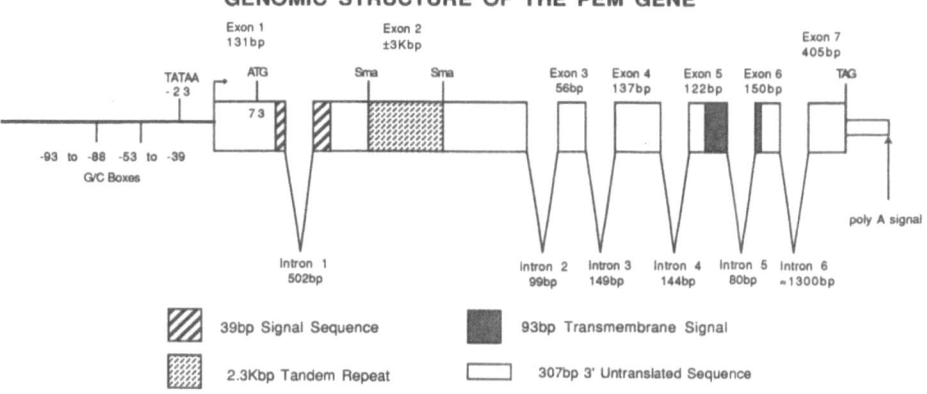

GENOMIC STRUCTURE OF THE PEM GENE

Fig. 3. Diagram of exon and intron organization of MUC 1 gene. Exons are designated by E and the size in bp is given inside the rectangles. Sizes of introns in bp are given below the lines. The tandem repeat, which varies in size from 2 to 6 kb, is designated as VNTR unit.

The genomic sequence is unusual both in its composition and structure. The overall genomic sequence is 65% G+C rich, whereas the tandem repeat region is 82% G+C, in marked contrast to the 40% G+C composition usually observed in mammalian genomic DNA. The 60 bp tandem repeats found in exon 2 vary in number and are extremely homogeneous in sequence, making this region an expressed variable number tandem repeat (VNTR) locus. Maintenance of precise repeats such as these at the DNA level is unusual, because of the abundance of CpG dinucleotides (6/60 bp) which have been shown in various studies to mutate at a significantly higher rate than other dimers (Savatier 1985; Cooper and Krawczak, 1989) (Fig. 4). It has been proposed that the primary cause of CpG deficiency in vertebrate DNA is deamination of 5mCpG to give TpG/CpA pairs (Bird 1980; 1985). Consequently, CpGs are not distributed randomly in the genome but are concentrated in CpG islands in which the DNA is unmethylated. Thus, we examined the methylation status of the tandem repeat to determine if the presence of the CpG dinucleotides could be attributed to a lack of methylation (Fig. 5). However, rather than the total lack of methylation in all tissues that is characteristic of a CpG island, the methylation pattern of the tandem repeat correlated with gene expression. Methylation was associated with lack of gene expression in peripheral blood lymphocyte DNA, whereas in tissues expressing MUC 1 (adenocarcinomas of the breast), the gene was not methylated.

GGC TCC ACC GCC CCC CCA GCC CAC GGT GTC
ACC TCG GCC CCG GAC ACC AGG CCG GCC CCG

Fig. 4.   DNA sequence of MUC 1 tandem repeat, showing GC richness of sequence and presence of 6 CpG dinucleotides in each 60 bp. The CpG dinucleotides are underlined.

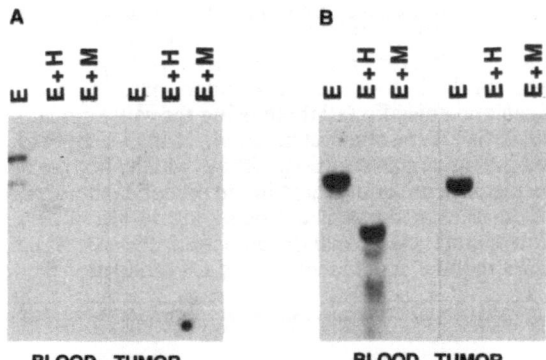

Fig. 5.   Methylation status of HpaII sites in the TR domain of the MUC 1 gene. Ten mg of genomic DNA prepared from PBL (blood) and breast carcinoma tissues (tumour) were digested with the restriction enzymes EcoRI (E), EcoRI and HpaII (E+H) (methylation sensitive) and EcoRI and MspI (E+M) (methylation insensitive) and electrophoresed through a 0.6% agarose gel. PEM alleles were detected by hybridization to a probe to the TR, pMUC7. A and B are representative autoradiographs from two individuals, one of whom is heterozygous for PEM (a) and one homozygous (b). Each 60 bp TR contains 3 sites for HpaII/MspI. Digestion by these enzymes results in loss of allelic bands. The presence of bands in the E+H lane of the blood samples indicates methylation of this DNA, whereas loss of hybridizing bands in E+H lane of tumor samples shows undermethylation of DNA.

SEQUENCES DETERMINING TISSUE-SPECIFIC EXPRESSION

It is of obvious interest to study the controlling sequences of a gene which is selectively expressed in defined epithelial tissues, is developmentally regulated and apparently upregulated in many carcinomas. Transgenic mice lines have been established utilizing the SacII band of genomic DNA containing the entire gene (manuscript in preparation). The band is approximately 12 kb in length with 1.5 kb of 5' flanking sequence and about 6 kb of 3' downstream sequences. Expression has been assessed using the monoclonal antibodies HMFG-1, -2 and SM-3 in indirect immunoperoxidase staining on methacarn-fixed tissues. The expression observed in the mouse lines correlates precisely with S. Zotter's elegant study (1989) on human adult tissues using 20 different Mabs. to the mucin. MUC 1 appears to be expressed in simple epithelial tissues (cells lining ducts or glands) which are secretory. Interestingly, the glycosylation pattern in the mouse appears similar to that in humans, with SM-3 showing minimal reactivity on normal tissues except for the lung where it shows a strong reactivity. It remains to be shown if the SM-3 epitope is revealed once tumors have been induced in the mice. It is surprising to find that the crucial regulatory sequences defining epithelial specificity are present either very close to or within the mucin gene. Experiments are in progress to define more precisely what the specific sequences are.

CELLULAR LOCALIZATION OF MUC 1

The mucin, MUC 1, has a very defined pattern of expression, being detected mainly on the apical surfaces of highly polarized epithelial tissues. This localization is intriguing, particularly since mammary epithelial cells are highly dependent upon the substratum on which cells are cultured in order to retain their tissue-specific phenotypes (Emerman et al., 1977). Yet it has been known for years that PEM exhibits an asymmetrical distribution on the plasma membranes of human breast cell lines in culture. This distribution is found even when the cells are grown on plastic without any exogenously added extracellular matrix components except for those present in fetal calf serum (Ormerod et al., 1981). Polarized expression is also evident in the majority of moderately- and well-differentiated carcinomas (Corcoran and Walker, 1990). This distribution appears to be mediated by interaction of the cytoplasmic domain of the mucin with components of the actin cytoskeleton (Parry et al., 1990). The functional interaction of PEM can be demonstrated by treating the breast cancer cells with cytochalasin D which causes depolymerization of the actin microfilaments, resulting in commensurate disruption of the apical expression of PEM (Fig. 6). Disruption of the keratin network with acrylamide or the microtubules with colchicine does not alter the localization of the mucin (data not shown). These results suggest that PEM is restricted to the apical cell surface by interactions with the microfilament network. It is possible that interactions between integral membrane proteins such as PEM and cytoplasmic structural proteins may be involved in the maintenance of the polarity of the plasma membrane and stabilization of the epithelial morphology. In undifferentiated malignancy, this interaction appears to be disturbed, resulting in increased expression of the mucin in the cytoplasm of the cells (Griffiths et al., 1987; Corcoran and Walker, 1990) which results in exposure of an epitope detected by the antibody SM-3. This epitope appears to be strongly associated with malignancy and may well be an internal epitope. Detailed studies of the interaction of the mucin cytoplasmic tail and characterization of the actin binding proteins with which it interacts may shed light on some of these events that occur in malignancy. Altered distribution within the cell of PEM may in turn lead to the altered glycosylation state of the molecule.

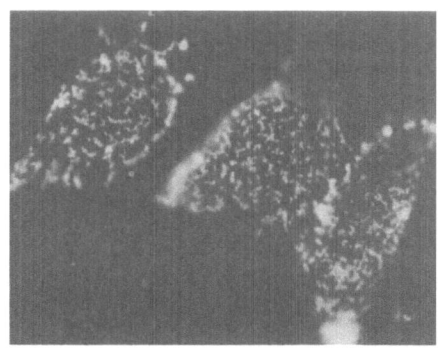 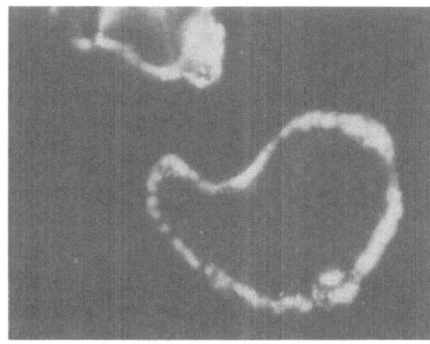

a   b

Fig. 6.   Interactions of the cytoplasmic tail with the cytoskeleton. Immunofluorescence localization of PEM on MCF-7 cells using polyclonal antiserum to PEM; a) untreated and b) following treatment with cytochalasin D at 1 mg/ml for two hours. After incubation the live cells were stained.

SPECIES CROSS-REACTIVITY

Periodic acid Schiff-treated silver stained gels identify high molecular weight glycosylated proteins present in the milk fat globule in a wide spectrum of mammals ranging from human to mouse (Patton et al., 1989). However, monoclonal and even polyclonal antibodies directed to the human mucin have shown cross-reactivity only with the mucin of higher primates (S. Patton, unpublished observations). Most of the antibodies react only with the tandem repeat. This lack of reactivity suggests that the sequence of the tandem repeat has altered during evolution. If the purpose of the tandem repeat is merely to provide an extended scaffold for O-glycosylation, then the sequence would not be important so long as the potential for O-glycosylation was maintained. In contrast, the functional interaction of the cytoplasmic tail with the actin cytoskeleton would suggest conservation; accordingly, these sequences would be most likely to cross-react with the mouse mucin.

The development of polyclonal antisera to two peptides within the 69 amino acid cytoplasmic portion which cross-react with the core protein of mice suggests functional conservation. Fig. 7 depicts the apical distribution in the lactating mammary gland, pancreas and lung, giving similar patterns of distribution to those seen in the human. To our knowledge, these antisera are the first to cross-react with the rodent protein and will allow us to do expression studies on the developing mouse embryo and mammary gland.

Although mucins have been assumed to have protective and/or lubrication roles in secretory epithelial tissues, the fact that many adenocarcinomas express high levels of this particular mucin may suggest additional functions. Indirect evidence has suggested mucins play a role in natural killer cell resistance (Bharathan et al., 1990), in evasion of immune response by tumor cells (Hanna, 1985) and may act as immune suppressors (Shimizu et al., 1990). The cytoskeletal interaction suggests the mucin may be involved in stabilizing cell morphology. As functional studies are much more readily performed in the mouse than in humans and in order to examine the evolutionary changes in the gene, we have cloned the mouse homologue of PEM.

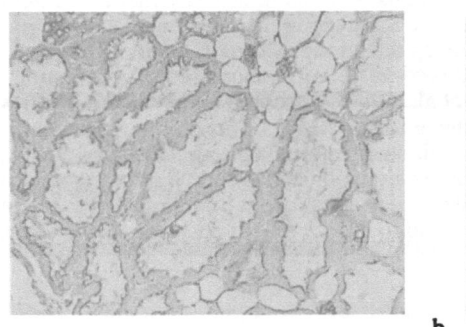 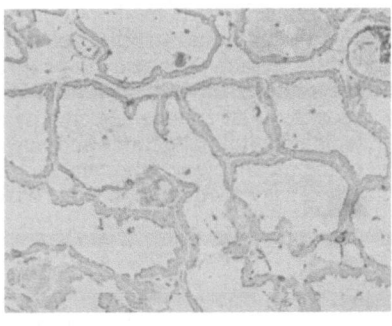

a   b

Fig. 7.   Mouse tissues stained with polyclonal antiserum CT-1 made to human cytoplasmic tail peptides. Staining shows species cross-reactivity. Staining was done using an indirect immunoperoxidase technique. a) mouse lactating mammary gland stained with CT-1 and b) with preimmune antiserum.

CLONING THE MOUSE HOMOLOGUE mMUC 1

Using the 3' sequence as a probe, the mouse mucin has been cloned (Spicer, Parry and Gendler, manuscript submitted). Interestingly, the gene has undergone substantial evolutionary changes while maintaining a high potential for O-glycosylation. The repeat domain of the mouse differs from that of the human in codon and amino acid sequence, the length and number of tandem repeats and the location of the repeats within the protein. The mouse sequence contains a variable 20-21 amino acid tandem repeat with amino acid composition similar to that of the human, although homology between the human and mouse repeats is less than 40% at the protein level (fig. 8). It was found that the polymorphism, so apparent in the human gene (Swallow et al., 1987; Gendler et al., 1988; 1990), has been lost in the mouse gene.

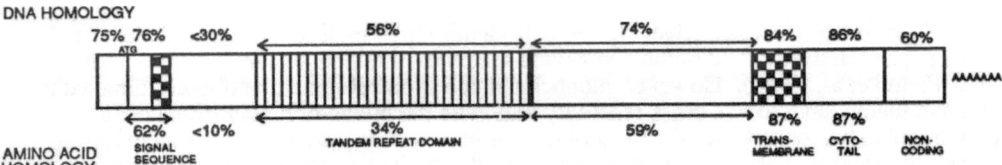

Fig. 8.   Cartoon of human and mouse mucin showing homology at DNA and amino acid level.

The consensus DNA and amino acid sequences are shown in fig. 9. It is notable that only 2 of the 11 repeats are identical in sequence in contrast to the human gene which contains largely identical repeats. Careful examination of the repeat sequences suggests that the present mouse gene was created from an unequal crossing-over event some time ago (Spicer and Gendler, manuscript submitted). Whether or not the ancient sequence was polymorphic is not known.

Human: G S T A P P A H G V T S A P D T R P A P

Mouse: D S T S S P V H S G T S S P A T R A P V
                                        S   E

Fig. 9.   Alignment of consensus amino acids in tandem repeats of human and mouse MUC 1. Note that the homology between the amino acids is only 45%.

POLYMORPHISM

The sequence homogeneity and variability in repeat unit number (from 20 to >125 repeats) may be a recent event occurring in large mammals and man and may be explained by unequal crossing over. Sequence similarities between repeats may cause chance misalignments in pairing, and the resultant crossover results in duplication of a series of repeats in one homologue or sister chromatid and their deletion in another. Thus, point mutations that have occurred will either spread quickly through the array or be removed. The extent of length polymorphism will depend on the rate of new allele production, with the most polymorphic VNTR loci (of which the human mucin is one) being the most homogeneous in sequence. Jeffreys and co-workers (1985) have postulated that the VNTR core may be an eukaryotic recombination signal. A key part of the hypothesis is that the Chi sequence (which is a signal for recombination in λ and E.coli), or a slight variation of Chi, is present in many VNTR loci; likewise, Chi or a 7/8 match is present in all the mucin genes so far identified. If Chi or some variation of Chi can serve as a signal for recombination, then a speculative model for the generation of tandem repeats can be invoked. The DNA duplex near the Chi sequence is nicked, repair synthesis and ligation of the nicked strand results in duplication of the Chi-like core sequence, which subsequently promotes mispairing and unequal exchange, leading to amplification to form a tandem repeat (Jarman and Wells, 1989).

However, even a single mismatch in sequences can lead to a dramatic decrease in recombination efficiency (Shen and Huang, 1986; Liskay et al., 1987). If point mutations were to accumulate within repeats at a VNTR locus at a rate greater than could be compensated for by recombination between alleles, then a threshold point can be imaged beyond which repeats would no longer share enough homology to recognize each other. Allele length would then become fixed and new alleles would fail to be generated. Significantly, the extensive polymorphism observed in humans and most other mammals has been found to be absent in the mouse, despite the 60-63 bp tandem repeat. The mouse mucin repeats appear to be degenerate repeats, although there is evidence of a previous polymorphsim or unequal crossover event. The lack of polymorphism in the mouse is most likely attributable to an accumulation of mismatches. Studies on the relative rates of nucleotide substitution in rodents and other mammals, such as man and bovine, reveal that rodents evolve significantly faster, primarily due to their shorter generation times (Wu and Li, 1985). This could be a possible reason for the observed lack of polymorphsim so far only noted in rodents.

Alternatively, the ancestral gene may not have been polymorphic. At some point during the divergence of the respective mammalian lineages, an unpredictable recombination event may have resulted in a duplication within the repeat-like domain. Once the initial duplication had occurred, chances for further unequal crossover events were increased, leading to the extensive polymorphism observed, and the subsequent high level of recombination between alleles in this region acted to maintain the repeat sequence in the face of point mutation. It will be of interest to clone the gene from other species, since evolution may be proceeding differently at this locus, and analysis of gene sequence data can lead to new insights into the evolutionary relationships between organisms.

REFERENCES

Abe, M. and Kufe, D., 1989, Sequence analysis of the 5' flanking region of the human DF3 breast carcinoma-associated antigen gene, Biochem. Biophys. Res. Commun., 165:644.

Bharathan, S., Moriarty, J., Moody, C.E. and Sherblom, A.P., 1990, Effect of tunicamycin on sialomucin and natural killer susceptibility of rat mammary tumor ascites cells, Cancer Res. 50: 5250.

Bhargava, A.K., Woitach, J.T., Davidson, E.A. and Bhavanandan, V.P., 1990, Cloning and cDNA sequence of a bovine submaxillary gland mucin-like protein containing two distinct domains, Proc. Natl. Acad. Sci. USA 87:6798.

Bird, A.P., Taggart, M., Frommer, M., Miller, O.J. and Macleod, D., 1985, A fraction of the mouse genome that is derived from islands of nonmethylated, CpG-Rich DNA, Cell, 40:91.

Burchell, J. and Taylor-Papadimitriou, J., 1989, Antibodies to human milk fat globule molecules, Cancer Invest., 17:53.

Burchell, J., Gendler, S., Taylor-Papadimitriou, J., Girling, A., Lewis, A., Millis, R. and Lamport, D., 1987, Development cna characterization of breast cancer reactive monoclonal antibodies directed to the core protein of the human milk mucin, Cancer Res., 47:5476.

Burchell, J., Taylor-Papadimitriou, J., Boshell, M., Gendler, S. and Duhig, T., 1989, A short sequence, within the amino acid tandem repeat of a cancer-associated mucin, contains immunodominant epitopes, Int. J. Ca., 44:691.

Cooper, D.N. and Krawczak, M., 1989, Cytosine methylation and the fate of CpG dinucleotides in vertebrate genomes, Hum. Genet., 83:181.

Corcoran, D. and Walker, R.A., 1990, Ultrastructural localization of milk fat globule membrane antigens in human breast carcinomas, J. Pathol. 161: 161.

Emerman, J.T., Enami, J., Pitelka, D.R. and Nandi, S., 1977, Maintenance and induction of morphological differentiation in dissociated mammary epithelium on floating collagen membranes, In Vitro, 13: 316.

Gendler, S.J., Burchell, J.M., Duhig, T., Lamport, D., White, R., Parker, M. and Taylor-Papadimitriou, J., 1987, Cloning of partial cDNA encoding differentiation and tumor-associated mucin glycoproteins expressed by human mammary epithelium, Proc. Natl. Acad. Sci. USA, 84:6060.

Gendler, S., Taylor-Papadimitriou, J., Duhig, T., Rothbard, J. and Burchell, J., 1988, A highly immunogenic region of a human polymorphic epithelial mucin expressed by carcinomas is made up of tandem repeats, J. Biol. Chem., 263:12820.

Gendler, S.J., Lancaster, C.A., Taylor-Papadimitriou, J., Duhig, T., Peat, N., Burchell, J., Pemberton, L., Lalani, E.-N. and Wilson, D., 1990, Molecular cloning and expression of human tumor-associated polymorphic epithelial mucin, J. Biol. Chem., 265:15286.

Griffiths, A.B., Burchell, J., Gendler, S., Lewis, A., Blight, K., Tilly, R. and Taylor-Papadimitriou, J., 1987, Immunological analysis of mucin molecules expressed by normal and malignant mammary epithelial cells, Int. J. Cancer, 40: 319.

Gum, J.R., Byrd, J.C., Hicks, J.W., Toribara, N.W., Lamport, D.T.A and Kim, Y.S., 1989, Molecular cloning of human intestinal mucin cDNAs, J. Biol. Chem., 264:6480.

Gum, J.R., Hicks, J.W., Swallow, D.M., Lagace, R.L., Byrd, J.C., Lamport, D.T.A., Siddiki, B. and Kim, Y.S., 1990, Molecular cloning of cDNAs derived from a novel human intestinal mucin gene, Biochem. Biophys. Res. Commun., 171:407.

Hanisch, F.-G., Uhlenbruck, G., Peter-Katalinic, J., Egge, H., Dabrowski, J. and Dabrowski, U., 1989, Structures of neutral O-linked polylactosaminoglycans on human skim milk mucins, J. Biol. Chem., 264: 872.

Hanna, N., 1985, The role of natural killer cells in the control of tumor growth and metastasis, Biochim. Biophys. Acta, 780: 231.

Hull, S., Bright, A., Carraway, K., Abe, M., Hayes, D. and Kufe, D., 1989, Oligosaccharide differences in the DF3 sialomucin antigen from normal human milk and the BT20 human breast carcinoma cell line, Cancer Comm., 1: 261.

Jarman, A.P. and Wells, R., 1989, Hypervariable minisatellites: recombinators or innocent bystanders?, TIBS 5: 367.

Jeffreys, A.J., Wilson, V. and Thein, S.L., 1985, Hypervariable 'minisatellite' regions in human DNA, Nature, 314: 67.

Khorrami, A., Lan, M.S., Metzgar, R.S. and Kaufman, B., 1989, Characteristics of a sulphated human pancreatic adenocarcinoma mucin glycoprotine, Glycoconjugate J., 6: 428.

Lan, M.S., Batra, S.K., Qi, W.-N., Metzgar, R.S. and Hollingswirth, M.A., 1990, Cloning and sequencing of a human pancreatic tumor mucin cDNA, J. Biol. Chem., 265: 15294.

Lancaster, C.A., Peat, N., Duhig, T., Wilson, T., Taylor-Papadimitriou, J. and Gendler, S.J., 1990, Structure and expression of the human polymorphic epithelial mucin gene: an expressed VNTR unit, manuscript submitted.

Ligtenberg, M.J.L., Vos, H.L., Gennissen, A.M.C. and Hilkens, J., 1990, Episialin, a carcinoma-associated mucin, is generated by a polymorphic gene encoding splic variants with alternative amino termini, J. Biol. Chem., 265: 5573.

Liskay, R.M., Letsopu, A. and Stachelek, J.L., 1987, Homology requirement for efficient gene conversion between duplicated chromosomal sequences in mammalian cells, Genet., 115:161.

Ormerod, M.G., Monaghan, P., Easty, D. and Easty, G.C., 1981, Asymmetrical distribution of epithelial membrane antigen on the plasma membranes of human breast cell lines in culture, Daig. Histopath., 4: 89.

Parry, G., Beck, J.C., Moss, L., Bartley, J. and Ojakian, G.K., 1990, Determination of apical membrane polarity in mammary epithelial cell cultures: the role of cell-cell, cell-substratum, and membrane-cytoskeleton interactions, Exp. Cell Res., 188: 302.

Patton, S., Huston, G.E., Jenness, R. and Vaucher, Y., 1989, Differences between individuals in high-molecular weight glycoproteins from mammary epithelia of several species, Biochim. Biophys. Acta, 980: 333.

Probst, J.C., Gertzen, E.-M. and Hoffmann, W., 1990, An integumentary mucin (FIM-B.1) from *Xenopus laevis* homologous with von Willebrand Factor, Biochem., 29: 6240.

Savatier, P., Trabuchet, G., Faure, C., Chebloune, Y., Gouy, M., Verdier, G. and Nigon, V.M., 1985, J. Mol. Biol., 182: 21.

Shimizu, M., Tanimoto, H., Azuma, N. and Yamauchi, K., 1990, Growth inhibition of Balb/c 3T3 cells by a high-molecular -weight mucin-like glycoprotein of human milk fat globule membrane, Biochem. Internatl., 20: 147.

Shen, P and Huang, H.V., 1986, Homologous recombination in *Escherichia coli*: dependence on substrate length and homology, Genet., 112: 441.

Siddiqui, J., Abe, M., Hayes, D., Shani, E., Yunis, E., Kufe, D., 1988, Isolation and sequencing of a cDNA coding for the human DF3 breast carcinoma-associated antigen, Proc. Natl. Acad. Sci. USA, 85: 2320.

Swallow, D.M., Gendler, S.J., Griffiths, B., Corneu, G., Taylor-Papadimitriou, J. and Bramwell, M.E., 1987, The human tumour-associated epithelial mucins are coded by an expressed hypervariable gene locus PUM, Nature, 327: 82.

Timpte, C.S., Eckhardt, A.E., Abernethy, J.L. and Hill, R.L., 1988, Porcine submaxillary gland apomucin contains tandemly repeated, identical sequences of 81 residues, J. Biol. Chem., 263: 1081.

Williams, C.J., Wreschner, D.H., Tanaka, A., Tsarfaty, I., Keydar, I and Dion, A.S., 1990, Multiple protein forms of the human breast tumor-associated epithelial membrane antigen (EMA) are generated by alternative splicing and induced by hormonal stimulation, Biochem. Biophys. Res. Commun., 170: 1331.

Wreschner, D.H., Hareuveni, M., Tsarfaty, I., Smorodinsky, N., Horev, J., Zaretsky, J., Kotkes, P., Weiss, M., Lathe, R., Dion, A and Keydar, I., 1990, Human epithelial tumor antigen cDNA sequences, Eur. J. Biochem., 189: 463.

Wu, C.-I. and Li, W.-H., 1985, Evidence for higher rates of nucleotide substitution in rodents than in man, Proc. Natl. Acad. Sci. USA, 82: 1741.

Zotter, S., Hageman, P.C., Lossnitzer, A., Mooi, W.J. and Hilgers, J., 1988, Tissue and tumor distribution of human polymorphic epithelial mucin, Cancer Rev., 11-12: 55.

STRUCTURE, PROCESSING, DIFFERENTIAL GLYCOSYLATION AND BIOLOGY OF EPISIALIN

J. Hilkens, M.J.L. Ligtenberg, S. Litvinov, H.L. Vos,
A.M.C. Gennissen, F. Buys, and Ph. Hageman

Department of tumor biology, The Netherlands Cancer Institute
(Antoni van Leeuwenhoek Huis), Plesmanlaan 121, Amsterdam,
The Netherlands

ABSTRACT

Episialin is a carcinoma associated antigen encoded by the MUC1 gene.
We have cloned genomic and cDNA coding for this protein and full length
cDNA has been sequenced. The mRNA can be differentially spliced depending
on a single nucleotide polymorphism in the second exon. The main part of
the coding domain of the MUC1 gene consist of repeated seqences of 60 bp.
Numerous allelic forms of the gene have been determined which differ in
the number of repeats. Each of the repeated sequences contains several
proline, threonine and serine residues; the latter are potential O-linked
glycosylation sites. The large amount of prolines and the O-linked glycosylation give the episialin molecule a very rigid structure pointing
into the extracellular space. Based on the predicted amino acid sequence,
episialin contains a transmembrane domain. However, the transmembrane
domain is absent in the molecule that is released from carcinoma cells.
At least three precursor forms of the molecule can be distinguished. The
most mature precursor form of episialin is undersialylated and is present
at the cell surface. This premature form is internalized and further sialylated. The glycosylation of episialin is variable both within one cell
line and among different cell lineages. This has resulted in the generation
of monoclonal antibodies that show a high tissue preference.
Episialin cDNA has been transfected into HBL-100 cells under the control
of a CMV promotor. Transfectants with high levels of episialin expression
showed decreased aggregation properties.

INTRODUCTION

Episialin (formerly called MAM-6) is a sialylated glycoprotein present
at the cell surface of most exocrine secretory epithelial cells. It is a
large molecule (M_r >400.000) consisting for more than 50% of carbohydrates
which are mainly O-linked. In carcinomas it is the major sialylated glycoprotein. In normal tissues the molecule is, with a few exceptions, only
present at the apical side of exocrine secretory epithelial cells. It was
originally identified by monoclonal antibodies (mAbs) raised against human
milk fat globule membranes (Taylor-Papadimitriou et al., 1981; Hilkens et
al., 1984). As determined with mAbs, the expression of the molecule is increased in many types of carcinoma cells relative to the corresponding normal tissues (Zotter et al., 1987). Episialin is a membrane bound glyco-

Breast Epithelial Antigens, Edited by R.L. Ceriani
Plenum Press, New York, 1991

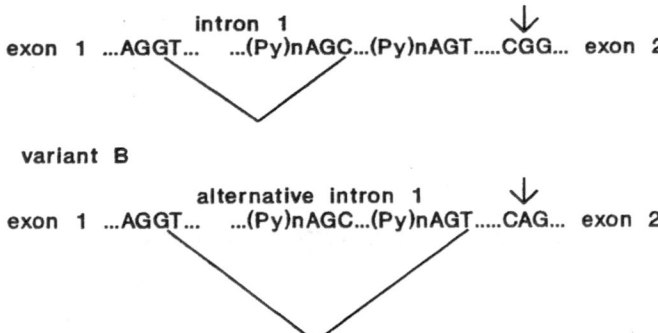

FIG. 1. **Alternative Splicing of Episialin mRNA.** The polymorphic nucleotide are marked with an arrow. (Py)nAG is the splice acceptor site.

protein but at least part of the molecules are released from the cell and appear in the medium of cultured cells and in the serum of breast cancer patients. The mechanism of release of episialin from the cell is not yet fully understood.

A sandwich assay has been developed (CA 15-3 assay) to determine the serum levels of episialin using a mAb developed in our group (115D8, Hilkens et al., 1984; Hilkens et al., 1986) and a mAb developed by Kufe and collegues (DF3, Kufe et al., 1984). The level of episialin in the serum of breast cancer patients correlates with the extent of the disease and serial measurements can be used to monitor the development of the disease in these patients.

Numerous mAbs have been raised against episialin by various other groups. Amongst these mAbs are HMFG-1 and 2 (Taylor-Papadimitriou et al., 1981), EMA (Ormerod et al., 1985), SM-3 (Burchell et al., 1988). The latter mAb has a preference for breast cancer cells. Many of the mAbs described in the literature recognize different glycoforms of episialin (Hilkens et al., 1989).

During the previous years we and others have extensively studied the episialin molecule since it is a major cell surface component that is often altered in carcinoma cells with respect to glycosylation, level of expression and cellular localization. Some of our most recent results will be summarized below.

RESULTS AND DISCUSSION

The Episialin Gene and the deduced Structure of the Molecule

We have cloned the genomic sequence of the episialin gene (designated MUC1) as well as the complete cDNA. Sequence analysis showed that the episialin gene consists of 7 exons (Ligtenberg et al., 1990; Williams et al., 1990). Exon 2 is the largest exon encoding the C-terminal part of the signal peptide and the repeat region of the molecule. Each repeat consist of 60 base pairs with an almost perfectly conserved sequence. The number of repeats varies among different individuals or cells lines

(Ligtenberg et al., 1990; Wreschner et al., 1990; Lan et al., 1990) leading to different molecular weights of episialin. In fact, episialin molecules of two main size classes can be recognized: Molecules encoded by alleles with 30-45 repeats and alleles with 60-90 repeats. Very few alleles were identified with an intermediate number of repeats. We have found two different splice variants which use alternative splice acceptor sites for exon 2. We have shown that the choice of the splice acceptor site is dependent on a single nucleotide polymorphism in the second exon eight nucleotides downstream of the second splice acceptor site; if a G is present at this position variant A is expressed whereas variant B is expressed when the nucleotide at this position is a A (fig. 1, Ligtenberg et al., in press). The single nucleotide polymorphism determining the choice of the splice acceptor site correlates in most cases with the length

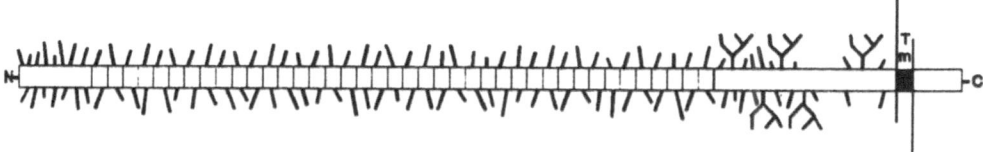

TM: Transmembrane domain

FIG. 2. **The episialin molecule: A model deduced from the cDNA sequence.**
In this model the protein backbone comprises 1264 amino acids and an array of 40 repeat units (800 amino acids), which is the approximate number of repeats in the short allele of T47D cells. The extracellular domain of the molecule is almost completely covered by O-linked glycans. Five N-linked glycans are present C-terminal of the repeat region. The C-terminal part of the molecule consists of a typical trans-membrane domain of at least 24 amino acids followed by a cytoplasmic tail of 69 amino acids (from Hilkens, in press).

polymorphism described above; most alleles containing the high number of repeats (60-90) have a G, whereas an A is present at this position in alleles with the relatively small number of repeats (30-45). The latter result indicates that unequal exchange between the repeat regions of the alleles seldomly occurs, suggesting that the major variation in length of the repeat region has most likely arisen by duplication of the repeat region, while the minor variations are due to slippage of the replication system. This result also clearly shows that splicing can depend on exon sequences. The splice variants encode molecules with different signal peptides possibly leading to mature molecules with different N-termini. It has not yet been established whether there are differences in biological function between both splice variants.

We have used the polymerase chain reaction using primers derived from exon sequences surrounding the splice sites in a search for other splice variants. In particular, we have looked for splice variants that lack a transmembrane domain as have been described by Wreschner et al. (1990) but we have failed to detect such splice variants in various episialin expressing cell lines.

We have deduced the structure of the molecule based on the predicted amino acid sequence. The episialin gene encodes a transmembrane molecule with a large extracellular domain, which mainly consists of repeated sequences of 20 amino acids, and a 69 amino acid cytoplasmic domain (Ligtenberg et al., 1990; Wreschner et al., 1990; Gendler et al., 1990; Lan et al., 1990). The repeats comprise more than half of the polypeptide backbone, even in the smallest allele detected thus far. Both the repeats and the surrounding sequences contain many potential attachment sites for O-linked glycans. The high percentage of proline residues and the extensive glycosylation of the molecule result in an extended and rigid structure. The deduced structure of episialin is shown in fig 2.

Processing, Intracellular Routing and Release of Episialin

We and others have shown by means of biosynthetic labeling followed by immunoprecipitation that episialin is synthesized as a single polypeptide chain of a relatively high molecular weight, in most cell lines approximately 200 kDa or more (Hilkens and Buys, 1988; Linsley et al., 1988). This precursor is rapidly converted to a second precursor of approximately 20 kDa lower molecular weight, probably by proteolysis. The main processing of episialin occurs by the addition of numerous O-linked glycans to this precursor, which leads to a premature form of episialin with an apparent molecular weight over 400 kDa on SDS-polyacrylamide gels. The final processing step involves sialylation and increases the mobility of the molecule on SDS-gels. All processing steps are schematically indicated in figure 3.

We have studied the final maturation step of episialin using a mAb, 201E9 preferentially reactive with premature episialin and almost non-reactive with mature or released episialin. Binding of this mAb to mature episialin is positively affected by neuraminidase treatment confirming the notion that the final maturation step involves sialylation. Neuraminidase protection experiments showed that cell surface episialin is internalized. Moreover, we have shown that cell surface exposed episialin acquires additional sialic acid residues following internalization and that the molecule is subsequently recycled to the cell surface. Evidence for internalization of episialin has been obtained before by our group using EM techniques (Calafat et al., 1988).

Despite the fact that the amino acid sequence predicts that episialin should be a transmembrane molecule, episialin can be detected in extracellular fluids. We have investigated whether the putative proteolytic cleavage of the early precursor that converts the 220 kDa precursor into a 200 kDa molecule (the small allele product in ZR-75-1 cells, described above) would release episialin from the transmembrane domain. For this purpose we have raised a rabbit antiserum against a peptide representing part of the cytoplasmic domain of episialin. Using this antiserum we have shown that the cytoplasmic domain of episialin was still present in all precursor molecules and in the mature molecule localized at the cell surface membrane, suggesting that the early proteolytic cleavage can not result in release of episialin from the membrane. However, the antiserum against the cytoplasmic domain of episialin does not react with episialin present in the medium of breast carcinoma cell lines in vitro suggesting that the release of episialin from the membrane is a late or post maturation event that involves proteolysis at or near the cell surface (litvinov et al., manuscript in preparation).

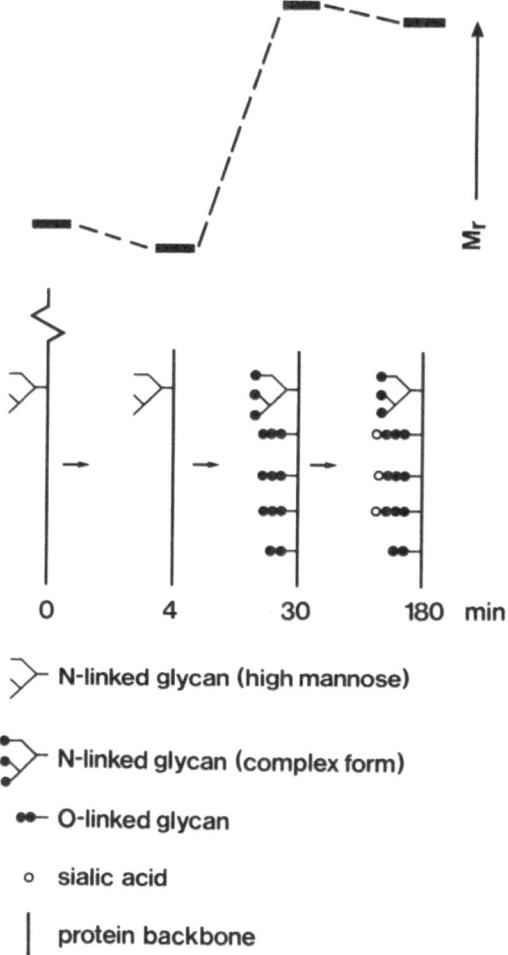

FIG. 3. Outline of the processing of episialin. The approximate time of
    appearance of the precursors is indicated (from Hilkens, in press).

Episialin Glycoforms and mAbs with Preferential Reactivity for Tumors or
certain Tissue Types

    The repetitive part of the protein backbone of episialin is very immuno-
genic (Gendler et al., 1988) and most mAbs raised against episialin bind
to this part of the molecule. Most of the mAbs tested reacted in ELISA
assays and on immunoblots with fragments of episialin encoded by the
repetitive part of the gene present in a bacterial expressed beta-galacto-
sidase fusion protein. For example, mAbs 139H2 and DF3 strongly react
with the protein backbone. However, mAb 115D8 only weakly reacts with the
protein backbone; full reactivity is dependent on the presence of carbohydrates, in particular sialic acid (Litvinov et al., in preparation).
Other mAbs, like Ca1 (Ashall et al., 1982) and 115G2, seem to be directed
against epitopes involving entirely or mainly carbohydrates since they do
not react with the bacterial fusion proteins and do not precipitate the

Table 1. Immunoperoxidase staining on formalin-fixed paraffin-embedded normal tissues

	201E9	201H4	139H2
breast			
epithelium	+	-/+	++
secretion product	-	-/+	++
Skin			
epidermis	-	-	-
sebaceous gland	++	++	++
sweat gland canaliculi	-	trace	++
lung			
pneumocytes type 1	-	-	-/+
pneumocytes type 2	-/+	-/+	++
bronchus epithelium	+	-	++
placenta			
villus epithelium	++	++	++
liver			
hepatocytes	-	-	-
bile ducts	-/+	-	++

early precursor forms of episialin which do not contain O-linked carbohydrates but only the fully glycosylated mature form of episialin. However,
it is difficult to prove that no amino acid residues are involved as long
the mAbs have not been tested on isolated carbohydrate structures. Other
investigators have established the peptide sequence of the minimal epitope
of some of these mAbs. Several epitopes partly overlap (Xing et al., 1989;
Burchell et al., 1989). Some epitopes include a potential glycosylation
site that is apparently not or not always used.

In addition to the mAbs raised against episialin by immunizing mice with
human milk fat globule membranes (Hilkens et al., 1984), we have recently
also raised mAbs against episialin present on breast cancer cells and
against non-glycosylated fusion proteins mentioned above (Litvinow et al.,
manuscript in preparation). We have determined the tissue specificity of
these new mAbs. Two mAbs, 175C5 and 175G7, raised against breast cancer
cells and mAb 201E9 raised against a bacterial fusion protein, have a
preference for breast carcinomas relative to normal breast epithelium.
The reactivity of mAbs 175C5 and 175G7 on other tissues was comparable to
the reactivity of standard mAbs directed against episialin, which are
reactive with almost all exocrine glandular epithelia. A similar mAb, SM-
3, has been produced by Burchell et al. (1987). Another mAb developed in
our group against fusion proteins (202H4) reacted with an epitope with a
limited distribution on normal tissues. In a few tissues, like sebaceous
gland, placental cells and certain cell types in embryonal carcinomas;
the reactivity was very strong, while in most other normal epithelia the
reactivity was much weaker than observed with other mAbs, like 139H2
(Table 1). Also the reaction pattern of MAb 201E9 deviated from the
standard anti-episialin mAbs. Moreover, mAb 201E9 reacted mainly with
intracellular and membrane bound episialin as was discussed above, while
another recently generated mAb, 202G3, preferentially reacted with secreted
episialin as determined by sandwich radioimmunoassays (not shown). These
results indicate that secreted episialin indeed contains different
epitopes compared with cell bound episialin.

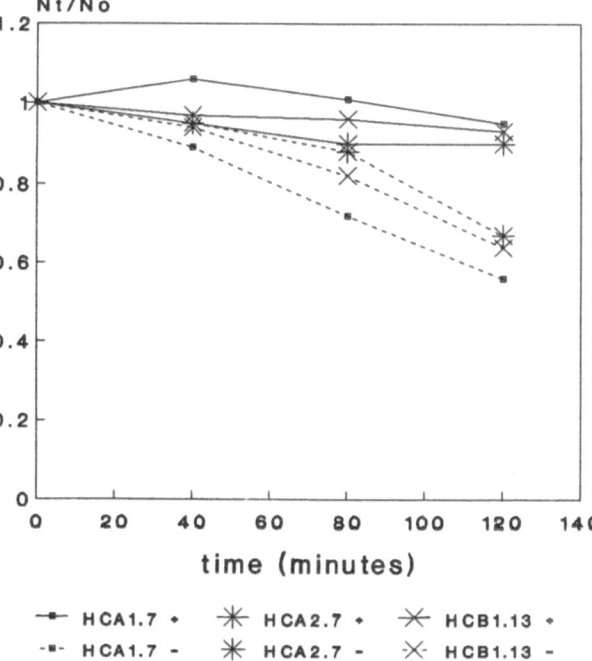

Fig. 4. **Aggregation of HBL-100 transfectants and revertants.** Cells were detached in with trypsin-EDTA and incubated for the time indicated on a rotary shaker at 37° C in a CO_2 incubator. The amount of aggregates was determined using a coulter counter. The relative decrease in particle number is indicated in the figure.

Previously, We have described differential reactivity of a large number of mAbs against episialin with adenolymphomas of the parotid gland (Zotter et al., 1988). In fact we were able to differentiate these mAbs in at least three groups based on the reactivity pattern on differentiated cells in this tumor; mAbs were reactive with basal-like cells, apical cells or with both cell types. Also Abe and Kufe (1987) reported that episialin consist of a family of molecules with different epitope composition, although they did not show tissue preference of these epitopes.

We have investigated the biochemical background of the preferential binding of the various mAbs against episialin to different tissues. For this purpose we used carcinoma cell lines originating from different tissues as models (Hilkens et al., 1989). All mAbs against the protein backbone precipitate the precursor molecule of episialin from all cell lines expressing this epithelial sialomucin. However, some of these mAbs precipitate mature molecules only from a restricted number of carcinoma cell lines. Even within a single cell line some mAbs precipitate only a subset of the mature episialin molecules. The difference in reactivity of these mAbs, which are directed against the protein moiety of episialin, is caused by alterations in accessibility of the epitopes for mAbs due to differential glycosylation of the molecule in different cell lines. The differentially glycosylated molecules, called glycoforms, each show a restricted expression pattern resulting in positive and negative cell lines. These results explain the differential reactivity of certain mAbs with carcinomas and normal epithelial cells and the differential reactivity of others with tissues of different histological origin. It is important

31

to realize that the differential reactivity is in most cases not an all or nothing phenomenon but a preferential reactivity with certain tissues or tumors. Even within one type of carcinoma the number of epitopes accessible for a given anti-episialin mAb may vary in individual tumors. Differential glycosylation could occur by linking O-linked glycans to different serines and threonines or could be due to differences in length or branching of the glycans.

The Biological Effect of Episialin Overexpression in Tumor Cells

The function of episialin is not yet known. The molecule may have a protective function in normal tissues. Episialin is normally expressed at the apical surface but in carcinoma cells polarisation is often lost which results in the presence of episialin at the plasma membrane facing the stroma or adjacent cells. Moreover, there is evidence that in carcinoma cells the expression of episialin is increased. Overexpression of this rigid and strongly negatively charged molecule (due to the abundance of sialic acid residues on the glycosidic side chains) may have a strong destabilizing influence on cell-substratum and cell-cell interactions including intercellular adhesion. To test this hypothesis we have transfected a SV-40 transformed human mammary epithelial cell line, HBL-100, which normally expresses a low level of episialin, with a vector containing episialin cDNA under the control of the cytomegalovirus immediate-early-promotor. Transfectants with different amounts of episialin on their cell membrane were selected. Transfectants with the highest levels of episialin had a decreased cellular aggregation capacity and showed aberrant adhesion properties; sometimes a large proportion of the cells were growing in suspension. Revertants of these transfectants which had lost episialin expression regained their normal aggregation properties and cell-matrix interactions (Fig. 4). These results indicate that episialin expression in the transfectants to levels also found on carcinoma cells reduces the cellular adhesion properties. The anti-adhesion and aggregation effect may be explained by the rigid and extended structure of the episialin molecule which points into the extracellular space. Above a certain density, this molecule may prevent intercellular interactions needed for cell-cell aggregation or cell-matrix interactions. This observation suggests that a high episialin expression may have a profound effect on the ability of the cells to participate in cellular interactions and on the metastatic potential of carcinoma cells.

REFERENCES

Abe, M., and D.W. Kufe (1987). Identification of a family of high molecular weight tumor-associated glycoproteins. J. Immunol. 139, 257-261.
Ashall, F., M.E. Bramwell, and H. Harris (1982). A new marker for human cancer cells. I. The Ca antigen and the Ca1 antibody. Lancet ii: 1-6.
Burchell, J., S. Gendler, J. Taylor-Papadimitriou, A. Girling, A. Lewis, R. Millis, and D. Lamport (1987). Development and characterization of breast cancer reactive monoclonal antibodies directed to the core protein of the human milk mucin. Cancer Res. 47, 5476-5482.
Burchell, J., J. Taylor-Papadimitriou, M. Boshell, S. Gendler, and T. Duhig (1989). A short sequence within the amino acid tandem repeat of a cancer associated mucin contains immunodominant epitopes. Int. J. Cancer 44, 691-696.
Calafat, J., Molthoff, C., Jansen, J., and Hilkens, J. (1988). Endocytosis and intracellular routing of an antibody ricin conjugate. Cancer Res., 45: 3822-3827.
Gendler, S.J., J.M. Burchell, T. Duhig, D. Lamport, R. White, M. Parker,

and J. Taylor-Papadimitriou (1987). Cloning of partial cDNA encoding
differentiation and tumor-associated mucin glycoproteins expressed by
human mammary epithelium. Proc. Natl. Acad. Sci. 84, 6060-6064.

Gendler, S.J., J. Taylor-Papadimitriou, T.Duhig, J. Rothbard, and J.M.
Burchell, (1988). A highly immunogenic region of a human polymorphic
epithelial mucin expressed by carcinomas is made up of tandem repeats.
J. Biol. Chem. 263, 12820-12823.

Gendler, S.J., C.A. Lancaster, J. Taylor-Papadimitriou, T.Duhig, N.
Peat, J. Burchell, L. Pemberton, E-N Lalani and D. Wilson (1990).
Molecular cloning of human tumor associated polymorphic epithelial
mucin. J. Biol. Chem. 265, 15286-15293.

Hilkens, J., Buijs, F., Hilgers, J., Hageman, Ph., Calafat, J., Sonnenberg,
A. and Van Der Valk, M. (1984) Monoclonal antibodies to human milkfat
globule membranes detecting differentiation antigens of the mammary
gland and mammary tumors. Int. J. Cancer 34, 197-206

Hilkens, J., V. Kroezen, J.M.G. Bonfrer, M. de Jong-Bakker and P.F.
Bruning. (1986) MAM-6 antigen, a new serum marker for breast cancer
monitoring. Cancer Res. 46, 2582-2587.

Hilkens, J. and Buys, F. Biosynthesis of MAM-6, an epithelial sialomucin;
evidence for the involvement of a rare proteolytic cleavage step in
the endoplasmic reticulum. (1988) J. Biol. Chem. 263: 4215-4222.

Hilkens, J, Buys F, and Ligtenberg M (1989). Complexity of MAM-6, an
epithelial sialomucin, associated with carcinomas. Cancer Res. 49:
786-793.

Hilkens, J. CA 15-3 assay for the detection of episialin; a serum marker
for breast cancer. In: Clinical Cancer Markers: Diagnosis, Prognosis and
Monitoring. S. Sell ed., The Humana Press. In press.

Kufe, D., G. Inghiriami, M. Abe, D. Hayes, H. Justi-Wheeler, and J.
Schlom (1984). Differential reactivity of a novel monoclonal antibody
(DF3) with human malignant versus benign breast tumors. Hybridoma 3,
223-232.

Lan, M.S., S.K. Batra, W-N. Qi, R.S. Metzger and M.A. Hollingsworth
(1990). Cloning and sequencing of a human pancreatic tumor mucin cDNA.
J. Biol. Chem. 265, 15294-15299.

Ligtenberg, M.J.L., H.L. Vos, A.M.C. Gennissen, and J. Hilkens (1990). A
carcinoma associated mucin is generated by a polymorphic gene encoding
splice variants with alternative amino termini. J. Biol. Chem. 265,
5573-5578.

Ligtenberg, M.L., Gennissen, A.M.C., Vos, H.L., Hilkens, J. (1991) A single
nucleotide polymorphism in an exon dictates allele dependent differential splicing of episialin mRNA. Nucl. Acid Res., in press.

Linsley, P.S, J.C. Kallestadt, and D. Horn (1988). Biosynthesis of a high
molecular weight cancer-associated mucin glycoprotein. J. Biol. Chem.
263, 8390-8397.

Ormerod MG, McIlhinney RA, Steel K and Shimizu M (1985) Mol. Immunol. 22,
265-269.

Swallow, D.M., S. Gendler, B. Griffiths, G. Corney, J. Taylor-Papadimitriou, and M.E. Bramwell (1987). The human tumor-associated epithelial
mucins are coded by an expressed hypervariable gene locus PUM. Nature
328, 82-84.

Taylor-Papadimitriou, J., J.A. Peterson, J. Arklie, J. Burchell, R.L.
Ceriani, and W.F. Bodmer (1981). Monoclonal antibodies to epithelium
specific components of the human milkfat globule membrane: Production
and reaction with cells in culture. Int. J. Cancer 28, 17-21.

Williams, C.J., Wreschner, D.H., Tanaka, A., Tsarfaty, I., Keydar, I.,
Dion, A.S. (1990). Multiple protein forms of the human breast tumor-
associated epithelial membrane antigen (EMA) are generated by alternative splicing and induced by hormonal stimulation. Biochem. Biophys.
Res. Comm. 170, 1331-1338.

Wreschner, D.H., M. Hareuveni, I. Tsarfaty, N. Smorodinsky, J. Horev, J.
Zaretsky, P. Kotkes, M. Weiss, R. Lathe, A. Dion, and I. Keydar (1990).
Eur. J. Biochem. 189, 463-473.

Xing, P.X., J.J. Tjandra, K.Reynold, P.J. McLaughlin, D.F.J. Purcell, and I.F.C. McKenzie (1989). Reactivity of anti-human milk fat globule antibodies with synthetic peptides J. Immunol. 142, 3503-3509.

Zotter, S., Lossnitzer, A., Hageman, Ph.C., Delemarre, J.F.M., Hilkens, J., and Hilgers, J. Immunohistochemical localization of the epithelial marker MAM-6 in invasive malignancies and highly dysplastic adenomas of the large intestine (1987). Lab. Invest. 57: 193-199.

Zotter, S., Hageman, Ph.C., Lossnitzer, A., van den Tweel, J., Hilkens, J., Mooi, W.J. and Hilgers, J. Monoclonal antibodies to epithelial sialo-mucins recognize epitopes of different cellular sides in adenolymphomas of the parotid gland. (1988) Int. J. Cancer. Supplement 3, 45-49.

MOLECULAR CLONING AND EXPRESSION OF BREAST MUCIN-ASSOCIATED ANTIGENS

D. Larocca, J.A. Peterson, R. Urrea, J. Kuniyoshi,
A. Bistrain, G. Walkup and R.L. Ceriani

John Muir Cancer and Aging Research Institute
2055 N. Broadway
Walnut Creek, CA

INTRODUCTION

The fat droplets in milk are encapsulated in a membrane that is derived from the apical surface of breast epithelial cells during lactation when the fat droplet buds off of the cell surface forming the milk fat globule. A variety of membrane proteins are found in the human milk fat globule (HMFG). In order to find breast specific components polyclonal antisera were originally raised against purified HMFG membrane and adsorbed with non-breast tissue. The adsorbed serum recognized 3 bands primarily, with molecular weights of 150 kDa, 70 kDa to 46 kDa[1]. It did not bind the high molecular weight mucin-like component that we initially termed NPGP since the mucin is also expressed in other tissues. Monoclonal antibodies (MoAbs) raised against HMFG membrane most often bind the mucin component, reflecting its high immunogenicity. However, we have also selected MoAbs that bind, 70,000 kDa and 46,000 kDa components. We have used MoAbs against the lower molecular weight components to isolate cDNAs that encode the antigens recognized by these MoAbs. We present here the characterization of the cDNAs, the genes that encode them and expression of recombinant HMFG proteins in E. Coli.

Cloning of a cDNA Encoding a 70KDa Antigen in HMFG

A cocktail of Mc13 and McR2 was used to immunoscreen a λgt11 cDNA expression library prepared from human breast tissue (Clontech, Palo Alto, CA). There were 3 out of about 900,000 positive phage plaques, one of which bound both monoclonal Abs and was further characterized. The sequence of a 297 bp cDNA, BA70-1, which we have previously described[2], was used to design oligonucleotide primers for anchor PCR[3]. Two DNA fragments that extended the BA70 sequence in the 3' direction were obtained that were approximately 400 and 800 base pairs long. The DNA sequence of the 800 base pair fragment and the deduced amino acid sequence are shown in Figure 1. The 400 and 800 base pair fragments differ at base 323, with the latter containing an additional T at this position. The amino acid sequence shown in Figure 2 is derived from the 800 base pair clone with the additional T at 323 deleted. The sequence shown contains a potential sulfated tyrosine at position 109 and is high in serine and threonine which could be glycosylated but contains no potentially n-linked glycosylation sites.

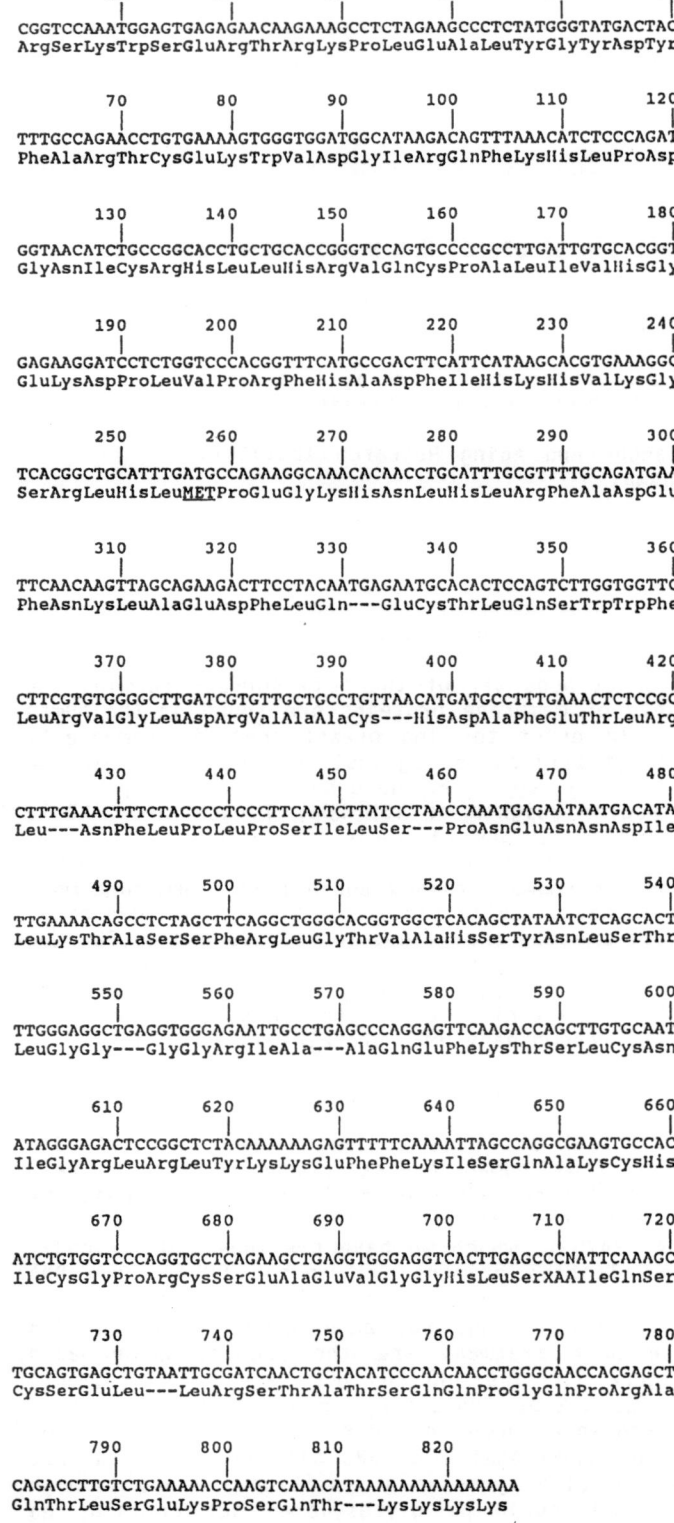

```
        10        20        30        40        50        60
         |         |         |         |         |         |
CGGTCCAAATGGAGTGAGAGAACAAGAAAGCCTCTAGAAGCCCTCTATGGGTATGACTAC
ArgSerLysTrpSerGluArgThrArgLysProLeuGluAlaLeuTyrGlyTyrAspTyr

        70        80        90       100       110       120
         |         |         |         |         |         |
TTTGCCAGAACCTGTGAAAAGTGGGTGGATGGCATAAGACAGTTTAAACATCTCCCAGAT
PheAlaArgThrCysGluLysTrpValAspGlyIleArgGlnPheLysHisLeuProAsp

       130       140       150       160       170       180
         |         |         |         |         |         |
GGTAACATCTGCCGGCACCTGCTGCACCGGGTCCAGTGCCCCGCCTTGATTGTGCACGGT
GlyAsnIleCysArgHisLeuLeuHisArgValGlnCysProAlaLeuIleValHisGly

       190       200       210       220       230       240
         |         |         |         |         |         |
GAGAAGGATCCTCTGGTCCCACGGTTTCATGCCGACTTCATTCATAAGCACGTGAAAGGC
GluLysAspProLeuValProArgPheHisAlaAspPheIleHisLysHisValLysGly

       250       260       270       280       290       300
         |         |         |         |         |         |
TCACGGCTGCATTTGATGCCAGAAGGCAAACACAACCTGCATTTGCGTTTTGCAGATGAA
SerArgLeuHisLeuMETProGluGlyLysHisAsnLeuHisLeuArgPheAlaAspGlu

       310       320       330       340       350       360
         |         |         |         |         |         |
TTCAACAAGTTAGCAGAAGACTTCCTACAATGAGAATGCACACTCCAGTCTTGGTGGTTC
PheAsnLysLeuAlaGluAspPheLeuGln---GluCysThrLeuGlnSerTrpTrpPhe

       370       380       390       400       410       420
         |         |         |         |         |         |
CTTCGTGTGGGGCTTGATCGTGTTGCTGCCTGTTAACATGATGCCTTTGAAACTCTCCGC
LeuArgValGlyLeuAspArgValAlaAlaCys---HisAspAlaPheGluThrLeuArg

       430       440       450       460       470       480
         |         |         |         |         |         |
CTTTGAAACTTTCTACCCCTCCCTTCAATCTTATCCTAACCAAATGAGAATAATGACATA
Leu---AsnPheLeuProLeuProSerIleLeuSer---ProAsnGluAsnAsnAspIle

       490       500       510       520       530       540
         |         |         |         |         |         |
TTGAAAACAGCCTCTAGCTTCAGGCTGGGCACGGTGGCTCACAGCTATAATCTCAGCACT
LeuLysThrAlaSerSerPheArgLeuGlyThrValAlaHisSerTyrAsnLeuSerThr

       550       560       570       580       590       600
         |         |         |         |         |         |
TTGGGAGGCTGAGGTGGGAGAATTGCCTGAGCCCAGGAGTTCAAGACCAGCTTGTGCAAT
LeuGlyGly---GlyGlyArgIleAla---AlaGlnGluPheLysThrSerLeuCysAsn

       610       620       630       640       650       660
         |         |         |         |         |         |
ATAGGGAGACTCCGGCTCTACAAAAAAGAGTTTTTCAAAATTAGCCAGGCGAAGTGCCAC
IleGlyArgLeuArgLeuTyrLysLysGluPhePheLysIleSerGlnAlaLysCysHis

       670       680       690       700       710       720
         |         |         |         |         |         |
ATCTGTGGTCCCAGGTGCTCAGAAGCTGAGGTGGGAGGTCACTTGAGCCCNATTCAAAGC
IleCysGlyProArgCysSerGluAlaGluValGlyGlyHisLeuSerXAAIleGlnSer

       730       740       750       760       770       780
         |         |         |         |         |         |
TGCAGTGAGCTGTAATTGCGATCAACTGCTACATCCCAACAACCTGGGCAACCACGAGCT
CysSerGluLeu---LeuArgSerThrAlaThrSerGlnGlnProGlyGlnProArgAla

       790       800       810       820
         |         |         |         |
CAGACCTTGTCTGAAAAACCAAGTCAAACATAAAAAAAAAAAAAAA
GlnThrLeuSerGluLysProSerGlnThr---LysLysLysLys
```

Figure 1.
The nucleotide sequence and amino acid translation of cDNA clone DL1-800. Dashed line indicates stop codons.

```
                 (P)           (P)
                  |             |
   1 Arg-Ser-Lys-Trp-Ser-Glu-Arg-Thr-Arg-Lys-Pro-Leu-Glu-Ala-Leu-

     (P)
      |
  16 Tyr-Gly-Tyr-Asp-Tyr-Phe-Ala-Arg-Thr-Cys-Glu-Lys-Trp-Val-Asp-

  31 Gly-Ile-Arg-Gln-Phe-Lys-His-Leu-Pro-Asp-Gly-Asn-Ile-Cys-Arg-

  46 His-Leu-Leu-His-Arg-Val-Gln-Cys-Pro-Ala-Leu-Ile-Val-His-Gly-

  61 Glu-Lys-Asp-Pro-Leu-Val-Pro-Arg-Phe-His-Ala-Asp-Phe-Ile-His-

  76 Lys-His-Val-Lys-Gly-Ser-Arg-Leu-His-Leu-Met-Pro-Glu-Gly-Lys-

  91 His-Asn-Leu-His-Leu-Arg-Phe-Ala-Asp-Glu-Phe-Asn-Lys-Leu-Ala-

                 (P) SO4                 _____
                  |   |
 106 Glu-Asp-Ser-Tyr-Asn-Glu-Asn-Ala-His-Ser-Ser-Leu-Gly-Gly-Ser-
     _____
 121 Phe-Val-Trp-Gly-Leu-Ile-Val-Leu-Leu-Pro-Val-Asn-Met-Met-Pro-

                      (P)
     _____|_
 136 Leu-Lys-Leu-Ser-Ala-Phe-Glu-Thr-Phe-Tyr-Pro-Ser-Leu-Gln-Ser-

 151 Tyr-Pro-Asn-Gln-Met-Arg-Ile-Met-Thr-Tyr
```

Figure 2. The derived amino acid sequence of DL1-800 with T deleted at base 323. Arrow indicated end of BA70 cDNA encoded sequence. (P) indicates potential phosphate group. (SO4) indicates potential sulfated tyrosine. Putative transmembrane domain is overlined.

A smoothed buried helix profile of this sequence using the RAOARGOS program of PCGENE [Intelligenetics, Palo Alto, CA]) indicates the presence of a potential transmembrane domain in the sequence extended by PCR. The same transmembrane domain is present in the 400 base pair sequence. However, if the T at base 323 is present in the 800 base pair sequence the reading frame ends prior to the transmembrane domain. We do not know if the additional T represents a polymorphism in the cells from which these sequences were obtained (Pan-Cl) or is a consequence of the PCR amplification (2 independent clones contained this additional T). We are currently screening a ZR75 cell cDNA library and a genomic library in an attempt to get the full length cDNA of the BA70 antigen. We have previously reported that the BA70 gene is polymorphic with respect to several different restriction enzyme fragment lengths[2].

The sequence obtained so far shows little homology to known sequences; however, it is of interest to note that a short region of homology to Von Willebrand Factor was found extending from residue 44 to 55. The 70 kDa component is associated with the breast mucin by disulfide bonds suggesting that it is a linker protein[4]. A link protein of the gastrointestinal mucin[5] also has homology to Von Willebrand factor.

We have studied the expression of the BA70 mRNA in carcinoma cell lines from breast and other tissues[2]. This data is summarized in table 1. The lowest level of expression was observed in Raji and 3 ER⁻

Table 1. Relative intensities and size of mRNAs in breast and nonbreast tumor cell lines detected with strand specific cloned HMFG cDNA probes

cDNA clone	BREAST EPITHELIAL									Cervix HeLa	Colon HT29	Panc-reas PanC1	Ovary SKOV3	Lung A549	B-Cell Raji	Message Length (Kb.)
	361[3]	331[3]	MCF7[3]	T47D[3]	ZR75[3]	BT20[4]	SKBR3[4]	E11G[4]	HS578T[1,4]							
11-2A	+	++	+	+++	+++	++	+	-	-	-	-	+/-	+	-	-	4.7-6.2
BA70	++	++	++	++	++	+/-	+/-	+/-	++	++	++	++	++	++	+/-	1.8
BA46	++	++	++	+	+	++	-	AMP[2]	AMP	+	++	+	AMP	AMP	-	2.2

1 Not epithelial   2 AMP = Amplified Expression   3 Estrogen Receptor Positive   4 Estrogen Receptor Negative

breast cell lines. The highest expression was in PanCl. Two mRNA species were detected with the BA70 cDNA probe that were 1.8 Kb and a 0.65 kb long. However, the 0.65 kb RNA was not specific for epithelial cells and is not long enough to encode at 70 kDa protein. It therefore most likely represents an altered splicing product or a separate but related gene product.

Cloning of a cDNA Encoding a 46KDa Antigen in HMFG

We used a cocktail of MoAbs that bound the 46 kDa component of HMFG to immunoscreen the human lactating breast λgt11 library described above. We obtained several cDNAs that produced fusion proteins that bound all the MoAbs in the cocktail except Mc3; therefore, Mc3 probably can not bind a peptide epitope or perhaps its epitope was not present in the library. We obtained 1365 base pairs of sequence information from 2 overlapping cDNA clones (figure 3). This sequence represents the 3' end of the complete cDNA since it ends in a poly A stretch preceded by the poly A signal sequence AATATA[6]. The mRNA length shown by northern blot analysis is ⁻2.2 kb. We are currently screening additional libraries in order to obtain the full length sequence. As indicated in figure 3, there are 4 potential N-linked glycosylation sequences in the deduced amino acid sequence. The sequence is asparagine and leucine rich. A comparison of the nucleotide sequence of BA46 with the Gen Bank data base revealed extended homology with human serum factors V and VIII[7] and with protein C. The deduced protein sequence however shares identity only with factors V and VIII (see figure 4) but not with protein C since most of this nucleotide homology lies in an intron[8]. There is also amino acid homology with MFG-E8[9] and discoidin[10].

The function of the 46 kDa component is unknown. The homology with factors V and VIII is in the Cl, C2 region. This region is thought to be involved in the phospholipid binding that is needed for activity of these blood clotting system components[11]. There is also homology with discoidin, a soluble lectin, involved in slime mold aggregation and also shown to bind phospholipid. Recently a 67 kDa component from mouse milkfat globule, MFG-E8, was cloned and sequenced and also shown to have a similar homology to blood clotting factors V and VIII. Our partial sequence of BA46 has 62% identity with MFG-E8 and is therefore likely to be evolutionarily related to the mouse protein.

We studied the expression of the BA46 mRNA in various carcinoma cell lines using labeled BA46 cDNA as probe. In contrast to the BA70 mRNA, BA46 mRNA (2.2Kb) levels varied greatly among the cell lines tested (see table 1). Cell lines HS578T (Breast), E11-G (Breast), A549 (Lung) and SKOV3 (ovary) had much greater BA46 RNA levels than the other cell lines. Raji, a lymphoid cell line, had very low levels of BA46 RNA. The significance of this result is at present unclear; however, over-expression of some genes (i.e., Her2/neu[12] and EGF receptor[13]) in breast and other carcinomas have been correlated with outcome of disease.

Southern blot analysis has shown that the BA46 gene was not amplified in the cell lines in which it was overexpressed. The higher RNA expression levels are therefore a result of increased transcription or stability of the BA46 mRNA. We did not observe any restriction fragment length polymorphisms in the BA46 gene in contrast to what was observed for the BA70 gene.

```
                    10            20            30            40            50
                    |             |             |             |             |
      GAT TTC ATC CAT GAT GTT AAT AAA AAA CAC AAG GAG TTT GTG GGT AAC TGG AAC
      Asp Phe Ile His Asp Val Asn Lys Lys His Lys Glu Phe Val Gly Asn Trp Asn

              60            70            80            90            100
              |             |             |             |             |
      AAA AAC GCG GTG CAT GTC AAC CTG TTT GAG ACC CCT GTG GAG GCT CAG TAC GTG
      Lys Asn Ala Val His Val Asn Leu Phe Glu Thr Pro Val Glu Ala Gln Tyr Val

      110           120           130           140           150           160
      |             |             |             |             |             |
      AGA TTG TAC CCC ACG AGC TGC CAC ACG GCC TGC ACT CTG CGC TTT GAG CTA CTG
      Arg Leu Tyr Pro Thr Ser Cys His Thr Ala Cys Thr Leu Arg Phe Glu Leu Leu

              170           180           190           200           210
              |             |             |             |             |
      GGC TGT GAG CTG AAC GGA TGC GCC AAT CCC CTG GGC CTG AAG AAT AAC AGC ATC
      Gly Cys Glu Leu Asn Gly Cys Ala Asn Pro Leu Gly Leu Lys Asn Asn Ser Ile

        220           230           240           250           260           270
        |             |             |             |             |             |
      CCT GAC AAG CAG ATC ACG GCC TCC AGC AGC TAC AAG ACC TGG GGC TTG CAT CTC
      Pro Asp Lys Gln Ile Thr Ala Ser Ser Ser Tyr Lys Thr Trp Gly Leu His Leu

                    280           290           300           310           320
                    |             |             |             |             |
      TTC AGC TGG AAC CCC TCC TAT GCA CGG CTG GAC AAG CAG GGC AAC TTC AAC GCC
      Phe Ser Trp Asn Pro Ser Tyr Ala Arg Leu Asp Lys Gln Gly Asn Phe Asn Ala

              330           340           350           360           370
              |             |             |             |             |
      TGG GTT GCG GGG AGC TAC GGT AAC GAT CAG TGG CTG CAG GTG GAC CTG GGC TCC
      Trp Val Ala Gly Ser Tyr Gly Asn Asp Gln Trp Leu Gln Val Asp Leu Gly Ser

      380           390           400           410           420           430
      |             |             |             |             |             |
      TCG AAG GAG GTG ACA GGC ATC ATC ACC CAG GGG GCC CGT AAC TTT GGC TCT GTC
      Ser Lys Glu Val Thr Gly Ile Ile Thr Gln Gly Ala Arg Asn Phe Gly Ser Val

              440           450           460           470           480
              |             |             |             |             |
      CAG TTT GTG GCA TCC TAC AAG GTT GCC TAC AGT AAT GAC AGT GCG AAC TGG ACT
      Gln Phe Val Ala Ser Tyr Lys Val Ala Tyr Ser Asn Asp Ser Ala Asn Trp Thr

        490           500           510           520           530           540
        |             |             |             |             |             |
      GAG TAC CAG GAC CCC AGG ACT GGC AGC AGT AAG ATC TTC CCT GGC AAC TGG GAC
      Glu Tyr Gln Asp Pro Arg Thr Gly Ser Ser Lys Ile Phe Pro Gly Asn Trp Asp

              550           560           570           580           590
              |             |             |             |             |
      AAC CAC TCC CAC AAG AAG AAC TTG TTT GAG ACG CCC ATC CTG GCT CGC TAT GTG
      Asn His Ser His Lys Lys Asn Leu Phe Glu Thr Pro Ile Leu Ala Arg Tyr Val

      600           610           620           630           640
      |             |             |             |             |
      CGC ATC CTG CCT GTA GCC TGG CAC AAC CGC ATC GCC CTG CGC CTG GAG CTG CTG
      Arg Ile Leu Pro Val Ala Trp His Asn Arg Ile Ala Leu Arg Leu Glu Leu Leu

  649 GGC TGT TAG TGG CCA CCT GCC ACC CCC AGG TCT TCC TGC TTT CCA TGG GCC CGC
      Gly Cys ---

  703 TGC CTC TTG GCT TCT CAG CCC CTT TAA ATC ACC ATA GGG CTG GGG ACT GGG GAA
  757 GGG GAG GGT GTT CAG AGG CAG CAC CAC CAC ACA GTC ACC CCT CCC TCC CTC TTT
  811 CCC ACC CTC CAC CTC TCA CGG GCC CTG CCC CAG CCC CTA AGC CCC GTC CCC TAA
  865 CCC CCA GTC CTC ACT GTC CTG TTT TCT TAG GCA CTG AGG GAT CTG AGT AGG TCT
  919 GGG ATG GAC AGG AAA GGG CAA AGT AGG GCG TGT GGT TTC CCT GCC CCT GTC CGG
  973 ACC GCC GAT CCC AGG TGC GTG TGT CTC TGT CTC TCC TAG CCC CTC TCT CAC ACA
 1027 TCA CAT TCC CAT GGT GGC CTC AAG AAA GGC CCG GAA GCC CCA GGC TGG AGA TAA
 1081 CAG CCT CTT GCC CGT CGG CCC TGC GTC GGC CCT GGG GTA CCA TGT GCC ACA ACT
 1135 GCT GTG GCC CCC TGT CCC CAA GAC ACT TCC CCT TGT CTC CCT GGT TGC CTC TCT
 1189 TGC CCC TTG TCC TGA AGC CCA GCG ACA CAG AAG GGG GTG GGG CGG GTC TAT GGG
 1243 GAG AAA GGG AGC GAG GTC AGA GGA GGG CAT GGG TTG GCA GGG TGG GCG TTT GGG
 1297 GCC CTC ATG CTG GCT TTT CAC CCC AGA GGA CAC AGG CAG CTT CCA AAA TAT ATT
 1351 TAT CTT CTT CAC GGG AAA AAA AAA AAA AAA ACC G
```

Figure 3. DNA sequence and derived amino acid sequence of BA46-1 cDNA. Potential n-linked glycosylation are underlined.

```
46kDa   F I H D V N K K K H K E F V G N W N K N A V H V N L
FAV     F K G N S T R N V M Y F N G N S D A S T I K E N N Q
FAVIII  Y R G N S T G T L M V F F G N V D S S G I K H N I

        F E T P V E A Q Y V R L Y P T S C H T A C T L R F E L L G
        F D P P I V A R Y I R I S P T R A Y N R P T L R L E L Q G
        F N P P I I A R Y I R L H P T H Y S I R S T L R M E L M G

        C E L N G C A N P L G L K N N S I P D K Q I T A S S S Y K
        C E V N G C S T P L G M E N G K I E N K Q I T A S S F K K
        C D L N S C S M P L G M E S K A I S D A Q I T A S S Y F T

        T W G L H L F S W N P S Y A R L D K Q G N F N A W V A G S
        S W W G D Y - - W E P F R A R L N A Q G R V N A W Q A K A
        N M F A T - - - W S P S K A R L H L Q G R S N A W R P Q V

        Y G N D Q W L Q V D L G S S K E V T G I I T Q G A R N F G
        N N N K Q W L E I D L L K I K K I T A I I T Q G C K S L S
        N N P K E W L Q V D F Q K T M K V T G V T T Q G V K S L L

        S V Q F V A S Y K V A Y S N D S A N W T E Y Q D P R T G S
        S E M Y V K S Y T I H Y S E Q G V E W K P Y R L K S S M V
        T S M Y V K E F L I S S S Q D G H Q W T L F F Q N - - G K

        S K I F P G N W D N H S H K K N L F E T P I L A R Y V R I
        D K I F E G N T N T K G H V K N F F N P P I L S R F I R V
        V K V F Q G N Q D S F T P V V N S L D P P L L T R Y L R I

        L P V A W H N R I A L R L E L L G C
        I P K T W N Q S I A L R L E L F G C D - - - I Y
        H P Q S W V H Q I A L R M E V L G C E A Q D L Y
```

Figure 4.   Comparison of the derived BA46-1 amino acid sequence with the C-terminal sequence of human serum factors V and VIII Arrow indicates junction of C1 and C2 repeats.

We have also examined the expression of BA46 mRNA in carcinoma cell lines that were maintained as tumors in nude mice. The northern blot analysis showed the same variation in expression in the tumors as we observed in the cell lines. These data will allow us to examine how the expression of the BA46 might correlate with effectiveness of experimental therapy in these nude mice using anti-46KDa component monoclonal antibodies.

We have compared the relative expression of the BA46 and BA70 genes with the breast mucin expression in the same carcinoma cell lines. The results are summarized in Table 1. The rank order of expression of these 3 genes in the cell lines tested are clearly different. For example, the mucin mRNA appears to be highest in ER^+ cell lines, where this is not the case for BA46 and BA70. These data indicate that the 3 major components of milk fat globule respond independently to regulatory signals in tumor cells. This is important for devising therapeutic and diagnostic strategies since in cases where one of these antigens is low or absent, one or both of the other 2 antigens could be high.

Expression of HMFG Antigens at High Levels in E. Coli

In order to produce purified HMFG antigens for immunizations (to make 2nd generation MoAbs) and for use in serum assays[14], we inserted cloned cDNAs encoding the BA70, BA46 and mucin antigens into the expression vector pEX2 as originally described by Stanley and Luzio[15]. The cDNAs inserted into pEX2 are fused with B-galactosidase. The fusion proteins produced are packaged into inclusion bodies that can be easily purified by methods described by Marston[16]. A typical fusion protein inclusion body preparation is shown in figure 5. The vast majority of the fusion protein remains in the pellet following repeated washes in 0.5% Triton X-100, while the remaining E. Coli proteins are found in the supernatant. Using this system we have routinely obtained over 50 mg of fusion protein per liter of starting culture.

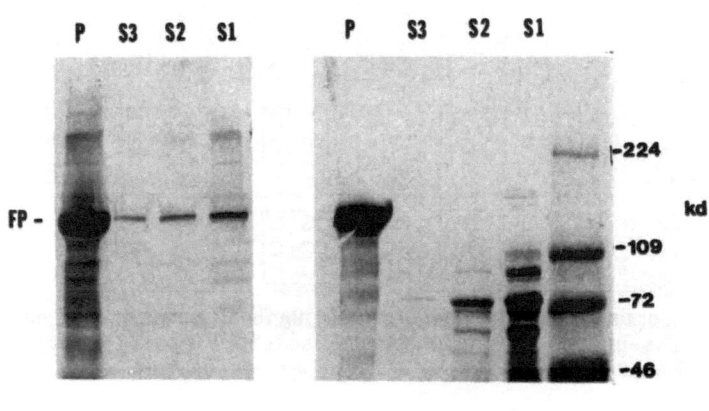

Figure 5.   Preparation of pEX/LB21 fusion protein. Western blot and coomassie stained PAGE analysis of pEXLB21 inclusion body preparation. Lysed cell pellet from 1 liter of induced (42C° 1 hour, 37C° 1 hour) N4830 cultures was centrifuged at 12,000 x g for 5 min. to obtain supernatant 1 (S1); repelleted and washed 2 x in 0.5% Triton X-100 to obtain supernatants 2 and 3 (S2 and S3); the pellet, (P), from S3 was suspended in 1x leammle sample buffer and loaded on the gel. FP = fusion protein. Molecular weight size standards are given in kilodaltons (kd) on right hand side. Yield from 1 liter of cells was > 50 mg. Similar results were obtained for pEX/DL-1 and pEX/NP5 fusion proteins.

Conclusions

We have successfully cloned partial cDNAs encoding 70KDa and 46KDa antigen components of HMFG. The mucin, BA70, and BA46 are encoded by 3 distinct genes that appear to be independently regulated in tumor cells. The BA46 gene is highly over-expressed in certain tumor cell lines relative to others, but this is not the result of gene amplification. We have obtained high level expression of the 3 major antigen components of HMFG as recombinant B-galactosidase fusion proteins in E. Coli. The fusion proteins are being used to make second generation antibodies in order to further characterize these antigens. We have also developed a novel assay for HMFG antigens in patient sera using the recombinant HMFG antigens[13]. Further studies are in progress to explore the use of BA70 and BA46 as tumor markers and targets for radioimmune therapies.

ACKNOWLEDGEMENTS

This work was funded by NIH Grant Nos. CA39932, CA42767

REFERENCES

1. R.L. Ceriani, K. Thompson, J.A. Peterson and S. Abraham. Surface differentiation antigens of human mammary epithelial cells carried on the human milk fat globule. Proc. Natl. Aca. Sci. USA, 74:582-586 (1977).

2. D. Larocca, J.A. Peterson, G. Walkup, R. Urrea, and R.L. Ceriani, Cloning and sequencing of a complementary DNA encoding a Mr 70,000 human breast epithelial mucin-associated antigen. Cancer Res., 58:5925-5930 (1990).

3. M.A. Frohman, PCR Protols: A guide to methods and applications, Academic Press, New York, pp. 28 (1990).

4. A.K. Duwe and R.L. Ceriani, Human milk fat globule membrane derived mucin is a disulfide-linked heteromer. Biochem. Biophys. Res Commun., 165:1305-1311 (1989).

5. A.M. Robertson, M. Mantle, R.E. Fahim, R.D. Specian, A. Bennick, S. Kawagishi, P. Sherman and J.F. Forstner. The putative "link" glycopeptide associated with mucus glycoproteins. Composition and properties of preparations from the gastrointestinal tracts of several mammals. Biochem. J., 261:637-647 (1989).

6. M.L. Birnstiel, M. Busslinger, and K. Strub, Transcription termination and 3' processing: The end is in site! Cell, 41:349-359 (1985).

7. W.H. Kane and E.W. Davie. Cloning of a cDNA coding for human factor V, a blood coagulation factor homologous to factor VIII and ceruloplasmin. Proc. Natl. Acad. Sci. USA, Vol. 83, pp6800-6804, Biochemistry (1986).

8. J. Plutzky, J. Joskins, G. Long and G.R. Crabtree. Evolution and organization of the human protein C gene. Proc. Natl. Acad. Sci. USA, Vol 83, pp. 546-550, Biochemistry (1986).

9. J.D. Stubbs, C. Lekutis, K.L. Singer, A. Bui, D. Yuzuki, U. Srinivasan, and G. Parry, cDNA cloning of a mouse mammary epithelial cell surface protein reveals the existence of epidermal growth factor-like domains linked to factor VIII-like sequences. Proc. Natl, Acad. Sci. USA, 87:8417-8421 (1990).

10. S. Poole, R.A. Firtel, E. Lamar and W. Rowekamp, Sequence and Expression of the Discoidin I gene family in Dictyostelium discoideum., J. Mol. Biol., 153: 273-289 (1981).

11. M. Arai, D. Scandella, and L.W. Hoyer. Molecular basis of factor VIII inhibition by human antibodies. Antibodies that bind to the factor VIII light chain prevent the interaction of factor VIII with phospholipid. J. Clin. Invest., 83:1978-1984 (1984).

12. D.J. Slamon, W. Godolphin, L.A. Jones, J.A. Holt, S.G. Wong, D.E. Keith, W.J. Levin, S.G. Stuart, J. Udove, A. Ullrich, and M.F. Studies of the HER-2/neu proto-oncogene in human breast and ovarian cancer. Press Science, 244: 707-712 (1989).

13. N.E. Davidson, Gelmann, E.P., Lippman, M.E., Dickson, R.B., Epidermal growth factor receptor gene expression in estrogen receptor-positive and negative human breast cancer cell lines. Mol. Endocrinol., 1, 216-223 (1987).

14. R.L. Ceriani, D. Larocca, J.A. Peterson, S. Enloe, R. Amiya, G. Walkup, and E.W. Blank. A novel serum assay using recombinant breast epithelial mucin antigen. (in this volume)

15. K. Stanley, and J. Luzio. Construction of a new family of high efficiency bacterial expression vectors: Identification of cDNA clones coding for human liver proteins. Embo J., 3, 1429-1434 (1984).

16. Marston, F.A.O. The purification of eucaryotic polypeptides expressed in E. Coli. In: DNA Cloning, IRL Press, England, Vol. 3:59 (1987).

MONOCLONAL ANTIBODIES REACTIVE WITH BREAST CANCER, MAMMARY MUCINS
AND SYNTHETIC PEPTIDES

Pei-xiang Xing and Ian F.C. McKenzie

Research Centre for Cancer and Transplantation
Department of Pathology, University of Melbourne
Parkville, Victoria, 3052, Australia

INTRODUCTION

Until recently most monoclonal antibodies to breast cancer were
made by injecting whole tumours, crude tumour extracts or human milk fat
globule membrane (HMFG) and selecting antibodies on the basis of a
preferential reaction with the tumour and lesser reactions on normal
tissues (1-5). A number of antibodies so produced in breast cancer have
been useful in development of serum tests, imaging and for therapy (6-
10). Nonetheless, the approach was crude, although a number of useful
reagents have been produced. By analysis of many different antibodies
produced around the world, it appears that for breast cancer, more than
90% of the antibodies react with mammary mucins (11). Further, it
appears that these antibodies could react either with carbohydrate
determinants in whole mucin (12), or on deglycosylated material, with the
protein core (13); some antibodies appear to require the presence of
both carbohydrate and protein (14,15,16). A major breakthrough in the
production of monoclonal antibodies to breast cancer came with the
isolation of cDNA clones coding for part of the protein core (17,18),
and later the whole cDNA sequence was obtained from different
laboratories (19,20). One feature of this work was the isolation of
small sequences in λgt11 expression libraries using monoclonal
antibodies - an unusual finding until it was realised that antibodies
were able to react with small peptides encoded by the small cDNA inserts
(see below). As discussed extensively elsewhere in this volume, a
remarkable feature of the protein core of the mammary mucin was the
presence of VNTR (variable number of tandem repeats), where a 60 bp
repeat sequence was repeated 40-80 times (19, 20). It is with peptides
in this sequence that most of the anti-protein core antibodies react.
The amino acid sequence of the N-terminal and C-terminal to the VNTR is
now known and studies are being performed to determine whether these
peptides are immunogenic; however, most of the current information
involves the 20 amino acids (encoded by the 60 bp of the VNTR) and these
will be discussed herein.

MATERIALS AND METHODS

Monoclonal antibodies were produced by standard techniques published
elsewhere (3); peptides were synthesised by two methods - using the
solid phase ABI synthesiser and, secondly, using the "pepscan" method
wherein 6-8-mer peptides are synthesised on pins (21). In the first
method the peptides are cleaved for subsequent testing; in the pin
method the peptide remains attached to the pin for subsequent ELISA
tests. Testing of antibodies with the peptides was based on ELISA tests
using peptides in four different configurations:

Breast Epithelial Antigens, Edited by R.L. Ceriani
Plenum Press, New York, 1991

(a) peptide in solid phase bound to the wells of a microtiter plate;
(b) small peptides conjugated to bovine serum albumin (BSA) with glutaraldehyde similarly adhere to the wells of the plate;
(c) peptides used in liquid phase for inhibition assays;
(d) peptides synthesised on the pins in solid phase.

It will be noted below that minor differences were obtained using the peptides in these different forms. The methods of performing the ELISA tests, immunohistochemistry are described elsewhere (3,22).

RESULTS AND DISCUSSION

1. <u>Antibodies react with small synthetic peptides</u>: These results have been published extensively elsewhere including in the preceding volume (23) and will only be presented briefly. Synthetic peptides of the 20-mer were produced and as it was not known what was the appropriate "starting" amino acid, a 24-mer was produced which runs into the next repeat. It is clear that the 24-mer of the sequence (PDTRPAPGSTAPPAHGVTSAPDTR) was reactive using a variety of techniques, and that the reactivity was found in the "right" 9 amino acids and not in the "left" 15 amino acids. However, the N-terminal 15 amino acids (P1-15) could be converted to reactivity by adding an additional N-terminal alanine (A-P1-15) and suggests that APDTR is part of the reactive epitope. This was confirmed by producing smaller peptides such as a synthetic 5-mer APDTR which was reactive with the antibodies, but only when bound to BSA or using the soluble phase inhibition method - presumably APDTR itself is too small to bind to plastic or the conformation changes when the peptide binds to plates.

2. <u>Antibodies react with 5 amino acids</u>: As indicated above, APDTR appeared to be the reactive epitope (22), and by using overlapping peptides synthesised on pins this was confirmed (Figure 1). In this method peptides are synthesised so that each new peptide moves along the 20-mer sequence, e.g. the first PDTRPA, second DTRPAP, the third TRPAPS and so on. The 5-mer APDTR and 4-mer PDTR were also made on the pins to confirm the minimum epitope recognized by mAb BC2, APDTR was reactive; PDTR was not (Figure 1). Using the pin method, it was clear that the reactive epitope was APDTR.

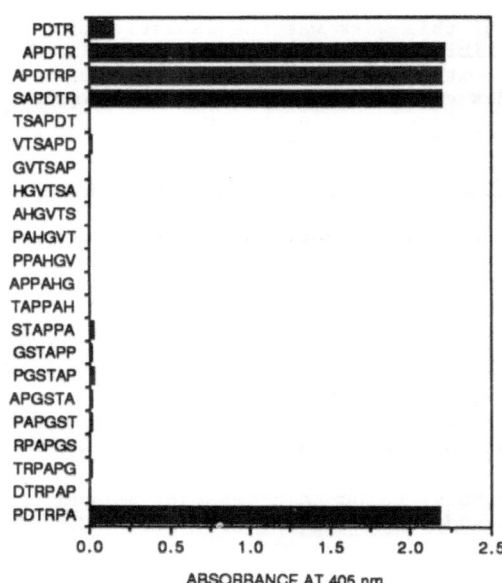

Figure 1. Six-mer overlapping peptides were synthesised on pins (starting PDTRPA - bottom left) and tested using mAb BC2 by ELISA; PDTR and APDTR were also made. Shown is the strong reaction with APDTR, APDTRP, SAPDTR and PDTRPA but not with PDTR.

3.   A single amino acid substitution alters antibody binding: It was
of interest to determine which amino acids in the APDTR sequence formed
a crucial part of the epitope. It was known from more classical studies
using well defined antigens such as lysozyme and myoglobin, that a
minimum of 3 amino acids were required to form an epitope but more
usually 5-8 amino acids - as indicated here with APDTR. It was of
interest, therefore, to determine the contribution of each amino acid to
the epitope. The pin method proved to be very useful where, in the 6-
mer SAPDTR, each amino acid could be substituted with 19 others making
6x20=120 different peptides. These were synthesised on the pins and
results shown (Figure 2). It was clear that both proline (P) and
arginine (R) were crucial amino acids in the epitope of BC2 for if they
were substituted by any other amino acid activity was totally lost.
This could indicate these are key contact residues, however, both have
substantial effects on folding (proline) or charge (arginine) which
could also alter folding, and it is not clear how these two amino acids
participate in the epitope. There was some flexibility with alanine (A)
which could be substituted with related amino acids P, S and with
aspartic acid (D) which could also be substituted with asparagine (N).
There was complete flexibility with threonine which could be substituted
with virtually any amino acid. Either threonine is not part of the
epitope at all - making APDTR "discontinuous"; or, a more likely
possibility, threonine is O-glycosylated and that the amino acid itself
does not participate in the epitope. It is also possible that the
attached carbohydrate is part of the epitope. At present we have no
means of confirming this as the precise sugars present are unknown.

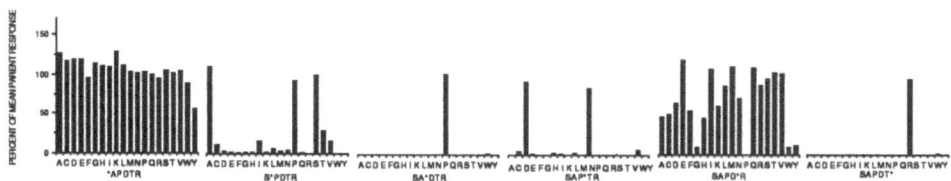

Figure 2. Each bar in the figure represents the ELISA result recorded
as the percent of mean parent response of peptides with mAb BC2 (i.e.
given by SAPDTR = 100%). Every amino acid in SAPDTR was substituted for
each peptide - indicated thus *.

In addition to substitution of a single amino acid the effect of
altering 3 or 4 "permissable" amino acids (S, A, D or T) was examined
(Figure 3). For example, the peptide VPPNQR retains only P and R of
SAPDTR, but gives the same reaction. By contrast, CPPDIR, also
retaining P and R, has no reactivity. It was noted that in most
peptides three substitutions could be made but in several, a combination
of peptides resulted, which could no longer bind antibody (e.g. ASPDER,
CPPDIR). It was apparent that fairly drastic alterations in the
structure could be made and activity retained (Figure 3).

4.   Conformation of the peptide epitope is important: In all of the
studies outlined above, the tacit assumption has been made that as the
linear peptide reacts with antibody, and then it is the primary sequence
which is important, and this included some discussion on whether the
epitopes were continuous (i.e. adjacent amino acids) or whether some
discontinuity was permissible. However, even small peptides are able to
form secondary and tertiary structures. With this in mind, we examined
the basic epitope APDTR either at the beginning of the peptide, eg. A-
P1-15; at the end, e.g. P1-24 or in the middle such as in P13-32
(PAHGVTSAPDTRPAPGSTAP) (Figure 4A and B). It was clear that by

comparing the binding of antibody to these peptides that significantly
more BC2 antibody could be bound when APDTR was in the middle of the
peptide, i.e. the position of APDTR is important, and from this we
conclude that the secondary structure developed alters antibody binding.
(We did not quantitatively measure the different amounts of these
peptides bound; it is possible that the differences seen were a
reflection of the different amounts of peptide bound in the plate). To
take this further, peptide TSA p1-24 was synthesised - it has two APDTR
epitopes and is a larger structure which could develop folding and
intrapeptide bonds. By chromatography it was shown to bind two
molecules of IgG antibody. However, of greater importance was the
demonstration in direct binding assays that significantly more antibody
bound this peptide than would be expected (Figure 4A). We conclude
albeit indirectly, that both the linear sequence, i.e. primary
structures of peptide and secondary and tertiary structures are also
important in the binding of antibody to the peptide.

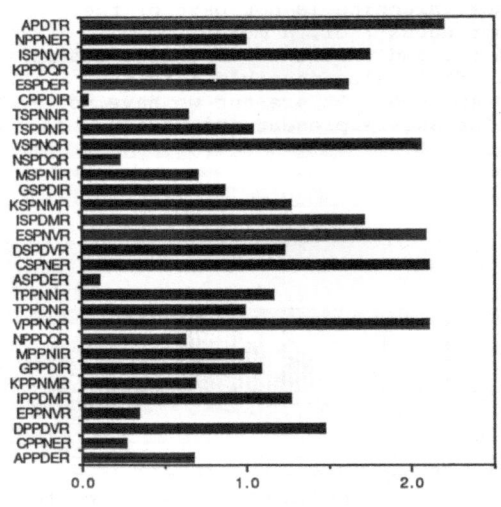

ABSORBANCE AT 405 nm

Figure 3. Twenty-nine 6-mer
peptides were synthesized on
the pins; these were
designed by choosing the
"permissible" amino acid
substitutions (S, A, D, T
but not P or R) of peptide
SAPDTR. ELISA tsting used
mAb BC2. APDTR was also
synthesized and tested as
positive control. As shown
many modified peptides
retained activity.

5.   In using peptides different methods give slightly different
results: As shown in Figures 1,4 and 5 the same peptide synthesised on
pins, or used in soluble form, or for inhibition studies give slightly
different results. Thus the sequence APDTR, when conjugated with BSA or
used in inhibition studies, is clearly the epitope reacting with the BC2
antibody. However, for the overlapping 6-mer peptides synthesised on
pins the epitope of mAb BC2 appears to be PDTR. The overlapping
peptides were examined; SAPDTR, APDTRP and PDTRPA were all equally
reactive (Figure 1) The common peptide is the 4-mer PDTR. However,
PDTR alone (Figure 1) is not reactive ~ nor was PDTR inhibitory. We
conclude that PDTR requires adjacent amino acids for its reactivity.
This is of no real significance, and it is important not to read too
much into the finer points of the results which could be merely due to
the different techniques used. For example, peptides on pins have one
end attached to the pin- the rest of the peptide is free to form

48

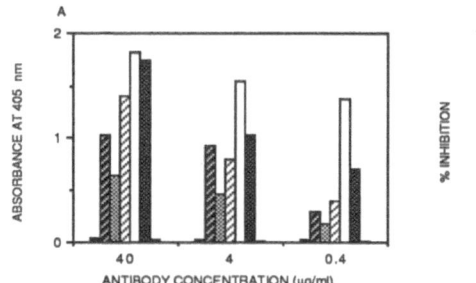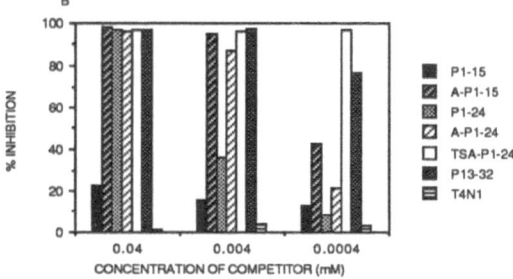

Figure 4. mAb BC2 reacting with VNTR peptides and T4N1 (negative control) using different concentrations of (A) antibody in a direct binding test using an ELISA based assay; (B) peptide in an inhibition assay (22), where the peptide in solution inhibited the BC2 binding to HMFG.

secondary structures - as opposed to peptides on plates where we assume that only a linear peptide is present.

6.   Are the Studies Unique? The results presented thus far are of interest and show that: (i) antibodies made against tumours can react with peptides; (ii) the length of the peptide can be quite small (3-5 amino acids); (iii) the alteration of a single amino acid can alter antibody binding; (iv) contact residues can be determined and others shown to be important in folding; and (v) the hydrophilic regions predicted to be immunogenic (by Hopp and Woods type analysis; 24) are indeed so. This is, indeed, a major advance in cancer immunobiology - from producing crude antibodies against crude mixtures of tumours to monoclonal antibodies which recognise a small epitope which can be easily defined and indeed used for further immunization (see below). However, several points should be made. Firstly, all of the above conclusions have previously been made with great clarity using defined antigens. None of the conclusions are, therefore, unique nor surprising. However, for those working in the cancer area with monoclonal antibodies, such a fine mapping of protein epitopes represents a real advance. The second point to make is that although synthetic peptides have been used, the role of carbohydrate in many of these reactions is mostly unknown. It is clear that with mucins many antibodies react solely with carbohydrate - as deglycosylating material leads to a total loss of reactivity. On the other hand, the anti-HMFG antibody BC2, while reacting with peptide, could give even better reactions if some carbohydrate was present on the peptide, for example, they react substantially better with HMFG than with peptide - although it is difficult to quantitate this because of the heterogeneous nature of HMFG. We are currently seeking means of artificially glycosylating the synthetic peptides to determine the role of carbohydrate in the reactions observed. What practical use can now be made of the data obtained thus far? The work has gone in several directions. Firstly, production of "second generation" anti-peptide antibodies. Secondly, examining the relationship between mucins in colon and breast cancer; and finally relating the antibody findings to those found with T cells.

7.   Second generation monoclonal antibodies: Synthetic peptides have now been used to produce another round of antibodies in two ways. Firstly, by immunising with HMFG and selecting by screening against peptides so that only anti-peptide antibodies are isolated. We have produced a number of these as characteristics have shown in Table 1. At this time, these antibodies do not appear to be substantially different from those produced by immunising and screening against HMFG and breast cancer and subsequently finding out that they reacted with peptides. In

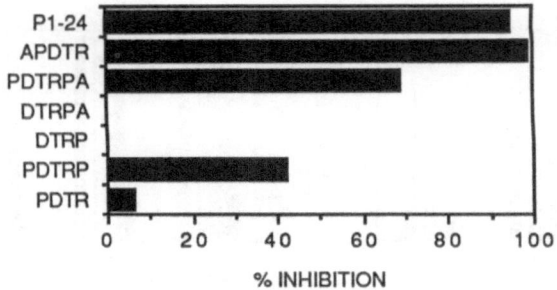

Figure 5. Six short peptides (consisting of 4-6 amino acids) were made to map the epitope of mAb BC2 using a liquid phase inhibition test. Peptides and BC2 were used at 0.1 mM and 1 µg/ml respectively.

the second approach, peptides have been used for immunisation – particularly the C-P13-32 peptide conjugated to Kehole limpet hemocyanin (KLH) with glutaraldehyde was found to be immunogenic in both mice and rabbits and led to the production of a series of monoclonal antibodies (Table 1) and rabbit polyclonal antisera. Again, at this early stage of testing, these antibodies do not appear to be substantially different to those produced by immunising with HMFG. Whether they will be superior reagents for diagnostic purposes (both *in vitro* and *in vivo*) or for therapeutic purposes remains to be seen.

8.   Relationship of colon and breast mucins: Recently cDNA clones coding for tandem repeats found in colon mucin have been isolated (25) and using the deduced amino acid sequence, we have produced monoclonal antibodies to synthetic peptides derived from this sequence. These antibodies are currently undergoing characterisation, but react in immunoperoxidase testing on colon cancer and give a differential staining pattern on normal tissue compared with a stronger reaction on formalin fixed carcinoma tissues. Of interest is the finding that these antibodies have no reaction with breast cancers. By contrast the anti-breast peptide antibodies described above react with both colon and breast cancer tissue. We can make a preliminary conclusion that mucins containing APDTR peptide found in breast tissue, are also found in colon cancers; by contrast one of the mucins found in colon cancer is not found in breast cancer. We would note further that the protein core of the breast cancer mucin has also been found in pancreatic and ovarian lesions (unpublished data) and the antibodies certainly react with other tissues such as lung and kidney. It remains to be seen whether there are such components as tissue specific mucins - perhaps the Gum *et al* (25) mucin is colon specific.

9.   The relationship of B cell and T cell epitopes: While the studies described herein are of interest, they basically demonstrate that the peptide APDTR is immunogenic in mice and has led to a series of interesting antibodies which could be useful for diagnostic or therapeutic purposes. However, it is important to note a recent link with the human immune response to breast cancer has been made by Dr Olivera Finn in Department of Immunology, Duke University. Dr Finn and colleagues have produced a series of T cell lines from the draining lymph nodes of patients with breast cancer (made by exposing the lines to different breast and pancreatic tumours as a source of antigen and stimulating continuously with Il-2) (26), and several T cell lines were produced which proliferate and are cytotoxic to breast cancer cells (apparently in the absence of HLA Class I restriction). What is of

Table 1

CHARACTERISTICS OF SECOND GENERATION ANTIBODIES

	HMPV[a]	BCP8[b]
IMMUNOGEN	HMFG	C-P13-32
p1-24	+	+
p13-32	+	+
HMFG	+	+
MI-29[c]	-	-
T4N1	-	-
Epitope	APDTR	DTR
Subclass	IgG1	IgG1
Normal Tissues		
breast	3/5[d]	7/9
kidney	6/6	2/2
pancreas	2/2	3/3
colon	0/7	1/6
lung	1/2	2/2
liver	0/1	0/2
spleen	0/1	0/1
stomach	4/4	4/4
brain	0/1	0/1
Lymphocyte	0/3	0/3
muscle	0/5	0/5
prostate	2/2	0/2
parathyroid	0/1	0/1
Carcinoma		
breast ca.	31/32	21/22
kidney ca	4/4	3/3
pancreas ca.	2/2	1/2
stomach ca.	4/4	3/4
colon ca.	4/7	7/14
lymphoma	0/1	0/1
neuroblastoma	0/1	0/1
nasopharyngeal ca.	0/1	0/1
Benign breast lesion	3/3	8/8

a. HMPV was produced by immunizing BALB/c mice, and their spleen cells boosted in vitro using the synthetic peptide C-p13-32 (Cysteine added to the N-terminal of p-13-32).

b. mAb BCP 8 was produced by immunizing BALB/c mice using C-p13-32-KLH (Cysteine was added to the N-terminal of p13-32) and screened using C-p13-32 in an ELISA test.

c. MI-29 is intestinal mucin peptide (KYPTTTPISTTTMVTPTPTPTGTQTPTTT) (24).

d. No. of positive/total tested.

interest, is that the antibodies of the type described herein can block the reaction between the cytotoxic T cell lines and the target cancer cell (26). Furthermore, Dr Finn has recently found that the synthetic APTDR containing peptides, in an appropriate antigen presentation model, can cause proliferation of the cell lines (O. Finn, personal communication). This is a finding of major importance for several reasons. Firstly, it demonstrates that the peptides which are immunogenic in mice are also immunogenic in patients. How or why patients should have such cells in the lymph nodes is not clear; nor is it clear whether these T cells have any influence on the occurrence or growth of the tumour is unknown, possibly killing some cells, or perhaps they stay in the lymph node and merely observe what is happening. Secondly, the findings in mice can now be used as a direct model of what is happening in humans as the same peptides appear to be immunogenic in both species. However, it will be noted that all of the studies presented herein refer to "B cell epitopes" wherein Dr Finn's work should refer to T cell epitopes. In this light one would expect amphipathic structures to be more likely to stimulate T cells (in the P1-24 sequence this is more likely to be PGSTAPPA than APDTR). Such studies are currently being performed by Dr Finn. However, the observations raise the intriguing possiblity that patients can be immunised against the mucins - particularly against the protein core and indeed the peptides described herein could form the basis of immunotherapeutic approach to breast cancer. However, it should be recalled that such peptides are indeed present in breast cancer cells which have been used in various vaccines, and it should also be noted that such peptides are present in normal tissues such as pancreas and kidney and could be sites for attack by either lymphocytes or antibodies if the patients were deliberately immunised.

SUMMARY

Most antibodies made to breast cancer are found to react with mammary mucins and on further examination the antibodies can be divided into those reacting predominantly with carbohydrate and those reacting predominantly with peptide. The anti-peptide antibodies can now be clearly defined following the cDNA cloning of the protein core of the mucins. Within this protein core are the VNTRs (variable number of tandem repeats) which contains the highly immunogenic sequence APDTR which appears to be the prime target of the anti-breast cancer antibodies which react with peptides. Taking this observation further, it is now possible to make "second generation" antibodies by immunising with synthetic peptides. Antibodies have been produced which give satisfactory reactions on breast cancer. Thus, the field has advanced considerably from having crude antibody mixtures made to crude mixtures of cancer cells, to immunising with peptides and making monoclonal anti-peptide antibodies. However, the value of these new reagents to patients with breast cancer remains to be seen. A recent advance demonstrating that patients have T lymphocytes which also react with the same peptide, indicates that the murine studies may pave the way for the development of important therapeutic reagents for patients with breast cancer.

REFERENCES

1.   S.A. Stacker, C.H. Thompson, S. Riglar and I.F.C. McKenzie. A new breast carcinoma antigen defined by a monoclonal antibody. J. Natl. Cancer Inst. 75:801-811 (1985).
2.   C.H. Thompson, S.L. Jones, R.H. Whitehead and I.F.C. McKenzie. A human breast tissue-associated antigen detected by a monoclonal antibody. J. Natl. Cancer Inst. 70:409-419 (1983).
3.   P.X. Xing, J.J. Tjandra, S.A. Stacker, J.G. Teh, C.H. Thompson, P.J. McLaughlin and I.F.C. McKenzie. Monoclonal antibodies reactive with mucin expressed in breast cancer. Immunol. Cell. Biol. 67:183-185 (1989).
4.   J. Taylor-Papadimitriou, J.A. Peterson, J. Burchell, J. Arklie, R.L. Ceriani and W.T. Bodmer. Monoclonal antibodies to epithelium specific components of the milk fat globule membrane: production and reactions with cells in culture. Int. J. Cancer 28:17-21 (1981).
5.   R.L. Ceriani, J.A. Petersen, J.Y. Lee, R. Moncada and E.W. Blank. Characterisation of cell surface antigens of human mammary epithelial cells with monoclonal antibodies prepared against human milk fat globule. Somat. Cell Genet. 9:415-427 (1983).
6.   R.L. Ceriani, M. Sasaki, H. Sussman, W.M. Wara and E.W. Blank. Circulating human mammary epithelial antigen in breast cancer. Proc. Natl. Acad. Sci. 79:5420-5424 (1982).

7. S.A. Stacker, N.P.M. Sacks., C.H. Thompson, C. Smart, R. Burton, J. Bishop, J. Golder, P.X. Xing and I.F.C. McKenzie. A serum test for the diagnosis and monitoring the progress of breast cancer. In: "Immunological Approaches to the Diagnosis and Therapy of Breast Cancer". R.L. Ceriani, ed., Plenum Press, New York. (1987).

8. D.F. Hayes, H. Sekine, T. Ohno, M. Aber, K. Keefe and D.W. Kufe. Use of a murine monoclonal antibody for detection of circulating plasma DF3 antigen levels in breast cancer patients. J. Clin. Invest. 75:1671-1678 (1985).

9. J.J. Tjandra, I.S. Russell, J.P. Collins, J.T. Andrew, M. Lichtenstein, D. Binns and I.F.C. McKenzie. Immunolymphoscintigraphy for the detection of lymph node metastases from breast cancer. Cancer Res. 49:1600-1608 (1989).

10. R.L. Ceriani and E.W. Blank. Experimental therapy of human breast tumours with ^{131}I-labelled monoclonal antibodies prepared against the human milk fat globule. Cancer Res. 48:4664-4672 (1988).

11. I.F.C. McKenzie and P.X. Xing. Mucins in breast cancer: Recent immunological advances. Cancer Cells, 2:75-78 (1990).

12. J. Hilkens, F. Buijs and M. Lightenberg. Complexity of MAM-6, an epithelial sialomucin associated with carcinomas. Cancer Res., 49:786-793 (1989).

13. J. Burchell, S. Gendler, J. Taylor-Papadimitriou, A. Girling, A. Lewis, R. Millis and D. Lamport. Development and characterization of breast cancer reactive monoclonal antibodies directed to the core protein of the human milk micin. Cancer Res., 47:5476-5482 (1987).

14 P.X. Xing, J.J. Tjandra, K. Reynolds, P.J. McLaughlin, D.F.J. Purcell and I.F.C. McKenzie. Reactivity of anti-human milk fat globule antibodies with synthetic peptides. J. Immunol. 142:3503-3509 (1989).

15. M. Abe and D. Kufe. Structural analysis of the DF3 human carcinoma-associated protein. Cancer Res., 49:2834-2839 (1989).

16. J. Burchell, J. Taylor-Papadimitriou, M. Boshell, S. Gendler and T. Duhig. A short sequence, within the amino acid tandem repeat of a cancer-associated mucin, contains immunodominant epitopes. Int. J. Cancer, 44:691-696 (1989).

17. S.J. Gendler, J.M. Burchell, T. Duhig, D. Lamport, R. White, M. Parker and J. Taylor-Papadimitriou. Cloning of partial cDNA coding differentiation and tumour-associated mucin glycoproteins expressed by human mammary epithelium. Proc. natl. AcAd. Sci. 84:6060-6064 (1987).

18. J. Siddiqui, M. Abe, D. Hayes, E. Shani, E. Yunis and D. Kufe. Isolation and sequencing of a cDNA coding for the human DF3 breast carcinoma associated antigen. Proc. Natl. Acad. Sci. USA 85:2320-2323 (1988).

19. D.H. Wreschner, M. Hareuveni, I. Tsarfaty, N. Smorodinsky, J. Horev, J. Zaretsky, P. Kotkes, M. Weiss, R. Lathe, A. Dion and I. Keydar. Human epithelial tumor antigen cDNA sequences - differential splicing may generate multiple protein forms. Eur. J. Biochem. 189:463-474 (1990).

20. M.J.L. Lightenberg, H.L. Vos, A.C.M. Gennissen and J. Hilkens. Episialin, a carcinoma-associated mucin, is generated by a polymorphic gene encoding splice variants with alternative amino-termini. J. Biol. Chem. 265:5573-5578 (1990).

21. H.M. Geyson, S.J. Rodda, T.J. Mason, G. Tribbick and P.G. Schoofs. Strategies for epitope analysis using peptide synthesis. J. Immunol. Meth. 102:259-274 (1987)

22. P.X. Xing, K. Reynolds, J.J. Tjandra, X.L. Tang and I.F.C. McKenzie. Synthetic peptides reactive with anti-human milk fat globule membrane. Cancer Res. 50:89-96 (1990).

23. P.X. Xing, K. Reynolds, J.J. Tjandra, X.L. Tang, D.F.J. Purcell and I.F.C. McKenzie. Reaction of antibodies to human milk fat globule (HMFG) with synthetic peptides. In: Breast Cancer

Immunodiagnosis and Immunotherapy. R.L. Ceriani,ed., Plenum Press, New York (1989).

24. T.P. Hopp and K.R. Woods. A computer program for predicting protein antigenic determinants. Mol. Immunol. 20:483-489 (1983).

25. J.R. Gum, J.C. Byrd, J.W. Hicks, N.W. Toribara, D.T.A. Lamport and Y.S. Kim. Molecular cloning of human intestinal mucin cDNAs: Sequence analysis and evidence for genetic polymorphism. J. Biol. Chem. 264:6480-6487 (1989).

26. K.R. Jerome, D.L. Barnd, C.M. Boyer, J. Taylor-Papadimitriou, I.F.C. McKenzie, R.C. Bast and O.J. Finn. Adenocarcinoma reactive cytotoxic T lymphocytes recognize an epitope present on the protein core of epithelial mucin molecules. In: "Cellular Immunity and the Immunotherapy of Cancer UCLA Symposia on Molecular and Cellular Biology, New Series". M.T. Lotze and O.J. Finn, ed., Willey-Liss, Inc., New York, NY (1990) (in press).

MOLECULAR ANALYSIS OF EPITOPIC HETEROGENEITY OF THE BREAST MUCIN

Jerry A. Peterson, David Larocca, Gary Walkup, Richard Amiya, and Roberto L. Ceriani

John Muir Cancer and Aging Research Institute, 2055 North Broadway, Walnut Creek, CA 94596

ABSTRACT

The breast mucin is a highly glycosylated large molecular weight glycoprotein on the surface of breast epithelial cells, in breast carcinoma cells, and is a major component of the human milk fat globule membrane. By epitope mapping we have determined that 5 different monoclonal antibodies (Mc1, Mc5, BrE1, BrE2, BrE3) raised against human milk fat globule and selected for breast specificity recognize 4 distinct but overlapping linear amino acid sequences encompassing the most hydrophilic region of the 20 amino acid tandem repeat that makes up a large part of polypeptide core of the breast mucin. Although these MoAbs bind to overlapping polypeptide epitopes they have different tissue and tumor specificities in histopathology, differ quantitatively in the binding to breast carcinoma cell lines by flow cytometry, and they have distinct competition patterns for binding to the native antigen on breast carcinoma cells. Even though BrE2 and BrE3 have the same polypeptide epitope, they differ in their relative binding to breast carcinoma cell as determined by flow cytometry and binding to fusion proteins containing mimotopes produced by cDNA clone isolated from a breast cDNA library. There is considerable variation in the breast mucin mRNA, but the expression of the different epitopes shows little correlation with mRNA levels. Surface expression of an epitope and detection of the epitope in secreted material is associated with its presence on a larger size molecular specie than the exclusively cytoplasmic expression of an epitope. This detailed analysis of the epitopic heterogeneity of an immunodominant region of the tandem repeat segment of the core polypeptide of the breast mucin, possibly involving altered glycosylation, reveals an epitopic heterogenity that may have functional significance and suggests approaches for preparing new monoclonal antibodies with improved qualities for breast cancer diagnosis and/or therapy.

INTRODUCTION

The first monoclonal antibodies (MoAbs) against breast epithelial cells were found to react with a high molecular weight mucin-like glycoprotein that had not previously been described on the surface of these cells (1,2). These were produced by using mice immunized with the

Breast Epithelial Antigens, Edited by R.L. Ceriani
Plenum Press, New York, 1991

human milk fat globule (HMFG) and breast epithelial cells isolated from human milk (1,2). Since then many more MoAbs have been produced using HMFG and various carcinoma cells and their membrane fractions for immunization. A large proportion of these MoAbs identified this mucin-like molecule (3), which we originally termed NPGP (2), but has also been referred to as sialomucin (4,5), PAS-O (6), polymorphic epithelial mucin (PEM) (7), MAM-6 (4), TAG-72 (8), Epithelial Membrane Antigen (EMA) (9), DF3 antigen (5), H23 antigen (10), and episialin (11). It has been called a normal epithelial antigen (2-4,6) as well as a tumor associated antigen (5,7-10). The entire coding sequence for the core protein of this epithelial cell associated mucin from breast cells (10-12) and pancreas (13) has been determined, and surprisingly, the sequences for breast and pancreas are virtually identical. The deduced amino acid sequence of the mucin polypeptide is made up of a variable number of 20 amino acid tandem repeats and unique 3' and 5' regions (10-13). At both ends of the tandem repeats are a few degenerate tandem repeats that increasingly deviate from the consensus repeat sequence the further they are away in either direction. The molecule exhibits restriction length polymorphism which is thought to be due to variations in the number of tandem repeats.

Even though a surprisingly large percentage of MoAbs prepared against breast and other epithelial cells have turned out to be against this epithelial mucin, they often have been described to have distinct properties. They have proved useful for immunodiagnosis in histopathology (14), development of serum assays for breast cancer (15), and in immunotherapy (16,17). Also, some MoAb against this molecule appear to be more tumor specific than others (18). If an MoAb recognizes the mucin on normal breast, it is always on the apical surface of the epithelial cells; however, in breast carcinomas it is often detected cytoplasmically as well as on the cell surface (3). It is thought that some of this heterogeneity is due to posttranslational changes, possibly due to alterations in glycosylation (3,18). In an attempt to better understand the nature of the apparent epitopic heterogeneity of the breast mucin, we have undertaken the detailed study of the epitopes of 5 different MoAbs, all resulting from immunizations with HMFG (2,3). These studies demonstrate that in spite that they all bind to an immunodominant region of the tandem repeat that encompasses 7 amino acids, they reveal a vast heterogeneity in the processing of this polymorphic molecule in different breast tumors and other tissues.

MATERIALS AND METHODS

Epitope mapping: Overlapping peptide hexamers were synthesized onto the ends of polyethylene pins using the Epitope Scanning Kit (Cambridge Research Biochemicals, Cambridge, UK), which is based on the method originally described by Geysen et al. (19). The polyethylene pins are arranged in a 8x12 configuration that fits into a 96 well microtiter dish. The pins are supplied with an alanine attached to the ends to which the amino acids are added consecutively using pentafluorophenyl active esters of fluorenylmethyloxycarbonyl (Fmoc)-L-amino acids. Each consecutive overlapping hexamer differs from the previous one by a single amino acid and enough are synthesized to span the entire sequence of the peptide to be tested so that every combination of hexamer is present. Each MoAb was tested for binding to the synthetic peptides using an ELISA method with horse radish peroxidase-conjugated goat anti-mouse IgG (Promega, Madison, WI) and color development with 2,2'azinobis (3-ethylbenzothiazoline-6-sulfonic acid (Sigma, St. Louis, MO).

Monoclonal antibodies: The preparation of monoclonal antibodies Mc1, Mc5, BrE1, BrE2, and BrE3 that recognize the breast mucin; Mc3 and Mc8 that recognize a 46 KDa glycoprotein; and McR2 and Mc13 that recognize a 70 KDa glycoprotein were all prepared by immunizing with HMFG and have been described previously (2,3). The MoAb DF3 that was prepared against breast carcinoma cells and also recognizes the breast mucin was kindly provide by Dr. Kaplan (20).

Binding to Synthetic peptides: The two polypeptides were synthesized by Multiple Peptide Systems (San Diego, CA). The peptides were bound to wells of microtiter plates (500 ng/well) using the methylated BSA method described previously (21). The MoAb binding was determined by a radioimmunobinding assay (21) using ^{125}I-labelled goat anti-mouse IgG.

Breast Cell Lines and Flow cytometry studies: Breast carcinoma cell lines were cultured in Dulbecco's Modified Eagle's medium plus 10% fetal bovine serum plus penicillin-streptomycin. Breast carcinoma cell lines ZR75, T47D MCF-7, and BT20 cells were obtained from the American Type Culture Collection; MDA-MB-331, MDA-MB-175, MDA-MB-361, and MDA- MB-435, and MDA-MB-231 from Dr. Relda Cailleau; and ELLG, MX1, and MX2 from Dr. Bogden as a transplantable tumor that we adapted to cell culture. For flow cytometry analysis the cells were removed from the culture vessels by trypsinization and cultured in methyl cellulose medium over agar medium for 48 hours to restore surface expression of the antigens as previously described (22). The cells were harvested as a single cell suspension then fixed with cold 70% ethanol for 30 minutes at 4°C which allowed antibody to stain both surface and cytoplasmic antigen. The cells were stained for 30 minutes at 4°C with FITC-conjugated MoAb dissolved in PBS + 1% calf serum. FITC-conjugation was carried out by the method of Kearney and Lawton (23). Total epitope content per cell was determined with a Coulter EPICS 753 flow cytometer equiped with an Argon laser operated with an excitation wavelength of 488 nm. Determination of percentage positive cells and relative staining intensity was determined with the IMMUNO program and quantitation and standardization was done using SIMPLY CELLULAR beads (Flow Cytometry Standards Corp., Research Triangle, NC) (24). The cell peak on forward vs 90° light scatter was gated to eliminate debris.

RESULTS

Epitope Mapping: Five different MoAbs (Mc1, Mc5, BrE1, BrE2 and BrE3), that have been prepared using the human milk fat globule (HMFG) for immunization, all identify epitopes on the highly glycosylated large molecular weight breast mucin. By immunohistochemistry they appear to recognize different epitopes since each had different tissue and tumor specificities (Table 1) (3). Each bound to a different spectrum of normal tissues and their specificities for different carcinomas were distinct; although, BrE2 and BrE3 were quite similar (Table 1). In addition, by screening breast λgt11 cDNA expression libraries with some of these MoAbs we isolated cDNA clones that produced fusion proteins that bound all of them, while other cDNA clones bound just some (25). This binding to the fusion proteins indicated that the epitopes for these 5 MoAbs all included the polypeptide portion of this glycoprotein. In order to confirm this, we tested for binding of these MoAbs to two synthetic polypeptide 20-mers (PDTRPAPGSTAPPAHGVTSA and APPAHGVTSAPDTRPAPGST) that spanned the tandem repeat consensus sequence (26,27), one starting at the beginning of the published 20 amino acid repeat (26) unit and the other starting in the middle. As shown in Table 2, all five MoAbs bound to both synthetic peptides, as did DF3, a MoAb produced by others (5) against breast carcinoma cells. Three other

MoAbs (2,3) against other components of the HMFG, that do not cross-react with the breast mucin, Mc13, against a 70 KDa glycoprotein, and Mc3 and Mc8, against a 46 KDa glycoprotein do not bind to these synthetic peptides (Table 2).

Table 1. Immunohistological specificity of anti-mucin MoAbs on normal tissues and various carcinomas.

Normal Tissue	Mc5	BrE1	BrE2	BrE3	Mc1 (HMFG2)
Breast	+ all	-	+ 6/11	+/-	+ all
Lung	alveoli ducts	alveoli	alveoli	alveoli	alveoli
Kidney	distal tubules	brush border	distal tubules	distal tubules	distal tubules
Pancreas	acini islets	-	acini ducts	acini	acini
Thyroid	+	-	-	-	-
Esophagus	+	-	-	-	-
Stomach	-	-	mucosa	mucosa	-
Carcinomas					
Breast	48/48	20/50	50/51	47/54	48/48
Lung	15/15	41/47	54/58	47/48	6/7
Pancreas	11/14	9/15	4/17	12/15	2/2
Liver	2/3	8/33	23/46	3/28	23/46
Kidney	5/15	0/15	6/14	6/13	2/8
Ovary	15/17	20/26	20/21	23/25	8/8
Colon	5/14	3/27	23/32	12/25	0/6
Thyroid	6/6	0/7	0/4	3/9	2/3

Table 2. MoAb binding to synthetic peptides of the tandem repeat of the breast mucin.

	Counts per minute	
MoAb	B47-15	B47-16
Mc1	15000	19400
Mc5	18100	21700
BrE1	19100	25120
BrE2	24200	29134
BrE3	21400	25230
DF3	27500	20700
Mc3	1100	1900
Mc8	1328	1640
Mc16	4300	3010

B47-15 synthetic peptide 20-mer: PDTRPAPGSTAPPAHGVTSA
B47-16 synthetic peptide 20-mer: APPAHGVTSAPDTRPAPGST

In order to determine the amino acid sequence of the tandem repeat that is essential for binding of each of these anti-mucin MoAbs, we synthesized a series of 20 overlapping amino acid hexamers on polyethylene pins that spanned the entire sequence of the tandem repeat. From the hexamers that each MoAb binds to it is possible to deduce the linear sequence of amino acids that are essential for binding of the particular MoAb to the peptide (19). As can be seen in Table 3, Mc1, Mc5, and BrE1 have distinct polypeptide epitopes, while BrE2 and BrE3 identify the same three amino acid sequence. This epitope mapping method identifies amino acid sequences that have strong binding for the MoAbs, but do not necessarily define the true epitope which can involve more amino acids and interactions with carbohydrate structures on the highly glycosylated and folded native molecule (28).

Table 3. Epitope mapping of MoAbs reactive with the tandem repeat sequence of the breast mucin.

MoAb	Reactive amino acid sequence	
	Tandem repeat sequence: TSAPDTRPAPGSTAPPAHGV	Mimotope:
Mc1 (HMFG2)	DTR	
Mc5	TRPAP	-DLRPGP- (NP5)
BrE1	DTRP	
BrE2	TRP	-HHTRPAL (NP4)
BrE3	TRP	-HHTRPAL (NP4)

Competition Studies: Since histological studies demonstrate that these 5 MoAbs have different tissue specificities (Table 1) (3) and thus appear to recognize different epitopes on the native mucin molecule in spite of their binding to overlapping peptide epitopes, we undertook competition studies for binding to T47D breast carcinoma cells that express the breast mucin antigen. Using radiolabelled Mc5 there was good competition with unlabelled Mc5; however, the other MoAbs against the breast mucin (Mc1, BrE1, BrE2, and BrE3) did not compete. If radiolabelled BrE2 was used there is good competition with unlabelled BrE2, partial competition with BrE3, but no competition with Mc1, Mc5, or BrE1. This demonstrates again that of the 5 MoAbs, BrE2 and BrE3 are the most similar. In neither assay was there competition with MoAbs against other components of the HMFG, such as the 70 KDa component (Mc13 and McR2) and the 46 KDa component (Mc8, Mc3), nor is there competition with a nonspecific immunoglobulin (MOPC).

Binding to fusion proteins and mimotopes: Further evidence that these MoAbs recognize distinct epitopes comes from studies of binding to fusion proteins produced by cDNA clones isolated from MCF-7 breast carcinoma λgt11 cDNA library (25). For example, the β-galactosidase fusion protein produced by the cDNA clone NP5 binds only Mc5, while the fusion protein produced by cDNA clone NP4 binds only BrE2 and BrE3 (Table 3). Both NP5 and NP4 appear to represent mimotopes resulting from screening of the λgt11 cDNA library and not the true antigen identified by these MoAbs (Table 3). NP5 cDNA identified a 0.9 kb mRNA species in breast cells (25). Sequencing of NP5 revealed that it corresponded to a fragment of the estrogen-responsive gene SP2 (29), but with a shifted reading frame causing it to produce a fusion protein different from the SP2 amino acid sequence (25). The deduced amino acid sequence of the NP5 fusion protein had a short region of homology with the 20 amino acid repeat of the breast mucin where 4 out of 6 amino acids were identical, -Asp-Leu-Arg-Pro-Gly-Pro- (the underlined amino acids are those that are identical) with the rest of the derived sequence being distinct. Epitope mapping of NP5 showed that only this hexamer bound Mc5 (Table 3). Even though NP5 may be an artifact of the selection in the MCF7 λgt11 cDNA library its fusion protein has proved very useful for constructing a radioimmunoassay that is superior in specificity and sensitivity to those using the native antigen (30). Upon sequencing of NP4 cDNA, it was found to have a short open reading frame of 7 amino acids containing a mimotope of the immunodominant region of the tandem repeat. The deduced sequence was His-His-<u>Thr-Arg-Pro-Ala</u>-Leu, where the underlined amino acids are identical to the tandem repeat consensus sequence. BrE2 binds to this fusion protein 4-fold greater than BrE3 at the same antibody concentration (1 µg/ml), while neither Mc1, Mc5, nor BrE1 bind at all. In contrast BrE2 and BrE3 bind equally well to the synthetic tandem repeat sequence (Table 2). Epitope mapping of this sequence showed that BrE2 and BrE3 bound to each hexamer containing the Thr-Arg-Pro (Table 3). The reduced binding of BrE3 to NP4 fusion protein indicates that amino acid substitutions adjacent to the essential amino acids of the epitope reduces its affinity for BrE3 compared to BrE2.

In order to further define the contribution of the polypeptide and oligosaccharide moieties to the epitope we have compared the affinity of Mc5 and BrE3 to the native breast mucin and to the fusion proteins. Monoclonal antibody Mc5 binds to the native breast mucin isolated from the HMFG with an affinity of 100-fold greater than to the fusion protein NP11-2 which contains the consensus sequence of the tandem repeat, and to the fusion protein NP5 containing the mimotope with an intermediate affinity (Table 4). In contrast, BrE3 has a similar affinity for both the native breast mucin and the fusion protein NP11-2 (Table 4). These results indicate that Mc5 requires a greater contribution of oligosaccharide moieties and/or the folded structure of the native mucin for its epitope than BrE3.

Table 4. Affinity constants for binding of monoclonal antibodies Mc5 and BrE3 to the native mucin (NPGP) and to fusion proteins containing the breast mucin tandem repeat (NP11-2) and a mimotope (NP-5).

Mc5	NPGP	$= 6.31 \times 10^{-9}$ M/1
	NP-5	$= 9.55 \times 10^{-8}$ M/1
	NP11-2	$= 4.00 \times 10^{-7}$ M/1
BrE3	NPGP	$= 1.32 \times 10^{-8}$ M/1
	NP11-2	$= 1.05 \times 10^{-8}$ M/1

Flow cytometry studies: Even though, BrE2 and BrE3 have the same essential amino acid trimer in their epitope they do not compete completely for binding to T47D cells, suggesting difference in their native epitopes. Immunohistochemical studies (3) and their binding to fusion proteins also suggest difference in their native epitope. Flow cytometry studies also indicate differences in their native epitopes, since as presented in Table 5 where the relative staining intensities are compared for binding of BrE2 and BrE3 to 12 different breast carcinoma cell lines, there are significant differences between these two MoAbs with 6 of the different cell lines. For 4 of the cell lines (ELL-G, ZR75, MX1, MX2) BrE2 stained at least 2-fold greated than BrE3. In contrast, for MDA-MB-331 and MDA-MB-231 BrE3 stained more intensely than BrE2 (Table 5). For all 4 cell lines the percent positive cells ranged from 90% - 100%.

60

Table 5. Flow cytometry analysis of total epitope content of MoAbs BrE2 and BrE3 on different breast cancinoma cell lines.

| Cell Line | Relative Epitope Content (mean channel No.) | |
	BrE2	BrE3
ELL-G	3246	715
MDA-MB-331	1633	3291
MDA-MB-435	29	30
MCF-7	356	292
MDA-MB-175	3769	3272
MDA-MB-231	658	1798
ZR75	1731	987
MDA-MB-361	404	98
BT-20	1092	1313
MX1	1633	198
T47D	1002	593
MX2	2233	729

Breast mucin mRNA levels and epitope expression: In order to investigate the regulation of expression of the various epitopes of the breast mucin we determined the relative levels of mRNA levels using Northern blot analysis with a specific cDNA probe (NP11-2) for the breast mucin in 6 different breast carcinoma cell lines (Table 6). Also, the relative total epitope content of 4 different MoAbs was determined with flow cytometry. Considerable variation in breast mucin mRNA levels was seen in the different breast cell lines; however, there was no direct correlation between the expression of the 4 different epitopes and mRNA levels (Table 6). For example, the cell lines MDA-MB-331 and T47D had equally high levels of breast mucin mRNA, but the expression of Mc5 epitope was 30-fold greater in MDA-MB-331 than T47D. The BrE1 epitope was similar in all cell lines in spite of the considerable variation in mRNA levels. The BrE2 epitope was highest in the cell line ELL-G which had the lowest mRNA level, while the epitope was 10-fold lower in MCF-7 which had higher mRNA levels. These results indicate that the expression of a given epitope requires not only mRNA and core protein synthesis but also posttranslational modification, probably involving glycosylation, and that each epitope exhibits an independent variability in expression.

Antigen size, surface expression and secreted breast mucin: Using Western blot analysis we have determined that in the cell line MDA-MB-331, Mc5 recognized two sized molecular species of the breast mucin (approximately 140 KDa and >200 KDa), while BrE3 recognized only the 140 KDa component (Table 7). Flow cytometry analysis revealed that the Mc5 epitope is expressed primarily on the cell surface in this cell line, while the BrE3 epitope was expressed mainly in the cytoplasm. Similar results were seen with MCF-7 breast cell line. In contrast, the ZR75 cells express both Mc5 and BrE3 epitopes on the cell surface and in this case BrE3 recognized also the >200 KDa component in these cells. In addition, the expression of the Mc5 epitope on the larger component and on the cell surface was associated with high levels of the Mc5 epitope on secreted material in both MDA-MB-331 and MCF-7 cells. Only in ZR75 cells where the BrE3 epitope was present on the larger molecule and on the cell surface was there significant expression of the BrE3 epitope on breast mucin material secreted from the cultured cells (Table 7). These results suggest that the breast mucin is processed differently in different cell lines and that although the breast mucin is probably secreted by all three cell lines, different epitopes can be expressed on the secreted material. Also, detection of an epitope in the secreted mucin is associated with its expression in a larger molecule and on the cell surface.

Table 6. Comparison of breast mucin mRNA and expression of different epitopes recognized by 4 MoAbs against the tandem repeat sequence in breast carcinoma cell lines measured by flow cytometry.

Cell line	Relative Epitope Content[1]				mRNA Level[2]
	Mc5	BrE1	BrE2	BrE3	
MDAMB331 (ER+)	6342	128	1102	3591	++++
T47D (ER+)	217	230	981	512	++++
ZR75 (ER+)	3577	290	1679	2847	++++
MCF-7 (ER+)	2315	126	341	471	+++
MDAMB361 (ER+)	317	102	404	98	+++
BT20 (ER-)	56	7104	1026	919	++
SKBR3 (ER-)	552	104	409	521	++
ELL-G (ER-)	390	307	3181	672	+/-

[1]Relative epitope content was determined by flow cytometry using FITC-conjugated MoAbs by multiplying the percentage positive cells with the mean relative channel number (normalized using SIMPLY CELLULAR beads).

[2]Relative mRNA content was determined using Northern blot with a probe for the tandem repeat (NP11-2).

Table 7. Comparison of surface and total epitope content, antigen size detected, and level of secreted mucin in different breast carcinoma cell lines using two different MoAbs against the breast mucin NPGP.

Cell line	MoAb	Epitope Content[1]		Antigen Size (MW)[2]	Level in Medium[3]
		Surface	Total		
MDAMB331	Mc5	2768	6342	>200 KDa 140 KDa	5.8 µg/ml
MDAMB331	BrE3	310	3591	140 KDa	0.01 µg/ml
MCF-7	Mc5	1083	2315	>200 KDa 140 KDa	8.1 µg/ml
MCF-7	BrE3	57	471	140 KDa	0
ZR75	Mc5	3672	3577	>200 KDa 140 KDa	0
ZR75	BrE3	2496	2847	>200 KDa 140 KDa	.37 ug/ml

[1]Relative epitope content was determined by flow cytometry using FITC-conjugated MoAbs by multiplying the percentage positive cells with the mean relative channel number (normalized using SIMPLY CELLULAR beads).

[2]Antigen size was determined using Western blot with MoAbs Mc5 and BrE3.

[3]Antigen content in media was determined by a competitive radioimmunoassay with either Mc5 or BrE3.

DISCUSSION

The present report demonstrates that 5 MoAbs, prepared by immunizing with human milk fat globule (HMFG) membranes and that were selected for breast specificity, identify polypeptide epitopes within an apparently immunodominant region of the 20 amino acid tandem repeat of the breast mucin. By epitope mapping Mc1, Mc5, and BrE1 have unique polypeptide epitopes while BrE2 and BrE3 have the same epitope, which is different from the former three. This region of the tandem repeat appears to be immunodominant since all MoAbs that have been isolated that bind to the polypeptide portion of the breast mucin and where the epitope has been characterized by binding to synthetic peptides, bind to the tandem repeat and more precisely to this same region (31-33). In fact, comparing the essential amino acids for binding determined by epitope mapping with overlapping hexamers or octomers, they all include a linear combination of amino acids within a 7 amino acid region between the serines that are expected glycosylation sites, one amino acid away on either side (Table 2). These include MoAbs prepared against HMFG and a variety of epithelial cell and membrane preparations. The MoAbs are SM-3 (33), Onc-M15 (33), LICR-LON-M8 (33), BC-1 (31), BC-2 (31), BC-3 (31), DF3 (32), and HMFG2 (7) [which is a different name for the MoAb Mc1 (2,3)]. The fact that there is a possible O-linked glycosylation site (threonine) in the middle of this region suggests that this site may not be glycosylated or its glycosylation does not inhibit the MoAb binding.

In spite of the fact that all 5 MoAbs have defined epitopes in a restricted immunodominant region of the tandem repeat sequence that overlap, the epitopes for them on the native breast mucin appear to be unique. They do not compete for binding to the native molecule, they have different tumor and tissue specificities, they are expressed at different quantitative levels on different breast cell lines as show by flow cytometry, they vary independently with breast mucin mRNA levels, they are expressed differently on different molecular species and on secreted breast mucin, they can be expressed on the cell surface or cytoplasmically, and their affinities can differ on different forms of the molecule. These results point to an epitopic heterogeneity that is derived from posttranslational modification of the breast mucin. Glycosylation probably plays an important role in the epitopic nature of the mature mucin, involving interaction of the MoAbs with oligosaccharide structures, but also resulting from the folding of the mature glycoprotein structure, possibly involving weaker interactions with other amino acids of the molecule than those shown to have strong binding by epitope mapping (28).

Clearcut differences in glycosylation, the length of the oligosacchardides, number of side changes, and their sugar composition, have been observed for this epithelial mucin. For example, the breast and pancreatic mucin have the same core protein sequence (12,13); however, the breast mucin is 50% carbohydrate (34) while the pancreatic mucin is 80% (35). The carbohydrate composition of the breast mucin in BT-20 breast carcinoma cells is considerably different that that isolated from the human milk fat globule (5), the latter representing the most mature form of the molecule. The expression of the breast mucin in the cytoplasm of breast carcinoma cells probably also is the result of the altered processing of the molecule. Therefore, this epithelial mucin is not only polymorphic with respect to the number of tandem repeats but also polymorphic in the forms of the processed molecule. It is known that tumor cells differ from normal in the levels of glycosyltransferases (36) and this can obviously affect the nature of glycoproteins that are synthesized; however, the significance of this in the neoplastic process is not known. The results showing the uniqueness of different epitopes on the breast mucin, for example the competition studies, suggest that different parts of the tandem repeat region are processed differently resulting in the presence of different epitopes in different regions of the molecule. How this occurs is not known since the specificity of O-linked glycosylation has not been defined. In contrast, N-linked glycosylation has been shown to require specific consensus sequences.

An alternative explanation for the uniqueness of different regions of the breast mucin with regard to expression of different epitopes could involve the observed degeneracy of the tandem repeat (10,12). At each end of the tandem repeat region there are degenerate repeats that increase in degeneracy the further they are away from the tandem repeats (10,12). Also, there could be amino acid substitutions in the repeat region. The observation that slight variations in the amino acid composition of the immunodominant region can result in greater affinity for some MoAbs, as represented by the mimotopes described here, could yield epitope heterogeneity of the tandem repeat region. In spite of the fact that the same concensus amino acid sequence has been described by several laboratories (10-13,26,27), no one has sequenced all the 20 to 80 tandem repeats in a single cDNA. The reason for this is that it is not possible using standard method of sequencing since any primers within the repeat would bind at multiple sites. When 14 of the tandem repeats were sequenced, Hareuveni et al. (37) found that there were some base substitutions that would result in variation in the amino acid sequences. Even though this shows that degenerate tandem repeat sequences can distinguish among these 5 MoAbs, their contribution to the binding specificity may be limited. The number of tandem repeats ranges from 20 to 80 or more and they clearly contain an immunodominant region. The degeneracy at either end of this region would represent no more than 6 repeats. Although only 14 consecutive repeats have been sequenced and revealed base substitutions, the immunodominant region appeared to be preserved since most of the base substitutions are in the non-immunogenic region and many do not lead to a change in the amino acid sequence. Also, the SmaI restriction endonuclease site in the latter region appears to be preserved since digestion with the latter enzyme breaks the tandem repeat region of the breast mucin into identical 60 base pairs fragments (26). Therefore, the hypothesis that degeneracy in the tandem repeat sequence is a major factor distinguishing the specificity of the different MoAbs for the breast mucin seems unlikely, although sequencing of more tandem repeats will be necessary to completely eliminate it.

These results demonstrate that by immunizing mice with native membrane material, be it derived from HMFG, normal breast epithelial cells or carcinoma cells from breast or other tissue origins, most MoAbs produced appear to recognize the breast mucin. A variety of MoAbs have been isolated that have distinct and different properties, but virtually all of them that recognize the polypeptide moiety bind to an immunodominant region of the tandem repeat. This is the case for 5 MoAbs that we describe here prepared against the HMFG membrane. In spite of their binding to overlapping linear amino acid sequences their epitopes on the native molecule are each unique. Their uniqueness probably arises from different folding configurations of the molecule and interaction with sugar moieties on this highly glycosylated molecule. This suggests that by modifying the native structure of the native molecule and/or by reconstructing it from the polypeptide backbone, such as by in vitro glycosylation of recombinant proteins or synthetic peptides it will be possible to design MoAbs that have preferred properties in diagnosis, tumor specificity, serum assays, immunohistopathology and/or therapy. With regard to the use of the breast mucin as marker for serum assays for breast cancer, different epitopes are prefentially expressed on secreted material. Our results demonstrate that Mc5 detects higher levels of secreted breast mucin in culture medium of breast epithelial cells, and also our unpublished data on detection of the breast mucin in breast cancer patients shows that Mc5 detects high serum levels while BrE3 epitope is very low or not detected in the same serum samples. This result suggests that Mc5 is a good MoAb for a serum assay while BrE3 may be better for radioimmunotherapy, since even though it is low in serum it is expressed highly in breast tumors. While radiolabelled Mc5 was poor in imaging tumors in breast cancer patients, successful imaging was obtained with BrE3 (unpublished data).

Although the function of this molecule is not clearly understood it is a major component of the cell surface of epithelial cells and exhibits different epitopes, probably due to different degrees of glycosylation, on different cell types and which are also altered in maligancy (32). Because of its mucin-like properties it has been considered to act as a lubricant on the cell surface of secretory epithelial cells, but also it may have other functions, possibility containing specific receptors. For example, it has been found to contain histocompatibility complex-restricted recognition sites by cytotoxic T-cells (38). Understanding the nature of its epitopic heterogeneity may shed more light on its function besides allowing the preparation of new generation MoAbs.

ACKNOWLEDGMENTS

This work is supported by NCI Grant Nos. CA39932, CA42767 and BRSG RR05929.

REFERENCES

1. Taylor-Papadimitriou, J., Peterson, J.A., Arklie, J., Burchell, J., Ceriani, R.L., and Bodmer, W.F. Monoclonal antibodies to epithelium-specific components of the human milk fat globule membrane: production and reaction with cells in culture. Int. J. Cancer, 28: 17-21, 1981.

2. Ceriani, R.L., Peterson, J.A., Lee, J.Y., Moncada, R., and Blank, E.W. Characterization of cell surface antigens of human mammary epithelial cells with monoclonal antibodies prepared against human milk fat globule. Somat. Cell Genet., 9: 415-427, 1983.

3. Peterson, J.A., Zava, D.T., Duwe, A.K., Blank, E.W., Battifora, H., and Ceriani, R.L. Biochemical and histological characterization of antigens preferentially expressed on the surface and cytoplasm of breast carcinoma cells identified by monoclonal antibodies against the human milk fat globule. Hybridoma, 9: 221-235, 1990.

4. Hilkens, J., Buijs, F., and Ligtenberg, M. Complexity of MAM-6, an epithelial sialomucin associated with carcinomas. Cancer Res., 49: 786-793, 1989.

5. Hull, S.R., Bright, A., Carraway, K.L., Abe, M., Hayes, D.F., and Kufe, D.W. Oligosaccharide differences in the DF3 sialomucin antigen from normal human milk and the BT-20 human breast carcinoma cell line. Cancer Comm., 1: 261-267, 1989.

6. Shimizu, M., Yamauchi, K., Miyauchi, Y., Sakurai, T., Tokugawa, K., and McIlhinney, R.A.J. High-Mr glycoprotein profiles in human milk serum and fat-globule membrane. Biochem. J., 233: 725-730, 1986.

7. Gendler, S., Taylor-Papadimitriou, J., Duhig, T., Rothbard, J., and Burchell, J. A highly immunogenic region of a human polymorphic epithelial mucin expressed by carcinomas is made up of tandem repeats. J. Biol. Chem., 263: 12820-12823, 1988.

8. Johnson, V.G., Schlom, J., Paterson, A.J., Bennett, J., Magnani, J.L., and Colcher, D. Analysis of a human tumor-associated glycoprotein (TAG-72) identified by monoclonal antibody B72.3. Cancer Res., 46: 850-857, 1986.

9. Heyderman, E., Strudley, I., Powell, G., Richardson, T.C., Cordell, J.L., and Mason, D.Y. A new monoclonal antibody to epithelial membrane antigen (EMA)-E29. A comparison of its immunocytochemical reactivity with polyclonal anti-EMA antibodies and with another monoclonal antibody, HMFG-2. Br. J. of Cancer, 52: 355-361, 1985.

10. Wreschner, D.H., Hareuveni, M., Tsarfaty, I., Smorodinsky, N., Horev, J., Zaretsky, J., Kotkes, P., Weiss, M., Lathe, R., Dion, A., and Keydar, I. Human epithelial tumor antigen CDNA sequences. Eur. J. Biochem., 189: 463-473, 1990.

11. Ligtenberg, M.J.L., Vos, H.L., Gennissen, A.M.C., and Hilkens, J. Episialin, a carcinoma-associated mucin, is generated by a polymorphic gene encoding splice variants with alternative amino termini. J. of Biol. Chem., 265: 5573-5578, 1990.

12. Gendler, S.J., Lancaster, C.A., Taylor-Papadimitriou, J., Duhig, T., Peat, N., Burchell, J., Pemberton, L., Lalani, E.N., and Wilson, D. Molecular cloning and expression of human tumor-associated polymorphic epithelial mucin. J. Biol. Chem., 265: 15286-15293, 1990.

13. Lan, M.S., Batra, S.K., Qi, W.N., Metzgar, R.S., and Hollingsworth, M.A. Cloning and sequencing of a human pancreatic tumor mucin cDNA. J. Biol. Chem., 265: 15294-15299, 1990.

14. Ceriani, R.L., Hill, D.L., Osvaldo, L., Kandell, C., and Blank, E.W. Immunohistochemical studies in breast cancer using monoclonal antibodies against breast epithelial cell components and with lectins. In: J. Russo (ed.), Immunocytochemistry in Tumor Diagnosis, pp. 233-263, Boston: Martinus Nijhoff Publications. 1985.

15. Ceriani, R.L., Sasaki, M., Sussman, H., Wara, W.M., and Blank, E.W. Circulating human mammary epithelial antigens in breast cancer. Proc. Natl. Acad. Sci. USA, 79: 5420-5424, 1982.

16. Ceriani, R.L., Blank, E.W., and Peterson, J.A. Experimental immunotheraphy of human breast carcinomas implanted in nude mice with a mixture of monoclonal antibodies against human milk fat globule components. Cancer Res., 47: 532-540, 1987.

17. Ceriani, R.L. and Blank, E.W. Experimental therapy of human breast tumors with 131I-labeled monoclonal antibodies prepared against the human milk fat globule [published erratum appears in Cancer Res 1989 Sep 15;49(18): 5236]. Cancer Res., 48: 4664-4672, 1988.

18. Burchell, J., Gendler, S., Taylor-Papadimitriou, J., Girling, A., Lewis, A., Millis, R., and Lamport, D. Development and characterization of breast cancer reactive monoclonal antibodies directed to the core protein of the human milk mucin. Cancer Res., 47: 5476-5482, 1987.

19. Geysen, H.M., Meloen, R.H., and Barteling, S.J. Use of peptide synthesis to probe vital antigens for epitopes to a resolution of a single amino acid. Proc. Natl. Acad. Sci. USA, 81: 3998-4002, 1984.

20. Abe, M. and Kufe, D.W. Identification of a family of high molecular weight tumor-associated glycoproteins. J. Immunol., 139: 257-261, 1987.

21. Ceriani, R.L. Solid phase identification and molecular weight determination of cell membrane antigens with monoclonal antibodies. In: K.B. Bechtol, T.J. McKern and R. Kennett (eds.), Monoclonal Antibodies and Functional Cell Lines. Progress and Applications, pp. 398-402, New York: Plenum Press. 1984.

22. Peterson, J.A., Bartholomew, J.C., Stampfer, M., and Ceriani, R.L. Analysis of expression of human mammary epithelial antigens in normal and malignant breast cells at the single cell level by flow cytofluorimetry. Exp. Cell Biol., 49: 1-14, 1981.

23. Kearney, J.F. and Lawton, A.R. B lymphocyte differentiation induced by lipopolysaccharide. J. Immunol., 115: 671-676, 1975.

24. Webb, N.R., Madoulet, C., Pierre-Francois, T., Broussard, D.R., Sneed, L., Nicolau, C., and Summers, M.D. Cell-surface expression and purification of human CD4 produced in baculovirus-infected insect cells. Proc. Natl. Acad. Sci. USA, 86: 7731-7735, 1989.

25. Larocca, D., Peterson, J.A., Walkup, G., and Ceriani, R.L. High level expression in E. coli of an alternate reading frame of pS2 mRNA that encodes a mimotope of human breast epithelial mucin tandem repeat. 1990. (Submitted for publication)

26. Gendler, S.J., Burchell, J.M., Duhig, T., Lamport, D., White, R., Parker, M., and Taylor-Papadimitriou, J. Cloning of partial cDNA encoding differentiation and tumor-associated mucin glycoproteins expressed by human mammary epithelium. Proc. Natl. Acad. Sci. USA, 84: 6060-6064, 1987.

27. Siddiqui, J., Abe, M., Hayes, D., Shani, E., Yunis, E., and Kufe, D.W. Isolation and sequencing of a cDNA coding for the human DF3 breast carcinoma-associated antigen. Proc. Natl. Acad. Sci. USA, 85: 2320-2323, 1988.

28. Laver, W.G., Air, G.M.G, Webster, R.G., and Smith-Gill, S.J. Epitopes on protein antigens: Misconceptions and Realities. Cell, 61: 553-556, 1990.

29. Jakowlew, S.B., Breathnach, R., Jeltsch, J.M., Masiakowski, P., and Chambon, P. Sequence of the pS2 mRNA induced by estrogen in the human breast cancer cell line MCF-7. Nucleic Acids Res., 12: 2861-2878, 1984.

30. Ceriani, R.L., Larocca, D., Peterson, J.A., Enloe, S., Amiya, R., Enloe, S., and Blank, E.W. A novel serum assay using recombinant breast epithelial mucin antigen. (see this volume)

31. Xing, P.X., Tjandra, J.J., Reynolds, K., McLaughlin, P.J., Purcell, D.F.J., and McKenzie, I.F.C. Reactivity of anti-human milk fat globule antibodies with synthetic peptides. J. Immunol., 142: 3503-3509, 1990.

32. Abe, M. and Kufe, D.W. Structural analysis of the DF3 human breast carcinoma-associated protein. Cancer Res., 49: 2834-2839, 1989.

33. Burchell, J., Taylor-Papadimitriou, J., Boshell, M., Gendler, S., and Duhig, T. A short sequence, within the amino acid tandem repeat of a cancer-associated mucin, contains immunodominant epitopes. Int. J. Cancer, 44: 691-696, 1989.

34. Shimizu, M. and Yamauchi, K. Isolation and characterization of mucin-like glycoprotein in human milk fat globule membrane. J. Biochem., 91: 515-524, 1982.

35. Lan, M.S., Khorrami, A., Kaufman, B., and Metzgar, R.S. Molecular characterization of a mucin-type antigen associated with human pancreatic cancer. The DU-PAN-2 antigen. J. Biol. Chem., 262: 12863-12870, 1987.

36. Parodi, A.J., Blank, E.W., Peterson, J.A., and Ceriani, R.L. Glycosyl transferases in mouse and human milk fat globule membranes. Mol. Cell Biochem., 58: 157-163, 1984.

37. Hareuveni, M., Tsarfaty, I., Zaretsky, J., Kotkes, P., Horev, J., Zrihan, S., Weiss, M., Green, S., Lathe, R., Keydar, I., and Wreschner, D.H. A transcribed gene, containing a variable number of tandem repeats, codes for a human epithelial tumor antigen. Eur. J. Biochem., 189: 475-486, 1990.

38. Barnd, D.L., Lan, M.S., Metzgar, R.S., and Finn, O.J. Specific, major histocompatibility complex-unrestricted recognition of tumor-associated mucins by human cytotoxic T cells. Proc. Natl. Acad. Sci. USA, 86: 7159-7163, 1989.

EXPRESSION AND PROGNOSTIC SIGNIFICANCE OF THE HER-2/*NEU* ONCOGENE DURING

THE EVOLUTIONARY PROGRESSION OF HUMAN BREAST CANCER

D. Craig Allred[1], Atul K. Tandon[2], Gary M. Clark[2],
and William L. McGuire[2]

Departments of Pathology[1] and Medicine/Oncology[2]
University of Texas Health Science Center
San Antonio, TX 78284

INTRODUCTION

About half of all breast cancer patients have metastatic disease in axillary lymph nodes when they are first seen by a physician. The untreated prognosis for these patients is very poor, and the decision to use adjuvant therapy (i.e., radiation, endocrine or chemotherapy) has become almost routine. The other half of patients, however, present without clinical-pathological evidence of metastatic disease, and appear to be cured by initial surgery. Unfortunately, the disease will recur in 20-30% of these patients within 5 years. The choice for adjuvant therapy in this setting is difficult and controversial. On the one hand, evidence from recent studies has shown that adjuvant endocrine and/or chemotherapy can significantly improve disease-free survival (DFS) in some patients with apparently localized breast cancer (i.e., axillary node-negative tumors) (1-3). This has resulted in an official recommendation by the National Cancer Institute that all patients with node-negative breast cancer should be considered for some form of adjuvant therapy (4). On the other hand, the disease does not recur in the majority (70-80%) of node-negative patients, supporting an alternative point of view that all of them should not receive potentially harmful adjuvant therapy (5). Proponents of both views would agree that patients at high-risk for recurrence should receive adjuvant therapy if there were reliable methods of identifying them.

The prospective recognition of high-risk, node-negative breast cancer patients has traditionally been based on an assessment of the size, hormone receptor status, histological grade and nuclear grade of the primary tumor (6). Recently, flow cytometric determinations of tumor DNA content and growth fraction have also been used to identify high-risk patients (7). However, consideration of all these factors has still been unable to predict the outcome of all patients, and there is an obvious need for more powerful markers to identify individual patients with poor prognosis who might benefit from adjuvant therapy.

Among the more promising candidates for new prognostic factors are activated proto-oncogenes, and one of the most intensely studied is HER-2/*neu*. Several recent studies have demonstrated that HER-2/*neu* is amplified or overexpressed in up to one third of human breast cancers. These manifestations of oncogene activation have been clearly shown to predict for poor outcome in patients with node-positive disease (8-15). However, results from studies evaluating the relationship of HER-2/*neu* to clinical outcome in node-negative patients have been far from conclusive (9-17). This chapter will briefly review the biology of HER-2/*neu*, followed by a more detailed discussion of the clinical implications of this oncogene in human breast cancer.

Breast Epithelial Antigens, Edited by R.L. Ceriani
Plenum Press, New York, 1991

The "*neu*" oncogene was first identified in the DNA of chemically-induced rat neuroblastomas using the NIH-3T3 transfection assay (18). Early studies showed that *neu* was highly homologous to the epidermal growth factor receptor (EGFR) gene and coded for a 185 kD trans-membrane protein (p185) (19-20). The human equivalent of *neu* was soon identified and cloned in 3 independent laboratories and called "HER-2" (21), "c-*erb*B-2" (22), and "*erb*B-related gene" (23). The human gene (refered to in this chapter as "HER-2/*neu*") also showed homology with the EGFR gene, and was localized to chromosome 17 (21-22, 24). Studies of the human oncoprotein showed that it was similar to EGFR with respect to amino acid sequence and intracellular tyrosine-kinase activity (21-22, 24). These structural and functional activities shared with EGFR, and results from animal studies in which antibodies to p185 inhibited the growth of xenografts containing activated HER-2/*neu* (25), suggested that the oncoprotein was a growth factor receptor. However, results from clinical studies attempting to correlate HER-2/*neu* activation with various measures of tumor proliferation rate have been conflicting, with some finding (26-28) and others not finding (29-30) a relationship between the two.

More recent studies have shown that transfection of HER-2/*neu* is not transforming in the NIH-3T3 assay if the gene is normal. The gene must be overexpressed to be transforming, which can occur by at least 2 mechanisms including amplification and mutation (31-33). In fact, the originally described rat *neu* gene was mutationally activated. Related investigations have shown that overexpression of HER-2/*neu* also occurs in short term cultures of normal breast epithelial cells (34), suggesting that overexpression per se does not always lead to malignant transformation.

Within the past 2 years, transgenic mice have been developed which carry an activated HER-2/*neu* gene, and overexpress p185 in several tissues (35). Of interest, these animals rapidly develop polyclonal adenocarcinomas in all breast tissue, but not in other tissues, suggesting that HER-2/*neu* activation induces malignant transformation by a single-step, tissue-restricted mechanism. However, more recent studies involving HER-2/*neu* transgenic mice have detected transgene expression in histologically normal mammary glands both before and during the asynchronous development of tumors, suggesting that expression is necessary but not sufficient to induce malignant transformation in mouse mammary tissue (36).

The biology of HER-2/*neu* in the context of both its normal and oncogenic functions is complex and incompletely understood. Studies of early human fetuses have shown prominent HER-2/*neu* expression in all 3 germ layers, providing convincing evidence that the normal gene plays a central role in embryogenesis (37-38). Other studies have shown a high incidence of HER-2/*neu* amplification and expression in several types of malignant human tumors, providing equally convincing evidence that the gene plays an important role in oncogenesis (39-42). The recent identification of a ligand for the HER-2/*neu* protein will hopefully increase our understanding of the functions of this gene and its product (43).

INCIDENCE OF HER-2/*NEU* ACTIVATION IN HUMAN BREAST CANCER

The first report of the involvement of the HER-2/*neu* oncogene in clinical breast cancer was by Slamon et al. (8), showing that HER-2/*neu* was amplified from 2 to 20 times in 30% of 189 samples of human breast carcinoma. Venter et al. (44) soon provided the first evidence of HER-2/*neu* overexpression in clinical breast tumors. These initial studies motivated many others within the past 3 years which have shown evidence of HER-2/*neu* activation (amplification and/or expression) in 10 - 40% of human breast cancers. This wide range of values is most likely a reflection of both technical and biological variability, and has been a serious impediment to our understanding of the clinical biology of this oncogene.

We have recently reviewed the literature regarding the results of various methods used to measure HER-2/*neu* activation in human breast cancer. Southern blotting was by far the most common method utilized to measure HER-2/*neu* amplification. The average incidence of amplification from 19 published studies using Southern blotting was about 22% in over 2000 total patients (8-10, 16, 26, 40, 44-50, 52-57). The range between values was large (10-34%), emphasizing several potential pitfalls in using this technique including an absolute necessity for uniformly and properly handled fresh-frozen specimens, a requirement for considerable technical expertise in conducting the test, and the nearly unavoidable dilutional effect of non-tumors cells contaminating the samples.

Several methods have been used to measure HER-2/*neu* protein expression in breast cancers and, among these, permanent-section immunohistochemistry was by far the most common. Figure 1 shows a typical example of a breast cancer overexpressing HER-2/*neu* as detected by immunostaining. The staining signal is localized to the surface membranes of tumor cells and most positive tumors are diffusely positive.

A review of 16 studies using immunohistochemistry on fixed tissue showed an average incidence of about 19% overexpression in more than 3600 patients (9, 14-15, 17, 29-30, 42, 57-65). These include our results showing 17.4% of tumors positive for HER-2/*neu* expression in a study involving 736 node-negative breast cancers (Table 1) (30). Similar to earlier findings by Gusterson et al. (61), we found expression to be restricted to the most common subtypes of ductal breast cancers. Several technical considerations are necessary to ensure accurate results when immunohistochemistry is being used to measure HER-2/*neu* expression. For example, high concentrations of anti-HER-2/*neu* antibodies can result in membrane and cytoplasmic staining of both benign and malignant cells. The reagents must be titered to react only with malignant cells, or the test will not discriminate between overexpression and baseline expression. The cellular location of the signal is also important as emphasized by recent in-vitro studies showing that development of the malignant phenotype in HER-2/*neu* transfected cells is strictly dependent on the presence of the oncoprotein at the cell surface (25, 66). De Potter et al. (63) have recently shown that the molecular weight of the protein associated with cytoplasmic staining (155 kD) is less than the protein associated with membrane staining

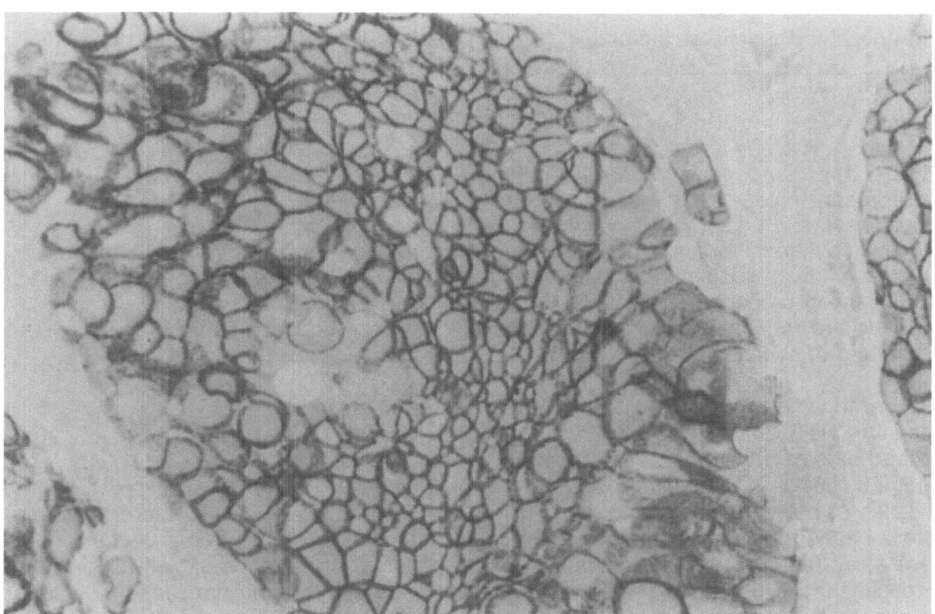

Fig. 1. Example of a breast carcinoma overexpressing the HER-2/neu oncoprotein as detected by permanent-section immunohistochemistry. Positive staining is localized to the surface membranes of malignant cells. (200X magnification)

TABLE 1. HER-2/*NEU* PROTEIN EXPRESSION IN HISTOLOGIC SUBTYPES
OF HUMAN BREAST CARCINOMAS

CATEGORY	POSITIVE/TOTAL CASES	% POSITIVE
All Carcinomas	128/736	17.4
All Invasive Carcinomas (± In-situ)	95/677	14.0
Infiltrating Ductal	95/649	14.6
Usual/NOS	91/601	15.1
Atypical Medullary	0/21	0
Medullary	0/11	0
Mucinous	3/11	27.3
Tubular	1/4	25.0
Papillary	0/1	0
Signet Ring Cell	0/4	0
Metaplastic	0/3	0
Infiltrating Lobular	0/21	0
All In-situ Carcinomas (± Invasive)	84/301	27.9
Ductal Carcinoma In-Situ	84/296	28.4
Comedo	61/90	67.8
Non-Comedo	23/206	11.2
Lobular Carcinoma In-Situ	0/5	0
Pure In-Situ Ductal Carcinomas	59	55.9
Combined In-Situ & Invasive Ductal Carcinomas	237	21.5
Pure Invasive Ductal Carcinomas	412	11.2

(185 kD), and more recent work by this group (67) suggests that the cytoplasmic signal may represent cross-reactivity with an uncharacterized mitochondrial protein. In addition, several studies have shown that the sensitivity of measuring HER-2/*neu* expression by immunohistochemistry is significantly less in formalin-fixed than in fresh-frozen tumor specimens (10, 29). Despite this problem, permanent section immunostaining probably detects most highly expressing tumors and, because fixed specimens are routinely prepared and relatively easy to obtain, will very likely become the "method-of-choice" for measuring this oncoprotein.

In addition to immunohistochemistry, Western blotting has been used successfully to measure HER-2/*neu* protein expression in breast cancers. The combined results from 4 published studies showed an average of 18.9% overexpression in 1135 breast tumors (12-13, 53, 56). These include results from our studies showing 17% expression in 728 cases of breast cancer analyzed by Western blotting (12). Like Southern blotting, Western blotting requires pampered fresh-frozen specimens, considerable technical expertise, and is subject to inaccuracies resulting from the presence of non-tumor cells in the specimens. When performed properly, however, Western blots provide a reliable, semi-quantitative assessment of HER-2/*neu* expression (Figure 2), and have been routinely performed in our laboratory on a large scale.

The incidence of HER-2/*neu* amplification and expression in clinical breast cancer appears to be nearly the same, suggesting that there is biological concordance of these phenomena. Consistent with this, several studies including ours (12) have directly examined the relationship between amplification and expression of HER-2/*neu* in the same series of tumors, and found a high correlation between the two (Figure 3). Other studies, however, have shown that, while expression and amplification are similar in late clinical stage tumors, expression is much more common than amplification in early stage tumors (56). In addition, the incidence of expression has been shown to be very high in in-situ carcinomas (30, 59, 61), while no evidence of amplification was found in the only study in which it was measured in non-invasive tumors (27). Our studies (30) have recently extended these findings by showing significant differences in HER-2/*neu* expression in tumors stratified on the basis of histological composition (Table 1). In an analysis of 736 tumors, we found a very high rate of (immunohistochemical) expression in pure in-situ carcinomas (56%), an intermediate incidence of expression in invasive carcinomas with a significant in-situ component (22%), and low levels in pure invasive carcinomas (11%). It can be argued that these histologically distinct categories represent sequential stages in tumor progression, and the decrease in oncogene expression observed during this (hypothetical) evolutionary sequence, combined with results from the studies cited above, suggest that HER-2/*neu* expression is a very early event in malignant transformation, and precedes oncogene amplification. Furthermore, to explain the relatively low incidence of expression in pure invasive carcinomas, one must hypothesize that individual tumors lose expression over time, and/or that many invasive carcinomas arise de novo by mechanisms not involving HER-2/*neu*. We have observed rare tumors in which the in-situ component was HER-2/*neu* positive, while the invasive component was negative, providing circumstantial evidence that some tumors lose expression. The observation that very small pure invasive carcinomas are common favors the hypothesis that these lesions may also arise de novo, independent of HER-2/*neu* activation.

To be meaningful, the interpretation and comparison of results between different studies must take into account both the techniques used to measure HER-2/*neu* activation, and the histological composition of the tumors being studied. Amplification and expression probably occur together in the "average" infiltrating carcinoma encountered in clinical practice, and the incidence of HER-2/*neu* activation in these lesions is likely to be somewhere between 20% and 30%. This assumption is consistent with recent results by Naber et al (68), in which amplification and expression were measured by several

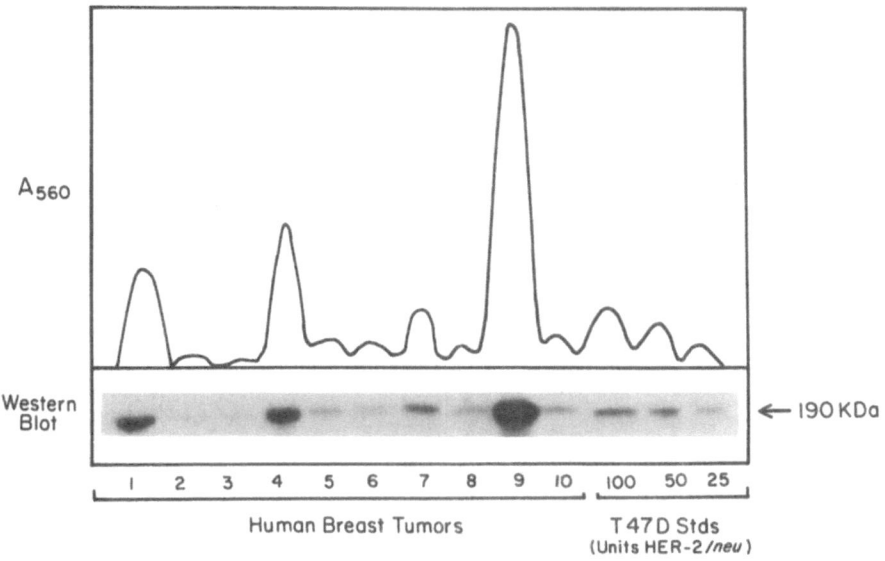

Fig. 2. Semiquantitation of HER-2/neu protein in tumor specimens by Western blot analysis. An SDS extract of human breast cancer cells (T47D) was used as our laboratory standard. The HER-2/neu bands at 190 kdD along with densitometric scan are shown. [With permission; from Tandon et al. (12)]

methods in the same series of fresh tumors specimens, and found to occur in the same tumors at a rate of 30%.

RELATIONSHIP OF HER-2/*NEU* WITH OTHER PROGNOSTIC FACTORS

One approach to studying the possible function of a potential prognostic factor is to evaluate its relationship with established indicators of clinical outcome. Many studies have looked at the relationship of HER-2/*neu* amplification and/or expression with other prognostically important clinical-pathological variables in human breast carcinoma. The most consistent findings that have emerged from these studies are that HER-2/*neu* activation is significantly associated with negative estrogen receptors (8-9, 12-13, 29-30, 49, 52,54-55, 63, 69), negative progesterone receptors (12-13, 29-30, 52, 54-55, 69), high histological grade (9, 27, 55, 64, 69), high nuclear grade (27, 30, 49, 60) and positive axillary lymph nodes (12-13, 46, 49, 51), all of which are indicators of poor prognosis.

Results from our studies (30) suggest that the relationship between HER-2/*neu* expression and other prognostic features is dependent on the histological composition and evolutionary stage of the tumors (Table 2). In an immunohistochemical analysis of 629 node-negative breast cancers, the subset of 221 tumors composed of both in-situ and invasive carcinoma showed significant correlations between HER-2/*neu* expression and younger patient age, premenopause, negative estrogen receptors, negative progesterone receptors, and high nuclear grade. In contrast, the subset of 408 pure invasive tumors showed a relationship with negative estrogen receptors only. There is currently no data available to explain the apparent independence that emerges between HER-2/*neu* expression and other prognostic features as lesions progress from "early" invasive carcinomas with a significant in-situ component, to "late" pure invasive lesions where, presumably, the in-situ component has been completely overgrown.

The relationship between HER-2/*neu* and the growth rate of tumors is of particular interest due to the hypothesized function of the oncoprotein as a growth factor receptor. The few studies that have addressed this relationship have found

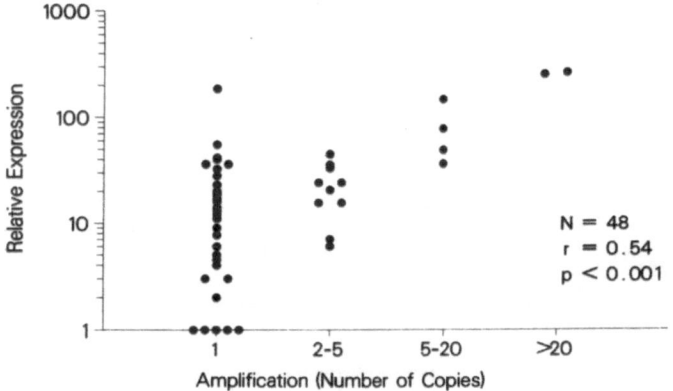

Fig. 3. Correlation between amplification and expression (as protein levels) of the HER-2/neu oncogene in human breast tumors. Amplification values are from Slamon et al. (8). (r = Spearman correlation coefficient). [With permission; from Tandon et al. (12)]

TABLE 2. ASSOCIATION OF HER-2/*NEU* PROTEIN EXPRESSION WITH CLINICAL-
PATHOLOGICAL FEATURES KNOWN TO HAVE PROGNOSTIC SIGNIFICANCE
IN HUMAN BREAST CANCER

CHARACTERISTIC	"EARLY" IDC[a] + DCIS[b]	"LATE" PURE IDC
No. of Patients	221	408
Age	p. = 0.03 (young)	ns[c]
Menopause	p = 0.02 (pre)	ns
Tumor Size	ns	ns
Estrogen Receptors	p = 0.002 (neg)	p = 0.01 (neg)
Progesterone Receptors	p = 0.003 (neg)	ns
Histologic Grade	ns	ns
Nuclear Grade	p = 0.005 (high)	ns
Ploidy	ns	ns
S Phase	ns	ns (p = 0.06)

Footnotes

IDC[a] = Infiltrating ductal carcinoma.

DCIS[b] = Ductal carcinoma in-situ.

ns[c] = not significant.

conflicting results. Heintz et al. (26) and Tsuda et al. (27) have reported a
significant association between HER-2/*neu* amplification and growth fraction as measured
by mitotic rate. Similarly, Bacus et al. (28) recently described a strong relationship
between HER-2/*neu* expression and proliferation rate in non-invasive carcinomas of the
breast. In contrast, Kommos et al. (29) found no correlation between HER-2/*neu*
expression and tumor growth fraction using Ki-67 immunostaining as a measure of cell
cycle activity. We found a near-significant trend ($p = 0.06$) in the relationship between
expression and S-phase as determined by flow cytometry (Table 2). These findings may
not be contradictory in the sense that, while HER-2/*neu* may be overexpressed, its ligand
may not be present in the same quantities in all tumors. Again, the recent identification of
a ligand for the HER-2/*neu* protein may help resolve this important issue (43).

ASSOCIATION OF HER-2/*NEU* WITH PATIENT OUTCOME

The first study regarding the prognostic significance of HER-2/neu was by Slamon
et al. (8), showing that amplification of the oncogene was associated with decreased
disease-free survival (DFS) and overall survival (OS) in 86 patients with node-positive
breast cancer. Since this initial report, there have been many others looking at the
prognostic significance of HER-2/*neu* activation in human breast cancer. At least 17
studies have been published which evaluated patients irrespective of clinical stage, with 9
finding (9, 11, 14, 17, 47-48, 59, 64-65) and 8 not finding (12, 15, 26, 46, 50, 60,
62, 70) a relationship between HER-2/*neu* and clinical outcome. The issue is much more
clear in studies analyzing only axillary node-positive patients, where 8 of 8 published
reports have shown a strong relationship between HER-2/*neu* activation and poor clinical
outcome (8-15). These include results from our laboratory involving 350 node-positive
patients in which HER-2/*neu* expression, as measured by Western blotting, was strongly
correlated with both early recurrence and decreased survival (Figure 4) (12).

In contrast, the relationship between prognosis and HER-2/*neu* activation is not
clear in node-negative breast cancer patients, which is unfortunate considering the
controversy regarding which of these patients should receive adjuvant chemotherapy. At
least 10 studies have addressed this issue, with 6 not finding (10-15) and 4 finding (9,
16-17, 71) correlations between HER-2/*neu* activation and clinical outcome (Table 3).

In 3 of the positive studies, relationships were observed between HER-2/*neu* activation and OS, but not with DFS (9, 16-17). In the positive study by Paik et al. (17) the correlation between oncogene activation and decreased OS was restricted to a subset of node-negative patients with low-risk tumors defined on the basis of having "good" nuclear grade tumors.

Results from our studies involving 453 untreated breast cancer patients with axillary node-negative tumors (71) failed to show a significant influence of HER-2/*neu* protein expression on clinical outcome (Figure 5A). We also looked at the prognostic significance of expression in subsets of this patient group stratified on the basis of the histological and evolutionary composition of their tumors. These analyses were negative with the exception of the subset of 179 low-risk untreated patients with "late" pure invasive carcinomas in which expression strongly correlated with early recurrence

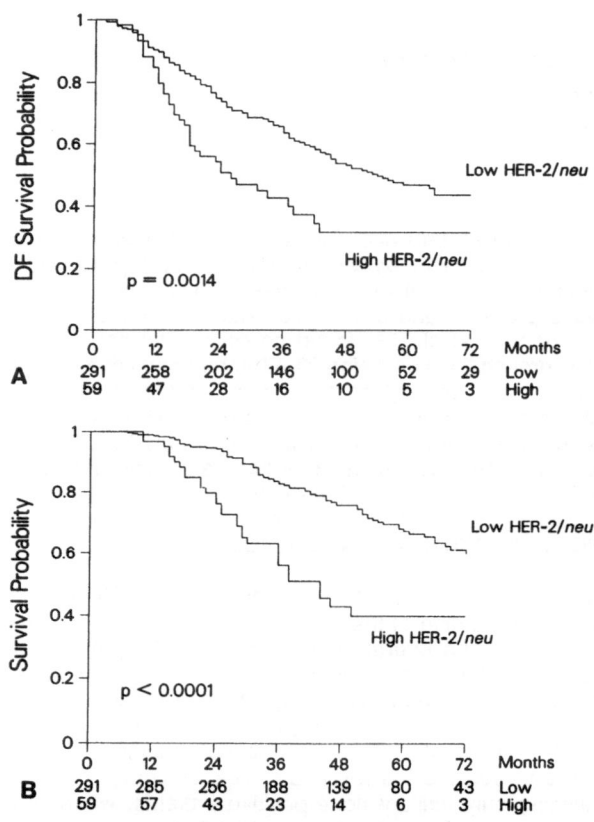

Fig. 4. Disease-free (A) and overall (B) survival curves in node-positive breast cancer patients as a function of HER-2/*neu* expression measured by Western blotting and densitometry. High HER-2/neu protein was defined as greater than 100 U. Median follow-up was 50 months. Values below the x-axis indicate the number of patients at risk at the interval shown. [With permission; from Tandon et al. (12)]

TABLE 3. PUBLISHED STUDIES WHICH HAVE EXAMINED THE PROGNOSTIC SIGNIFICANCE OF HER-2/*NEU* ACTIVATION IN PATIENTS WITH AXILLARY NODE-NEGATIVE BREAST CANCER

Reference[a]	No. of Patients	Recurrence[b]	Survival[c]
Negative Studies			
Slamon et al. (10)	181	ns[d]	ns
Tsuda et al. (11)	73	ns	ns
Tandon et al. (12)	378	ns	ns
Thor et al. (15)	141	ns	ns
Borg et al. (13)	192	ns	ns
Lovekin et al. (14)	250	ns	ns
Positive Studies			
Wright et al. (9)	44	ns	0.05
Ro et al. (16)	66	ns	0.02
Paik et al. (17)	94[e]	ns	0.0004
Allred et al. (71)	179[f]	0.0001	0.0001

Footnotes

References[a] are listed by first author in chronological order.

Recurrence[b] = associated with early recurrence indicated by univariate "p" value.

Survival[c] = associated with early death indicated by univariate "p" value.

ns[d] = not significant.

94[e] = low-risk subset of 183 total patients defined as having "good" nuclear grade tumors.

179[f] = low risk subset of 453 total untreated patients defined as having small (< 3 cm) ER-positive pure invasive tumors.

(Figure 5B). A similar correlation (p = 0.003) was observed between HER-2/*neu* expression and OS in this group of patients. The latter findings are consistent with the results of Paik et al. (17) in the sense that both studies identified relatively small groups of node-negative, otherwise low-risk patients which have very poor prognosis on the basis of HER-2/*neu* expression.

The reason for our observation that HER-2*neu* expression influences prognosis in pure invasive carcinomas is unknown. This difference may reflect associations between HER-2/*neu* and other prognostic factors which assume importance only in pure invasive lesions. For example, Bacus et al. (28) have recently shown a positive correlation between the growth rate of tumors and HER-2/*neu* expression in in-situ carcinomas of the breast. They further demonstrated that invasive HER-2/*neu* positive tumors were not associated with rapid proliferation unless they coexpressed epidermal growth factor receptor and, in this sense, perhaps invasive carcinomas have accumulated multiple activated oncogenes, or lost tumor suppressor genes, which cummulatively contribute to poor clinical outcome. Consistent with this idea, several recent studies have described finding simultaneous activation of multiple oncogenes in human breast tumors (46-47, 55). Other possible explanations for the emergence of prognostic significance associated with HER-2/*neu* expression in pure invasive carcinomas may involve the ligand for HER-2/*neu* . For example, an autocrine loop may exist between the ligand and oncoprotein in the small subset of "late" pure invasive tumors which have retained HER-2/*neu* expression, and this interaction may result in uncontrolled growth leading to poor prognosis.

SUMMARY

HER-2/*neu* is a proto-oncogene which is highly homologous to the epidermal growth factor receptor gene. It codes for a 185 kD trans-membrane protein with

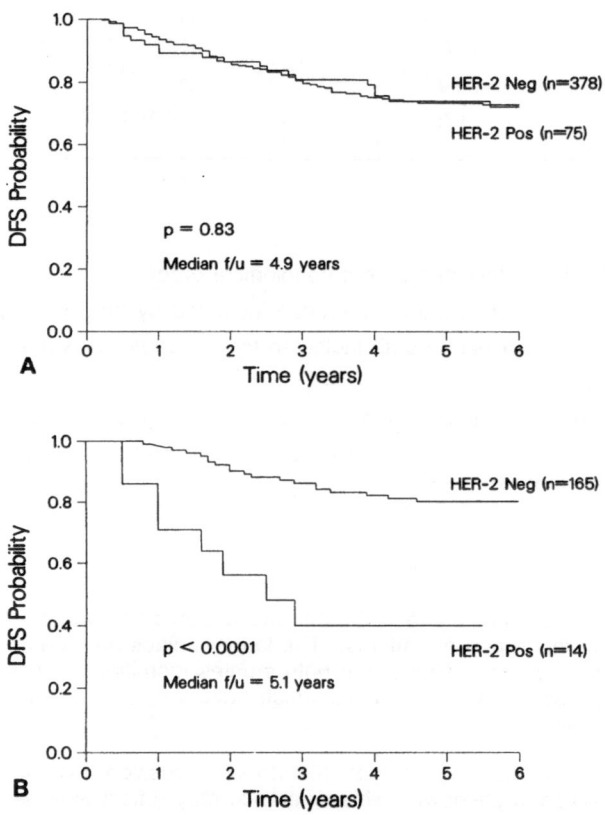

Fig. 5. (A). Disease-free survival curves for all untreated node-negative breast cancer patients stratified on the basis of HER-2/<u>neu</u> protein expression. (B) Disease-free survival curves for the subset of untreated node-negative patients with low-risk, pure invasive carcinomas stratified on the basis of HER-2/<u>neu</u> expression. Low risk was defined as small (< 3 cm), estrogen receptor positive tumors. [From Allred et al. (71)]

intracellular tyrosine-kinase activity, and is thought to function as a growth factor receptor. Amplification and/or overexpression of the HER-2/neu oncogene is present in up to 30% of invasive human breast cancers. The incidence of expression is nearly twice as high in pure in-situ carcinomas of the breast, suggesting that HER-2/neu plays a particularly important role in the early stages of tumor development. The observation that expression dramatically declines as tumors progress from non-invasive to pure invasive lesions, suggests that individual tumors lose expression over time, and/or that many invasive carcinomas arise de novo by mechanisms not involving HER-2/neu. Of clinical importance, amplificaton and expression of HER-2/neu have been shown to be strongly associated with poor prognosis in breast cancer patients with positive axillary lymph nodes. HER-2/neu activation also has a negative influence on prognosis in node-negative disease, but this relationship appears to be restricted to subsets of these patients.

REFERENCES

1. Fisher B, Constantino J, Redmond C, et al. A randomized clinical trial evaluating tamoxifen in the treatment of patients with node-negative breast cancer who have estrogen-receptor-positive tumors. N Engl J Med 1989; 320:479-84.
2. Fisher B, Redmond C, Nikolay V, et al. A randomized clinical trial evaluating sequential methotrexate and fluorouracil in the treatment of patients with node-negative breast cancer who have estrogen-receptor-negative tumors. N Engl J Med 1989; 320:473-8.
3. Mansour EG, Gray R, Shatila AH, et al. Efficacy of adjuvant chemotherapy in high-risk node-negative breast cancer. N Engl J Med 1989; 320:485-90.
4. Glick JH. Meeting highlights: adjuvant therapy for breast cancer. J Natl Cancer Inst 1988; 80:471-5
5. McGuire WL. Adjuvant therapy of node-negative breast cancer (Editorial). N Engl J Med 1989; 320:525-7.
6. McGuire WL, Tandon AK, Allred DC, Chamness GC, Clark GM. How to use prognostic factors in axillary node-negative breast cancer patients. J Natl Cancer Inst 1990; 82:1006-15.
7. Clark GM, Dressler LG, Owens MA, et al. Prediction of relapse or survival in patients with node-negative breast cancer by DNA flow cytometry. N Engl J Med 1989; 320:627-33.
8. Slamon DJ, Clark GM, Wong SG, Levin WJ, Ullrich A, McGuire WL. Human breast cancer: correlation of relapse and survival with amplification of the HER-2/neu oncogene. Science 1987; 235:177-82.
9. Wright C, Angus B, Nicholson S, et al. Expression of c-erbB-2 oncoprotein: a prognostic indicator in human breast cancer. Cancer Res 1989; 49:2087-90.
10. Slamon DJ, Godolphin W, Jones LA, et al. Studies of the HER-2/neu proto-oncogene in human breast and ovarian cancer. Science 1989; 244: 707-12.
11. Tsuda H, Hirohashi S, Shimosato Y, et al. Correlation between long-term survival in breast cancer patients and amplification of two putative oncogene-coamplification units: hst-1/int-2 and c-erbB2/ear-1. Cancer Res 1989; 49:3104-8.
12. Tandon AK, Clark GM, Chamness GC, Ullrich A, McGuire WL. HER-2/neu oncogene protein and prognosis in breast cancer. J Clin Oncol 1989; 7:1120-8.
13. Borg A, Tandon AK, Sigurdsson H, et al. HER-2/neu amplification predicts poor survival in node-positive breast cancer. Cancer Res 1990; 50:4332-7.
14. Lovekin C, Ellis IO, Locker A, et al. c-erbB-2 oncoprotein expression in primary and advanced breast cancer. Br J Cancer (In Press)
15. Thor AD, Schwartz LH, Koerner FC, et al. Analysis of c-erbB-2 expression in breast carcinomas with clinical follow-up. Cancer Res 1989; 49:7147-52.
16. Ro J, El-Naggar A, Ro JY, et al. C-erbB-2 amplification in node-negative human breast cancer. Cancer Res 1989; 49:6941-4.
17. Paik S, Hazan R, Fisher ER, et al. Pathological findings from the National Surgical Adjuvant Breast Project (protocol B-06): prognostic significance of erbB-2 protein overexpression in primary breast cancer. J Clin Oncol 1990; 8:103-12.

18. Shih C, Padhy LC, Murray M, Weinberg RA. Transforming genes of carcinomas and neuroblastomas introduced into mouse fibroblasts. Nature 1981; 290:261-4.

19. Padhy LC, Shih C, Cowing D, Finkelstien R, Weinberg RA. Identification of a phosphoprotein specifically induced by the transforming DNA of rat neuroblastomas. Cell 1982; 28:865-71.

20. Schecter AL, Stern DF, Vaidyanathan L, et al. The c-*neu* oncogene: an *erb*B-related gene encoding a 185,000 MW tumour antigen. Nature 1984; 312:513-6.

21. Coussens L, Yang-Fen TL, Liao YC, et al. Tyrosine kinase receptor with extensive homology to EGF receptor shares chromosomal location with *neu* oncogene. Science 1985; 230:1132-9.

22. Semba K, Kamata N, Toyoshima K, Yamamoto T. A v-*erb* related proto-oncogene, c-*erb*B-2, is distinct from c-*erb*B-1/epidermal growth factor receptor gene and is amplified in a human salivary gland adenocarcinoma. Proc Natl Acad Sci USA 1985; 82:6497-501.

23. King CR, Kraus MH, Aaronson SA. Amplification of a novel v-*erb*B-related gene in human mammary carcinoma. Science 1985; 229:974-6.

24. Schecter AL, Hung MC, Vaidyanathan L, et al. The c-*neu* gene: an *erb*B-homologous gene distinct from and unlinked to genes encoding the EGF receptor. Science 1985; 229:976-8.

25. Drebin JA, Link VC, Weinberg RA, Greene MI. Inhibition of tumor growth by a monoclonal antibody reactive with an oncogene-encoded tumor antigen. Proc Natl Acad Sci USA 1986; 83:9129-33.

26. Heintz NH, Leslie KO, Rogers LA, Howard PL. Amplification of the c-*erb*B-2 oncogene and prognosis of breast adenocarcinoma. Arch Pathol Lab Med 1990; 114:160-3.

27. Tsuda H, Hiroshashi S, Shimosato Y, et al. Correlation between histologic grade of malignancy and copy number of c-*erb*B-2 gene in breast carcinoma. Cancer 1990; 65:1794-1800.

28. Bacus SS, Ruby SG, Weinberg DS, Chin D, Ortiz R, Bacus JW. HER-2/*neu* oncogene expression and proliferation in breast cancers. Am J Pathol 1990; 137:103-11.

29. Kommoss F, Colley M, Hart CE, Franklin WA. Insitu distribution of oncogene products and growth factor receptors in breast carcinoma: c-*erb*B-2 oncoprotein, EGFR and PDGFR-b-subunit. Mol Cell Probes 1990; 4:11-23.

30. Allred DC, Clark GM, Molina R, et al. HER-2/*neu* oncogene expression during the evolutionary progression of human breast carcinoma. (Submitted).

31. Di Fiore PP, Pierce JH, Kraus OS, King CR, Aaronson SA. *Erb*B-2 is a potent oncogene when overexpressed in NIH/3T3 cells. Science 1987; 237:178-82.

32. Bargmann CI, Weinberg RA. Oncogenic activation of the *neu*-encoded receptor protein by point mutation and deletion. EMBO J 1988; 7:2043-52.

33. Di Marco E, Pierce JH, Knicley CL, Di Fiore PP. Transformation of NIH 3T3 cells by overexpression of the normal coding sequence of the rat *neu* gene. Mol Cell Biol 1990; 10:3247-52.

34. Benz CC, Scott GK, Santos GF, Smith HS. Expression of c-*myc*, c-Ha-*ras*1, and c-*erb*B-2 proto-oncogenes in normal and malignant human breast epithelial cells. J Natl Cancer Inst 1989; 81:1704-9.

35. Muller WJ, Sinn E, Pattengale PK, Wallace R, Leder P. Single-step induction of mammary adenocarcinoma in transgenic mice bearing the activated c-*neu* oncogene. 1988; 54:105-15.

36. Bouchard L, Lamarre L, Tremblay PJ, Jolicoeur P. Stochastic appearance of mammary tumors in transgenic mice carrying the MMTV/c-*neu* oncogene. Cell 1989; 57:931-6.

37. Mori S, Akiyama T, Yamada Y, et al. C-*erb*B-2 gene product, a membrane protein commonly expressed on human fetal epithelial cells. Lab Invest 1989; 61:93-7.

38. Quirke P, Pickles A, Tuzi NL, Mohamdee O, Gullick WJ. Pattern of expression of c-*erb*B-2 oncoprotein in human fetuses. Br J Cancer 1989; 60:64-9.

39. Yokota J, Yamamoto T, Toyoshima K, et al. Amplification of c-*erb*B-2 oncogene in human adenocarcinomas in vivo. The Lancet 1986; II:765-7.

40. Tal M, Wetzler M, Josefberg Z, et al. Sporadic amplification of the HER-2/*neu* protooncogene in adenocarcinomas of various tissues. Cancer Res 1988; 48:1517-20.

41. Gutman M, Ravia Y, Assaf D, Yamamoto Y, Rozin R, Shiloh Y. Amplification of c-*myc* and c-*erb*B-2 proto-oncogenes in human solid tumors: frequency and clinical significance. Int J Cancer 1989; 44:802-5.

42. McCann A, Dervan PA, Johnston PA, Gullick WJ, Carney DN. C-*erb*B-2 oncoprotein expression in primary human tumors. Cancer 1990; 65:88-92.

43. Lupu R, Colomer R, Zugmaier G, et al. Direct interaction of a ligand for the *erb*B-2 oncogene product with the EGF receptor and p185^{erb}B-2. Science 1990; 249:1552-5.

44. Venter DJ, Kumar S, Tuzi NL, Gullick WJ. Overexpression of the c-*erb*B-2 oncoprotein in human breast carcinomas; immunohistological assessment correlates with gene amplification. Lancet II 1987; 69-72.

45. van de Vijver M, van de Bersselaar R, Devilee P, Cornelisse C, Peterse J, Nusse R. Amplification of the *neu* (c-*erb*B-2) oncogene in human mammary tumors is relatively frequent and is often accompanied by amplification of the linked c-*erb*A oncogene. Mol Cell Biol 1987; 7:2019-23.

46. Cline MJ, Battifora H, Yokota J. Proto-oncogene abnormalities in human breast cancer: correlations with anatomic features and clinical course of disease. J Clin Oncol 1987; 5:999-1006.

47. Varley JM, Swallow JE, Brammar WJ, Whittaker JL, Walker RA. Alterations to either c-*erb*B-2 (*neu*) or c-*myc* proto-oncogenes in breast carcinomas correlate with poor short-term prognosis. Oncogene 1987; 1 :423-30.

48. Zhou D, Battifora H, Yokota J, Yamamoto T, Cline MJ. Association of multiple copies of the c-*erb*B2 oncogene with spread of breast cancer. Cancer Res 1987; 47:6123-5.

49. Berger MS, Locher GW, Saurer S, et al. Correlation of c-*erb*B-2 gene amplification and protein expression in human breast carcinoma with nodal status and nuclear grading. Cancer Res 1988; 48:1238-43.

50. Ali IU, Campbell G, Lidereau R, Callahan R. Amplification of c-*erb*B-2 and aggressive human breast tumors. Science 1988; 240:1795-6.

51. Tavassoli M, Quirke P, Farzaneh F, Lock NJ, Mayne LV, Kirkham N. C-*erb*B-2/c-*erb*A co-amplification indicative of lymph node metastasis, and c-*myc* amplification of high tumour grade in human breast carcinoma. Br J Cancer 1989; 60:505-10.

52. Zeillinger R, Kury F, Czerwenka K, et al. HER-2 amplification, steroid receptors and epidermal growth factor receptor in primary breast cancer. Oncogene 1989: 34:109-14.

53. Lacroix H, Iglehart JD, Skinner MA, Kraus MH. Overexpression of *erb*-2 or EGF receptor proteins present in early stage mammary carcinoma is detected simultaneously in matched primary tumors and regional metastases. Oncogene 1989; 4:145-51.

54. Adnane J, Gaudray P, Simon MP, Simony-Lafontaine J, Jcanteur P, Theille, C. Proto-oncogene amplification and human breast tumor phenotype. Oncogene 1989; 4:1389-95.

55. Garcia I, Dietrich PY, Aapro M, et al. Genetic alterations of c-*myc*, c-*erb*B-2, and c-Ha*ras* protooncogenes and clinical associations in human breast carcinomas. Cancer Res 1989; 49:6675-9.

56. Yamada Y, Yoshimoto M, Murayama Y, et al. Association of elevated expression of the c-*erb*B-2 protein with spread of breast cancer. Jpn J Cancer Res 1989; 80:1192-8.

57. Hanna W, Kahn HJ, Andrulis I, Pawson T. Distribution and patterns of staining of *neu* oncogene product in benign and malignant breast diseases. Modern Pathol 1990; 3:455-61.

58. Gusterson BA, Gullick WJ, Venter DJ, et al. Immunohistochemical localization of c-*erb*B-2 in human breast carcinomas. Mol Cellular Probes 1988; 2: 383-91.

59. van de Vijver MJ, Peterse JL, Mooi WJ, et al. *Neu*-protein overexpression in breast cancer. Association with comedo-type ductal carcinoma in situ and limited prognostic value in stage II breast cancer. N Engl J Med 1988; 319:1239-45.

60. Barnes DM, Lammie GA, Millis RR, Gullick WL, Allen DS, Altman DG. An immunohistochemical evaluation of c-*erb*B-2 expression in human breast carcinoma. Br J Cancer 1988; 58:448-52.

61. Gusterson BA, Machin LG, Gullick WJ, et al. Immunohistochemical distribution of

c-*erb*/b-2 in infiltrating and in-situ breast cancer. Int J Cancer 1988b; 42:842-5.

62. Gusterson BA, Machin LG, Gullick WJ, et al. C-*erb*B-2 expression in benign and malignant breast disease. Br J Cancer 1988; 58:453-7.

63. De Potter CR , Van Dalle S, van de Vijver MJ, et al. The expression of the *neu* oncogene product in breast lesions and in normal fetal and adult human tissues. Histopathol 1989; 15:351-62.

64. Walker RA, Gullick WJ, Varley JM. An evaluation for immunoreactivity for c-*erb*B-2 protein as a marker of poor short-term prognosis in breast cancer. Br J Cancer 1989; 60:426-9.

65. Gullick WJ, Love SB, Wright C, et al. c-*erb*B-2 protein overexpression in breast cancer is a risk factor in patients with involved and uninvolved lymph nodes. Br J Cancer (In Press).

66. Drebin JA, V C, Stern DF, Weinberg RA, Greene, MI. Down-modulation of an oncogene protein product and reversion of the transformed phenotype by monoclonal antibodies. Cell 1985; 41, 695-706.

67. De Potter CR, Quatacker J, Maertens G, et al. The subcellular localization of the *neu* protein in human normal and neoplastic cells. Int J Cancer 1989; 44:969-74.

68. Naber SP, Tsutsumi Y, Yin S, Zolnay SA, et al. Strategies for the analysis of oncogene overexpression. Studies of the *neu* oncogene in breast carcinoma. Am J Clin Pathol 1990; 94:125-36.

69. Marx D, Schauer A, Reiche C, et al. C-*erb*B-2 expression in correlation to other biological parameters of breast cancer. 1990; 116:15-20.

70. Zhou DJ, Ahuja H, Cline MJ. Proto-oncogene abnormalities in human breast cancer: c-*erb*B-2 amplification does not correlate with recurrence of disease. Oncogene 1989; 4:105-8.

71. Allred DC, Clark GM, Tandon AK, et al. (submitted). HER-2/*neu* expression identifies a restricted subset of axillary node-negative breast cancer patients with poor prognosis.

STRUCTURE AND FUNCTION OF MFG-E8: A NOVEL APICAL

MEMBRANE PROTEIN OF MOUSE MAMMARY EPITHELIAL CELLS

Gordon Parry, Christine Lekutis, Karen Singer, Anhthu Bui, Dale Yuzuki [#] and John Stubbs [#]

Cell and Molecular Biology Division
Lawrence Berkeley Laboratory
University of California
Berkeley, CA 94720

The milk fat globule membrane (MFGM) has proven to be an excellent source of cell surface glycoproteins that are useful markers for breast tumor diagnosis (1-3). Most studies have focused on the large mucin-like protein PEM, that is a dominant antigen on human milk fat globule membranes and the plasma membrane of breast tumor cells (4-6). Because of the remarkable immogenicity of PEM, little attention has been placed on other components of the milk fat globule membrane, although early studies demonstrated that some of the other, smaller glycoproteins had considerable potential as markers for tumor diagnosis (7).

The long term objective of the work discussed here is to build up a molecular picture of the apical surface of mammary epithelial cells, to predict the functions of the major components, and to understand the process of membrane assembly and secretion. With such basic information available, potential tumor markers could be identified by informed consideration of the structure and dynamics of the membrane. Many previous studies have demonstrated the value of the mouse for experimental studies of mammary epithelial cell function and prompted us to use the mouse as a model system for analysis of MFGM glycoproteins that may be important tumor markers.

Recently, we demonstrated that the mouse MFGM contained proteins with molecular weights of 220 kD, 150 kD, 92 kD, 66 kD, and 53 kD, in addition to PEM, which has a molecular weight > 300 kD (8). The major protein is the 66 kD molecule, with the 53 kD and 92 kD molecular weight components being present in small quantities. We have cloned and sequenced a full length cDNA coding for the 66 kD molecule and have named this protein MFG-E8 (8). In this manuscript we review the major structural features of the protein and predict its means of association with the plasma membrane. We also present immunocytochemical data demonstrating that MFG-E8 is located exclusively on the apical surface of the epithelium and suggest that it is preferentially incorporated into the budding fat droplet.

EXPERIMENTAL PROCEDURES

cDNA Cloning and Sequencing: The selection, cloning and sequencing of a cDNA molecule coding for MFG-E8 is described by Stubbs et al., (8). Computer analysis of the sequence and searches of Genbank and National Biomedical Research Foundation Protein Identification Resource were carried out using a software package from the University of California, San Francisco. Data base searches used a program devised by Pearson and Lipman (9). Sequence alignments were performed using the MALIGN and GENALIGN programs (10).

[#] Present address San Francisco State University, San Francisco CA 94132

Breast Epithelial Antigens, Edited by R.L. Ceriani
Plenum Press, New York, 1991

```
                                                         Met Gln Val Ser Arg Val Leu Ala
           GAATTCCGCATCAGAGCGCGTGGACCTTTTCCCGCGTCCCGCAGC    ATG CAG GTC TCC CGT GTG CTG GCC

 -14   Ala Leu Cys Gly Met Leu Leu Cys Ala Ser Gly Leu Phe Ala Ala Ser Gly Asp Phe Cys
       GCG CTG TGC GGC ATG CTA CTC TGC GCC TCT GGC CTC TTC GCC GCG TCT GGT GAC TTC TGT

   7   Asp Ser Ser Leu Cys Leu Asn Gly Gly Thr Cys Leu Thr Gly Gln Asp Asn Asp Ile Tyr
       GAC TCC AGC CTG TGC CTG AAC GGT GGC ACC TGC TTG ACG GGC CAA GAC AAT GAC ATC TAC

  27   Cys Leu Cys Pro Glu Gly Phe Thr Gly Leu Val Cys Asn Glu Thr Glu Arg Gly Pro Cys
       TGC CTC TGC CCT GAA GGC TTC ACA GGC CTT GTG TGC AAT GAG ACT GAG AGA GGA CCA TGC

  47   Ser Pro Asn Pro Cys Tyr Asn Asp Ala Lys Cys Leu Val Thr Leu Asp Thr Gln Arg Gly
       TCC CCA AAC CCT TGC TAC AAT GAT GCC AAA TGT CTG GTG ACT TTG GAC ACA CAG CGT GGG

  67   Asp Ile Phe Thr Glu Tyr Ile Cys Gln Cys Pro Val Gly Tyr Ser Gly Ile His Cys Glu
       GAC ATC TTC ACC GAA TAC ATC TGC CAG TGC CCT GTG GGC TAC TCG GGC ATC CAC TGT GAA

  87   Thr Glu Thr Asn Tyr Tyr Asn Leu Asp Gly Glu Tyr Met Phe Thr Thr Ala Val Pro Asn
       ACC GAG ACC AAC TAC TAC AAC CTG GAT GGA GAA TAC ATG TTC ACC ACA GCC GTC CCC AAT

 107   Thr Ala Val Pro Thr Pro Ala Pro Thr Pro Asp Leu Ser Asn Asn Leu Ala Ser Arg Cys
       ACT GCC GTC CCC ACC CCG GCC CCC ACC CCC GAT CTT TCC AAC AAC CTA GCC TCC CGT TGT

 127   Ser Thr Gln Leu Gly Met Glu Gly Gly Ala Ile Ala Asp Ser Gln Ile Ser Ala Ser Tyr
       TCT ACA CAG CTG GGC ATG GAA GGG GGC GCC ATT GCT GAT TCA CAG ATT TCC GCC TCG TAT

 147   Val Tyr Met Gly Phe Met Gly Leu Gln Arg Trp Gly Pro Glu Leu Ala Arg Leu Tyr Arg
       GTG TAT ATG GGT TTC ATG GGC TTG CAG CGC TGG GGC CCG GAG CTG GCT CGT CTG TAC CGC

 167   Thr Gly Ile Val Asn Ala Trp His Ala Ser Asn Tyr Asp Ser Lys Pro Trp Ile Gln Val
       ACA GGG ATC GTC AAT GCC TGG CAC GCC AGC AAC TAT GAT AGC AAG CCC TGG ATC CAG GTG

 187   Asn Leu Leu Arg Lys Met Arg Val Ser Gly Val Met Thr Gln Gly Ala Ser Arg Ala Gly
       AAC CTT CTG CGG AAG ATG CGG GTA TCA GGT GTG ATG ACG CAG GGT GCC AGC CGT GCC GGG

 207   Arg Ala Glu Tyr Leu Lys Thr Phe Lys Val Ala Tyr Ser Leu Asp Gly Arg Lys Phe Glu
       AGG GCG GAG TAC CTG AAG ACC TTC AAG GTG GCT TAC AGC CTC GAC GGA CGC AAG TTT GAG

 227   Phe Ile Gln Asp Glu Ser Gly Gly Asp Lys Glu Phe Leu Gly Asn Leu Asp Asn Asn Ser
       TTC ATC CAG GAT GAA AGC GGT GGA GAC AAG GAG TTT TTG GGT AAC CTG GAC AAC AAC AGC

 247   Leu Lys Val Asn Met Phe Asn Pro Thr Leu Glu Ala Gln Tyr Ile Arg Leu Tyr Pro Val
       CTG AAG GTT AAC ATG TTC AAC CCG ACT CTG GAG GCA CAG TAC ATA AGG CTG TAC CCT GTT

 267   Ser Cys His Arg Gly Cys Thr Leu Arg Phe Glu Leu Leu Gly Cys Glu Leu His Gly Cys
       TCG TGC CAC CGC GGC TGC ACC CTC CGC TTC GAG CTC CTG GGC TGT GAG TTG CAC GGA TGT

 287   Leu Glu Pro Leu Gly Leu Lys Asn Asn Thr Ile Pro Asp Ser Gln Met Ser Ala Ser Ser
       CTC GAG CCC CTG GGC CTG AAG AAT AAC ACA ATT CCT GAC AGC CAG ATG TCA GCC TCC AGC

 307   Ser Tyr Lys Thr Trp Asn Leu Arg Ala Phe Gly Trp Tyr Pro His Leu Gly Arg Leu Asp
       AGC TAC AAG ACA TGG AAC CTG CGT GCT TTT GGC TGG TAC CCC CAC TTG GGA AGG CTG GAT

 327   Asn Gln Gly Lys Ile Asn Ala Trp Thr Ala Gln Ser Asn Ser Ala Lys Glu Trp Leu Gln
       AAT CAG GGC AAG ATC AAT GCC TGG ACG GCT CAG AGC AAC AGT GCC AAG GAA TGG CTG CAG

 347   Val Asp Leu Gly Thr Gln Arg Gln Val Thr Gly Ile Ile Thr Gln Gly Ala Arg Asp Phe
       GTT GAC CTG GGC ACT CAG AGG CAA GTG ACA GGA ATC ATC ACC CAG GGG GCC CGT GAC TTT

 367   Gly His Ile Gln Tyr Val Glu Ser Tyr Lys Val Ala His Ser Asp Asp Gly Val Gln Trp
       GGC CAC ATC CAG TAT GTG GAG TCC TAC AAG GTA GCC CAC AGT GAT GAT GGT GTG CAG TGG

 387   Thr Val Tyr Glu Glu Gln Gly Ser Ser Lys Val Phe Gln Gly Asn Leu Asp Asn Asn Ser
       ACT GTA TAT GAG GAG CAA GGA AGC AGC AAG GTC TTC CAG GGC AAC TTG GAC AAC AAC TCC

 407   His Lys Lys Asn Ile Phe Glu Lys Pro Phe Met Ala Arg Tyr Val Arg Val Leu Pro Val
       CAC AAG AAG AAC ATC TTC GAG AAA CCC TTC ATG GCT CGC TAC GTG CGT GTC CTT CCA GTG

 427   Ser Trp His Asn Arg Ile Thr Leu Arg Leu Glu Leu Leu Gly Cys OC
       TCC TGG CAT AAC CGC ATC ACC CTG CGC CTG GAG CTG CTG GGC TGT TAA  TGCTCAGTCCTGCCA
```

```
GCCCAAACGATGAGGATGGCCAGAGGCTGAGGGGCCTCCTGGCCCTGCCTCCCAGGCCCTGCTGCCTTCTGTGGCTGAC
GACCTTCTCGGCCTTCCCTCCTGATTGTACTGGGGCTGGAGGCAGGAAGGGCCAGGGGATTTCAGAGTTGCCCCTTCACC
CTTTCCCTCACCCTGCAGCCCCCACAGGCCTCCTGCTAGCCCCCTTCTCTCAGGCATTCTGGGGGAGTTGGACAGGTCT
GAGATGAATAGAGAAGAAGAGTGAAGTTGGGGTATGTGGGCTATCTGTACCAACCACCCCAAGTCCTAAACTTCCTGCC
AGGGCTTGACTCAGGACTGAAGGGAGCCCCTGACTGCCCATCCCTCTCTGCACACCACACATTCCTCCATGTTCCATTC
CGGGAAGGAGAGGCCCACGTCCGCTTGCTGTGTCCCTTGGGTCACCAGGTCCTGCCTCCTATCTCCTGAGACGCCTCTTGA
CCCTTGCACTGGAGCCTCAGTTGACAAGGAGACTGGCGGGTCTGGAGAGGTCGGTGGCTCTGGGTGGTTGACAGGTTGG
CTGTGGGACCTCTGCTGGCTTGCTACCCAAGTTAACAAGCAGATTCCAAAATACATTCGTGTTCTCCACTGGAAAAAAA
AAAAAAAAGGAATTC
```

Figure 1. Nucleotide sequence and predicted amino acid sequence of the 2.1 kb cDNA coding for MFG-E8. The translated sequence is a protein with a molecular weight of 51,473. The underlined sequences denote possible N-glycosylation sites.

<u>Peptide synthesis and Production of Antibodies:</u> Peptides were synthesised at the Microchemical Facility at the University of California at Berkeley. Antibodies were produced in New Zealand white rabbits using 600 ug of peptides injected in an unconjugated form. Animals were immunized at 3 and 5 weeks after the first set of injections and bled 10 days after the third injection. Sera were screened by ELISA procedures using peptides that had been absorbed to the surface of Immunolon plates.

<u>Western Blotting Procedures:</u> SDS polyacrylamide gel electrophoresis and Western blotting were carried out as previously described (11). Typically peptide antisera were diluted 1:1000 for use. Alkaline phosphatase conjugated second antibodies were used to detect reactive bands.

<u>Immunocytochemical Procedures:</u> Mammary tissue from lactating mice was fixed in 3.5% formaldehyde and frozen in liquid nitrogen. Sections were cut on a cryostat at -30°C, post fixed in acetone at -20°C, and stained with rabbit antibodies against peptide GP-2. Primary antibodies were detected with Texas red conjugated goat anti rabbit IgG .

RESULTS AND DISCUSSION

<u>Structural Features of MFG-E8</u>

A full length cDNA, 2.1 kb long was shown to code for the complete sequence of MFG-E8 (fig.1), (8). The open reading frame in this sequence codes for a protein with a molecular weight of 51.5 kD. This is significantly smaller than the size of the protein as determined by SDS polyacrylamide gel electrophoresis and probably reflects post-translational modification of the core. The sequence contains four potential N-glycosylation sites and is rich in serine and threonine, and thus may be glycosylated at either N- or O-linkages (fig. 1).

Computer assisted analysis of the protein sequence reveals a striking organization of structural domains (fig. 2). A characteristic cleavable signal sequence of 22 aa is found at the N terminus and this is followed by an 86 amino acid cysteine rich domain. A short sequence containing 5 proline residues links the cysteine rich domain to two interally repeated sequences, referred to as C1 and C2. The C1 and C2 regions of the sequence together constitute 65% of the molecule and exhibit a 47% amino acid sequence identity. Residues 139 to 281 constitute C1 and residues 299-441 constitute C2. Certain sequences in these two regions are extremely well conserved, notably a stretch of 10 amino acids from 272-281 and 432-441. It is possible that these sequences are functionally important for MFG-E8.

Hydropathicity analysis of the complete sequence identifies the N-terminal sequence as the only stretch of hydrophobic amino acids long enough to constitute a transmembrane domain (figure 3). In general, the molecule is hydrophilic with occasional pockets of 4-7 hydrophobic amino acids. An additional sequence of 8 amino acids at the C-terminal is hydrophobic, and terminates in a Cys residue.

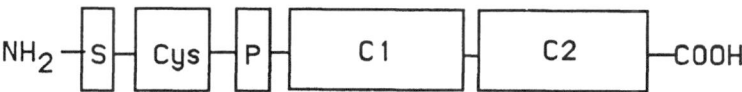

Figure 2. Principal structural features of MFG-E8 sequence. (S), putative cleavable signal sequence; (Cys), cysteine rich region; (P), proline rich sequence; (C1 and C2), internally repeated sequences of the molecule. The boundaries of each of these regions are defined in the text.

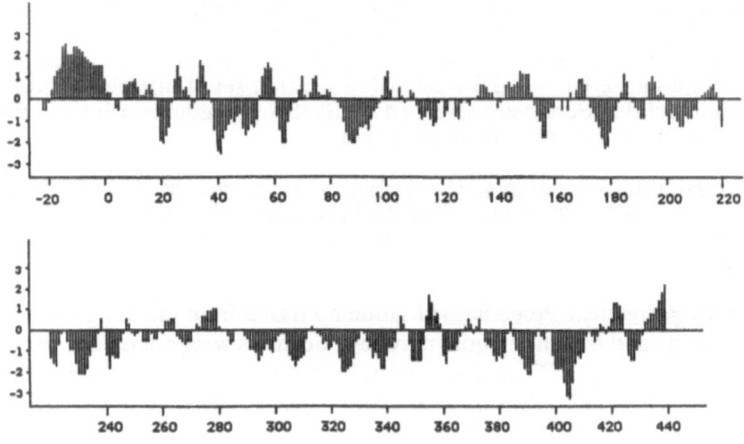

Figure 3. Hydropathicity plot of the complete amino acid sequence of MFG-E8. Hydropathicity index was calculated according to Kyte and Doolittle (12).

Homology of MFG-E8 to other Proteins: Implications for Predicting MFG-E8 Structure and its Interactions with the Plasma Membrane

As reported in detail previously, (8) the Cys rich A domain is composed of a tandem repeat of an EGF-like sequence with the organization $[CX_4CX_5CX_nCXCX_8C]_2$. Assuming that the disulfide bonds in this sequence are paired in the same way as for EGF, then a secondary structure for this domain can be established (13). The proposed structure is found to take on a conformation that is totally compatible with a secondary structure predicted by a Chou/Fassman analysis of the sequence (data not shown).

A second homology that is clear in this sequence is between the C_1 and C_2 domains of MFG-E8 and the C_1 and C_2 domains of coagulation factors V and VIII. These two coagulation factors are significantly larger than MFG-E8 and the region of similarity maps to the light chain region that has been implicated in the association of factor VIII with the surfaces of platelets. In particular, this region is thought to bind of factor VIII to phosphatidyl serine (14-16). A rigorous analysis of the specific regions of factor VIII that bind to phosphatidyl serine using synthetic peptides to block binding, revealed that the factor VIII binding site was close to the C terminal of the protein (17). Significantly this portion of the sequence of factor VIII is very similar to the highly conserved sequence of the C_1 and C_2 domains, that was discussed previously, and is also highly conserved between mouse MFG-E8 and human factor VIII. The sequence constitutes the terminal 22 amino acids of MFG-E8. Based on this information we speculate that MFG-E8 is also a phospholipid binding protein with specificity for phosphatidyl serine, and that the sequences of amino acids from 260 to 281, and from 420 to 441 constitute binding sites. A significant consideration in evaluating this hypothesis is that phosphatidyl serine is only a minor component of the outer leaflet of the phospholipid bilayer of many cells. While the assymmetry of the phospholipid bilayer of the milk fat globule membrane has not been determined it is possible that the quantity of membrane phosphatidyl serine may restrict the capacity of the membrane to bind MFG-E8.

Utilizing both information on the secondary structure of the cysteine rich domain and the possibility that MFG-E8, like factor VIII may bind phospholipids, we propose the structure shown in figure 4, as a model for how MFG-E8 may interact with the plasma membrane. In this model, the C1 and C2 domains are viewed as globular regions of the molecule that interact with the outer phospholipid leaflet of the bilayer, with the terminal ten amino acids of the C1 and C2 sequences each constituting a binding site. The proline rich sequence then functions as a link region that permits the EGF-like domain to be projected away from the surface of the bilayer. While this model is consistent with the structural data its validity clearly needs to be established experimentally.

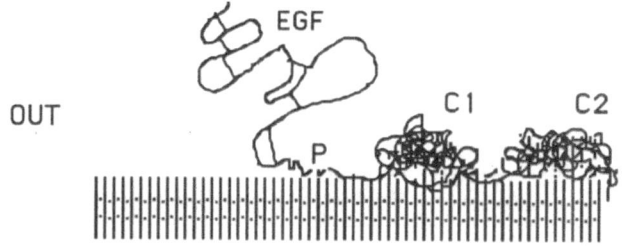

EGF

OUT

C1   C2

P

Figure 4. Proposed mechanism of interaction of MFG-E8 with the phospholipid bilayer.

MFG-E8 May Exist in Multiple Forms in the MFGM

We have raised two antibodies using peptides identical with portions of the MFG-E8 sequence: anti GP-1 that recognize the sequence of amino acids 389-410, and anti GP-2 that recognizes amino acids 58-75. Western blotting experiments using anti GP-1 (fig. 5) demonstrate that MFG-E8 is expressed not just as a 66 kD form but also as a 53 kD form. It is possible that the 53 kD form is a degradation product of the 66 kD molecule. However, N-terminal sequence analysis revealed that the 53 kD form contained the predicted N-terminal amino acids (8), and the Western blot shown above demonstrates the presence of sequences up to amino acid 410. Loss of the C-terminal amino acids from 411-441 by proteolysis could not bring about a reduction in size of the 66 kD molecule to a 53 kD form. It is most likely then that the difference in size is due to differential post-translational processing (possibly in glycosylation) of the two forms. Preliminary experiments using antibodies raised against GP-1, have also detected multiple forms of MFG-E8 in other epithelial tissues, including lung and stomach, (data not shown). This issue, however, clearly needs to be investigated futher as differential glycosylation of a single protein core to give rise to two distinct glycosylated forms in the same cell has not been extensively documented. It is possible that a post-translational modification other than glycosylation is involved.

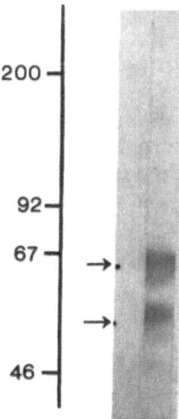

200 —

92 —
67 —
→ .

→ .

46 —

Figure 5. Western blot analysis of milk fat globule membrane proteins recognized by an antibody raised against the peptide Gp-1. Gp-1 is a hydrophilic segment of the predicted amino acid sequence shown in figure 1, including amino acids 389-410. The antibody recognizes the 66 kD band and also the 53 kD band. (Left lane: pre-immune serum; right lane, immune serum). Another affinity purified antibody eluted from purified 66 kD protein also gave the same pattern of reactivity, binding to both 66 and 53 kD proteins (8).

Immunocytochemical Staining with anti GP-2 Antibodies

Of several polyclonal antibodies generated against peptides GP-1 and GP-2, only one reacted with frozen sections of mammary tissue from lactating mice. The staining pattern seen with this antibody is shown in figure 6. The protein MFG-E8 is clearly associated only with the apical membrane and is totally absent from the basal and lateral membranes. It is also present on the membranes of the fat droplets that are seen in the lumen of the alveoli (the distorted shape of these droplets is a result of acetone fixation of the sections). A noticeable feature of the staining is that it is punctate, rather that being evenly distributed over the apical surface. While further analysis of this needs to be made, it is possible that MFG-E8 may be sequestered to localized regions of the apical membrane, such as to regions of budding fat droplets.

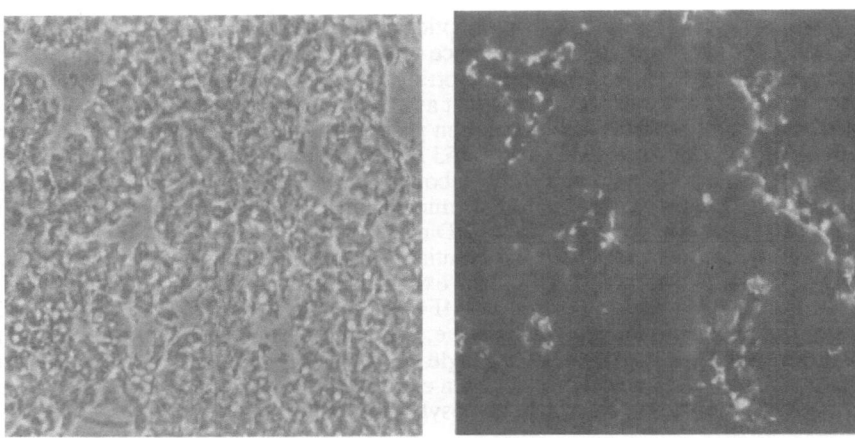

Figure 6. Immunofluorescence localization of MFG-E8 on the surface of mammary epithelia of a lactating gland. Left photograph: phase view; right photograph: fluorescence view.

CONCLUSIONS AND FUTURE PERSPECTIVES

MFG-E8 constitutes a novel component of the mammary epithelial cell surface. Structural analysis predicts it to be a peripheral membrane protein that is found exclusively on the apical surface of the cell. While the function of the protein is unknown, homology of the amino acid sequence to coaggulation factor VIII and to EGF-like proteins indicates directions of research that may lead to defining the function of the molecule. In particular, the possible phospholipid binding properties of MFG-E8, and protein-protein interactions that are typically associated with EGF-like domains (18) are important directions to pursue in a search for MFG-E8 function.

While the value of MFG-E8 as a tumor marker remains to be determined, preliminary work has demonstrated that several radiation induced mammary tumors express mRNA for MFG-E8, (manuscript in preparation). Additionally, the structural analysis presented here identifies several features of the molecule that are of interest in considering its value in tumor detection. Notably the absence of a transmembrane domain and the peripheral association of the molecule with the plasma membrane may permit MFG-E8 to be easily shed from the cell surface. It is possible then, that tumor associated MFG-E8 will be shed into the serum of tumor bearing animals and that antibodies reactive with MFG-E8 will be useful for monitoring the presence of tumor derived MFG-E8 in serum.

REFERENCES

1. Ceriani, R.L., Thompson, K., Peterson, J.A., and Abraham, S., Proc. Natl. Acad. Sci. (USA), 74, 582-586 (1977).

2. Taylor-Papadimitriou, J., Peterson, J.A., Arklie, J., Burchell, J., Ceriani, R.L., and Bodmer, W.F., Int. J. Cancer, 28, 17-21 (1981).

3. Ceriani, R.L. and Blank, E.W., Cancer Res. 48, 4664-4672 (1988).

4. Shimizu, M. and Yamauchi, K., J. Biochem. (Tokyo). 91, 515-524 (1982).

5. Abe, M., and Kufe, D., J. Immunol. 139, 257-261 (1987).

6. Gendler, S.J., Burchell, J.M., Duhig, T., Lamport D., White, R., Parker, M. and Taylor-Papadimitriou, J., Proc. Natl. Acad. Sci. (USA). 84, 6060-6064 (1987).

7. Ceriani, R.L., Sasaki, M., Sussman, H., Wara, W.M., and Blank, E.W., Proc. Natl. Acad. Sci. (USA).79, 5420-5424 (1982).

8. Stubbs, J. D., Lekutis, C., Singer, K.L. Bui, A., Yuzuki, D., Srinivasan, U., and Parry, G., Proc. Natl. Acad. Sci. (USA). 87, 8417-8421 (1990).

9. Pearson, W.R., and Lipman, D.J., Proc. Natl. Acad. Sci.(USA). 85, 2444-2448 (1988).

10. Sobel, E., and Martinez, H.M., Nucleic Acids Res. 14, 363-374 (1986).

11. Moss, L., Greenwalt, D.E., Cullen, B., Dinh, N., Ranken, R., and Parry, G., J. Cell Physiol. 137, 310-320 (1988).

12. Kyte, J., and Doolittle, R.F., J. Mol. Biol. 157, 105-132 (1982).

13. Montelione, G.T., Wuthrich, K., Nice, E.C., Burgess, A.W. and Scheraga, H.A., Proc. Natl. Acad. Sci.(USA). 84, 5226-5230 (1987).

14. Kane, W. H., and Davie, E. W., Blood 71, 539-555 (1988).

15. Ganz, P.R., Tackaberry, E.S., Rock, G., Thromb. Haemostasis 58, 222-228 (1987).

16. Arai, M., Scandella, D., and Hoyer, L.W., J. Clin. Invest. 83. 1978-1984 (1989).

17. Foster, P. A., Fulcher, C.A., Houghten, R.A., and Zimmerman T.S., Blood 75, 1990-2004 (1990).

18. Davis G.C., The New Biologist. 2, 410-419 (1990).

TARGETED LYSIS OF HUMAN BREAST CANCER CELLS BY HUMAN EFFECTOR CELLS ARMED WITH BISPECIFIC ANTIBODY 2B1 (ANTI-c-erbB-2 / ANTI-Fcγ RECEPTOR III)

David B. Ring,* Tim Shi,* Sylvia T. Hsieh-Ma,* John Reeder,* Audrey Eaton,* and Jeffrey Flatgaard**

* Department of Immunology
* Department of Purification Process Development
Cetus Corporation
1400 53rd Street
Emeryville, CA 94068

ABSTRACT

Hybridomas 520C9 (murine IgG_1 recognizing the proto-oncogene product c-erbB-2) and 3G8 (murine IgG_1 recognizing human FcγR III or CD16) were stained with vital fluorescent dyes and fused with PEG; hybrid hybridomas were isolated by dual fluorescence on a cell sorter. Cultures whose supernatants promoted lysis of SK-Br-3 breast cancer cells by human total mononuclear cells were chosen for subcloning. The immunoglobulin species produced by clone 2B1 could be resolved on anion exchange chromatography into parental antibody 520C9, a middle peak containing bispecific antibody, and parental antibody 3G8. The bispecific peak further resolved into major and minor peaks on cation exchange chromatography. The major peak appeared bispecific on the basis of SDS PAGE, isoelectric focusing, immunofluorescent binding to both tumor and effector cells, and ability to promote targeted cytolysis. The minor peak also appeared bispecific by gel criteria, but lacked tumor cell binding and cytolytic targeting ability. We tentatively suggest that the minor species of bispecific antibody is more heavily glycosylated at the hinge region in a manner that sterically interferes with activity of the 520C9 but not the 3G8 binding site. No evidence for immunoglobulin species containing heterologous light / heavy chain pairing was observed. Purified 2B1 bispecific antibody (active form) promoted lysis of erbB-2 positive breast, ovarian and lung cancer cells by human LGLs (large granular lymphocytes) and macrophages. Half maximal targeting activity was usually observed at antibody concentrations between 10 and 100 ng/ml. Target cell lysis was not blocked by the level of irrelevant immunoglobulin found in 100% autologous human serum. Lysis by human LGLs was somewhat reduced in the presence of high numbers of human PMNs, but substantial targeted killing due to 2B1 could be observed in unfractionated human blood.

INTRODUCTION

Cell-mediated cytolysis is a natural mechanism for the destruction of tumor cells or virally infected cells in vivo. Target cells may be bound and lysed by cytotoxic T cells whose T cell receptors specifically recognize processed antigen presented on HLA molecules. Alternatively, target cells may be recognized by specific antibodies and lysed by Fc receptor bearing effector cells including large granular lymphocytes, monocytes or activated macrophages. The latter process is referred to as antibody-dependent cell-mediated cytotoxicity, or ADCC.

Bispecific antibodies offer the prospect of deliberately targeting cytolytic effector cells recognized by one antibody binding site against chosen target cells recognized by another (Karpovsky et al., 1984). Such bispecific antibodies offer at least two advantages. First, a monospecific antibody must bind effector cells through its Fc portion, and must compete for such binding with other irrelevant antibody molecules. Bispecific antibodies can avoid such competition by including an Fc receptor binding site that does not compete with Fc binding. Second, monospecific antibodies cannot recruit cytotoxic T cells, which lack Fc receptors, but bispecific antibodies can be constructed to contain binding sites directed against the T cell receptor, and can then enable T cells to lyse a chosen target regardless of the T cells' original specificity. Bispecific antibodies can be constructed by chemical conjugation of whole antibodies (Karpovsky et al., 1984), by asymmetric derivatization and reassembly of antibody Fab' fragments (Brennan et al., 1985), and by generation of a hybrid hybridoma that assembles antibody molecules with one binding site derived from each parent antibody (Reading, 1984). Bispecific antibodies generated by the last route are particularly attractive because their preparation requires fewer and simpler steps once a suitable hybrid hybridoma is established.

In this paper we describe general results and problems experienced in fusing seven hybridomas that recognize human breast tumors and other carcinomas with 3G8, a hybridoma recognizing human Fcγ receptor III on human LGLs, PMNs and activated macrophages (Unkeless, 1979). We then describe in more detail the purification and characterization of a bispecific antibody recognizing the c-erbB-2 oncogene product and human Fc receptor III.

MATERIALS AND METHODS

Hybridomas and cell lines: Murine hybridomas 34F2, 113F1, 317G5 and 520C9 were made as previously described (Frankel et al., 1985; Ring et al., 1989). Murine hybridomas 15D3 and 17F9 against human P-glycoprotein were made at Cetus (J. Wrin and D. Liu, unpublished results). Hybridoma 3G8 was obtained from Dr. Jay Unkeless (Unkeless, 1979), and OVB3 was obtained from Dr. Ira Pastan (Willingham, Fitsgerald and Pastan, 1987). Human cancer cell lines BT-474, MDA-MB-175 VII and MDA-MB-361 were obtained from the Mason Research Institute; SK-Br-3 was a gift from Dr. J. Fogh; CaLu-3 was obtained from the American Type Culture Collection, and normal human foreskin fibroblast line CC95 was a gift from David Buck. All cell lines were kept in Iscove's modified Dulbecco's medium (IMDM) plus 2 mM glutamine and 10-15% heat-inactivated fetal bovine serum (FBS).

Immunofluorescent cell sorting of fused hybridomas: 520C9 cells were labeled by incubation in Hank's balanced saline solution (HBSS) containing 4 µg/ml hydroethidine (PolySciences, Inc., Warrington, PA) for 20 min at room temperature, then washed three times in growth medium without FBS. 3G8 cells were labeled by incubation for 10 min at 37 °C in complete medium with 0.4 µg/ml rhodamine 123 (Eastman Kodak, Rochester, NY) and 20 µM verapamil (Sigma, St. Louis, MO). The cells were then washed three times with serum-free medium containing 20 µM verapamil. Ten million hybridoma cells of each line were fused in 1 ml of 40% (w/v) PEG (J. T. Baker, Phillipsburg, NJ) in IMDM and diluted in 70 ml medium to recover for at least 4 hours at 37 °C before cell sorting. Flow cytometric experiments were performed using an Epics V cell sorter (Coulter Electronics, Hialeah, FL) coupled with an Innove 90 laser (Coherent Laser, Palo Alto, CA). The excitation wavelength used was 488 nm. A 550 nm dichroic mirror was used in combination with a 525 nm bandpass filter or a 610 nm long wavelength pass filter to measure fluorescence from rhodamine 123 and ethidium respectively. Data on forward angle light scatter and log 90° light scatter were collected simultaneously. The contribution of rhodamine 123 to the red fluorescence signal was subtracted using a built-in dual fluorescence compensation circuit when necessary. Doubly fluorescent cells from PEG fusions were either cloned directly in a 96-well round bottom plate at 1 cell/well or sorted into a single well of a 24 well culture plate (up to 5,000 cells/well), using a Coulter autoclone. Cells sorted into a 24-well plate were cloned manually 24 hours later at 0.5 cell per well in 96 well round bottom plates.

DNA content of hybrid hybridomas: One to four million cells were washed twice in phosphate buffered saline (PBS) and fixed by adding cold methanol to 70% for 60 min. After washing in PBS, the cells were incubated at least 30 min at room temperature with 20 µg/ml chromomycin A3 (Sigma) in PBS containing 15 mM Mg^{++}. Stained cells were analyzed on the EPICS V cell sorter, exciting at 457 nm and measuring fluorescence above 515 nm.

<u>Immunofluorescent analysis of antibody cell binding:</u> One million cells were washed with PBS / 1% bovine serum albumin (Sigma). The cells were incubated with first antibody at 5 μg/ml for 30 min at 4 °C. After washing three times in PBS/BSA, the cells were incubated with 10 μg/ml fluorescein isothiocyanate (FITC)-conjugated goat F(ab')$_2$ anti-mouse IgG Fc portion (Jackson ImmunoResearch Labs, West Grove, PA) for another 30 min at 4 °C. Propidium iodide (Molecular Probes, Inc., Eugene, OR) was added at 5 μg/ml in the last wash to stain dead cells. The samples were analyzed on the EPICS V cell sorter with an excitation wavelength of 488 nm and a 525 nm bandpass filter. Dead cells and cell debris were excluded based on forward angle light scatter and red fluorescence.

<u>Immunochemistry:</u> Anion exchange chromatography was carried out on a 0.5 x 5 cm 10 micron Mono Q column (Pharmacia) at a flow rate of 1 ml/min. Cation exchange chromatography was performed on a 0.5 x 5 cm 10 micron Mono S column (Pharmacia), also at 1 ml/min. Sodium dodecyl sulfate polyacrylamide gel electrophoresis (SDS PAGE) and isoelectric focusing (IEF) were carried out on a Pharmacia Phast gel apparatus using precast 20% acrylamide gels for SDS PAGE and precast 5-8 gradient gels for IEF. For Fab fragment production, papain (Sigma) was coupled to CNBr activated sepharose 4B (Pharmacia) at a ratio of 2 mg papain per ml beads, according to the Pharmacia protocol. Antibody at 10 mg/ml in PBS was added to papain-sepharose beads at 10 mg antibody / ml beads, and 20 mM cysteine HCL and 5 mM EDTA were also added. The mixture was rotated at 37 °C for 4 hours, the beads spun down, and the supernatant dialysed to 100 mM tris, pH 8.5. The dialysate was then run over a protein A sepharose column to remove undigested antibody and Fc fragments. Appropriate fractions (by SDS PAGE) were pooled, concentrated, and run on a Zorbax gpc column to resolve Fab from larger fragments. For Fab'2 fragment production, antibody was concentrated to 5-10 mg/ml and dialysed to sodium citrate, pH 3.5. Soluble pepsin was added at a pepsin : antibody ratio of 1:50 (w/w) and the mixture was rotated at 37 °C for 2.5 hours (520C9) or 3.5 hours (2B1). Digestion was stopped by neutralizing the mixture through dropwise addition of 1 M tris, pH 8.5. The digest was run on a Zorbax G-250 gpc column to remove small fragments, and after dialysis to tris pH 8.5, was run over a protein A sepharose column to remove undigested antibody and Fc fragments.

<u>Isolation and culture of human monocytes:</u> Human effector cells were isolated from donor buffy coats obtained from Stanford University Blood Bank (Palo Alto, Ca) and separated by Ficoll-Hypaque differential centrifugation. To isolate adherent cells, total mononuclear cells were plated in 24 well plates and allowed to adhere for 30 min at 37 °C, 5% CO2. Nonadherent cells were washed off with warm HBSS containing 50 μg/ml gentamicin. Adherent cells were cultured in AIM V medium (Gibco BRL, Gaithersburg, MD) containing 100 ng/ml M-CSF (Cetus Corp., Emeryville, CA) for 3 days before use in ^{3}H release assays.

<u>Preparation of human ovarian cancer ascites cells:</u> Samples of fresh human ovarian ascites were obtained from Dr. John Donohue, Mayo Clinic, Rochester, MN. Red blood cells and dead cells were removed by Ficoll-Hypaque differential centrifugation. Viable cells were plated out in 24 well tissue culture plates in AIM V medium with or without 10% heat inactivated human type AB serum (Gibco), and cultured from 0-2 days before use in ^{3}H release assays.

^{51}Cr release assays: Human PBLs were prepared by Ficoll-Hypaque the same day or prepared the previous day and cultured overnight in teflon bottles with 10 units/ml IL-2 (Cetus). In experiments using PMNs, PMNs were isolated by 1.5% dextran sedimentation after Ficoll-Hypaque differential centrifugation, and mixed with PBLs from the same donor at PMN:PBL ratios of around 2:1. Antibodies were diluted in IMDM with 2 mM glutamine, 50 μg/ml gentamycin, 5% heat inactivated FBS or 5-100% heat inactivated serum from the effector cell donor ("autologous serum"). Target cells (about 2 x 10^7 cells from one T150 flask) were prelabeled for 1 hour with 100 μCi sodium [^{51}Cr] chromate, washed extensively, and plated at 20,000 per V bottom microtiter well. Antibody dilutions were then added for 1 hour at 37 °C and removed before effector cells were added at an effector to target ratio of around 20:1. After 3 hours incubation at 37 °C, 100 μl supernatant was collected from each well to count chromium release. Each assay was done in triplicate. Percent specific lysis was calculated as (mean sample release - spontaneous release) / (maximum release - spontaneous release). Spontaneous release was measured from cells in medium alone, and maximum release was measured after lysis in 0.5% SDS.

^{3}H release assays: One day before an ADCC assay, target cells in T75 flasks (50% confluent) were labeled with 62.5 µCi of ^{3}H thymidine (6.7 µCi/mmole, from Dupont NEN, Boston, Mass) in 25 ml growth medium. After 30 hours, the cells were trypsinized off the flasks and washed 3 times. Twenty thousand labeled target cells were used per well in the ADCC assays. Medium used throughout the assay for diluting effector cells, antibodies and lymphokines was AIM V with 8 mM glutamine and 50 µg/ml gentamicin, and with or without 10% heat inactivated autologous serum or human type AB serum. PBLs from healthy donors were isolated by Ficoll-Hypaque and plated at 100 µl/well in 96 well round bottom tissue culture plates. Five units/ml IL-2 was added and effector cells were preincubated for 2-3 days before the addition of antibodies and ^{3}H labeled target cells. Tritium release in the supernatant was measured after 3 days, with Cytoscint scintillation fluid (ICN Biomedicals, Costa Mesa, CA). Each sample was tested in 4-6 replicates in each experiment.

RESULTS

Hybrid hybridoma 2B1 was produced from mouse IgG$_1$ κ monoclonal antibodies 520C9, recognizing c-erbB-2 (Ring et al., 1989; D. Ring, R. Clark and A. Saxena, manuscript in preparation), and 3G8, recognizing CD16 or human Fcγ receptor III (Unkeless, 1979). 520C9 was labeled with hydroethidine, which is converted to ethidium inside the cell and intercalates into DNA producing a red fluorescence (Bucana, Saiki and Nayar, 1986). 3G8 was labeled with rhodamine 123, which produces a green fluorescence (Johnson et al., 1981); verapamil was used to block the excretion of rhodamine 123 via P glycoprotein (Neyfakh, Dmitrevskaya and Serpinskaya, 1988). Concentrations of dyes were chosen not to produce toxic effects in the two hybridoma cell lines.

Labeled cells were mixed with or without PEG fusion and analysed by cell sorter. Figure 1 shows that PEG fusion produced a large population of cells showing both red and green fluorescence. These doubly fluorescent cells were cloned and their culture supernatants were screened for cytotoxic targeting ability, using a 3 hour ^{51}Cr release assay with SK-Br-3 breast cancer cells as targets and human PBLs as effectors. Several clones that caused specific lysis of SK-Br-3 were found and subcloned for further testing. These clones contained approximately twice as much DNA as parental hybridomas 520C9 and 3G8, suggesting that they were in fact hybrid hybridomas.

Subclone 2B1, derived from 5A5, was chosen for bispecific antibody purification. When 2B1 serum-free culture supernatant was chromatographed on a mono Q column, three protein peaks were observed (Figure 2A). The first peak eluted at the same position as 520C9 parental antibody, and showed a reducing SDS PAGE light chain mobility and isoelectric focusing position characteristic of 520C9 (Figure 3). Similarly, the third peak eluted at the expected position for 3G8, and showed the reducing SDS PAGE light chain mobility and isoelectric point expected for 3G8. The second peak showed an intermediate elution position, showed equal amounts of light chains with 520C9 and 3G8 mobilities on SDS PAGE, and showed an isoelectric focusing pattern of intermediate mobility. Pooled fractions from this second peak were concentrated, diafiltered into appropriate starting buffer, and chromatographed on a mono S column. Figure 2B shows that two protein peaks were observed, one eluting almost immediately and one eluting in the salt gradient. Both peaks contained equal amounts of 520C9 and 3G8 light chain and showed IEF patterns with mobility intermediate between 520C9 and 3G8, although the IEF pattern of the first peak was slightly more acidic than that of the second (Figure 3). Essentially identical chromatographic results were observed when a low pressure DEAE sepharose column was used instead of mono Q, or when low pressure S sepharose was substituted for mono S (data not shown).

Figure 4 shows the cytotoxic targeting activity of 2B1 fractions resolved on DEAE and S sepharose chromatography. DEAE peak 1 (520C9) was essentially inactive and DEAE peak 3 (3G8) contained four logs less activity than the middle peak, presumed to be bispecific antibody. When this middle peak was run on S sepharose, only the second S sepharose peak had cytotoxic targeting ability. In flow cytometry experiments (not shown), the second S sepharose peak bound both human PMNs (positive for Fc receptor III) and SK-Br-3 cells (positive for erbB-2), while the first S sepharose peak bound only PMNs. Thus, it appears that the first S sepharose peak fails to mediate cytotoxicity because it lacks an active 520C9 antibody binding site.

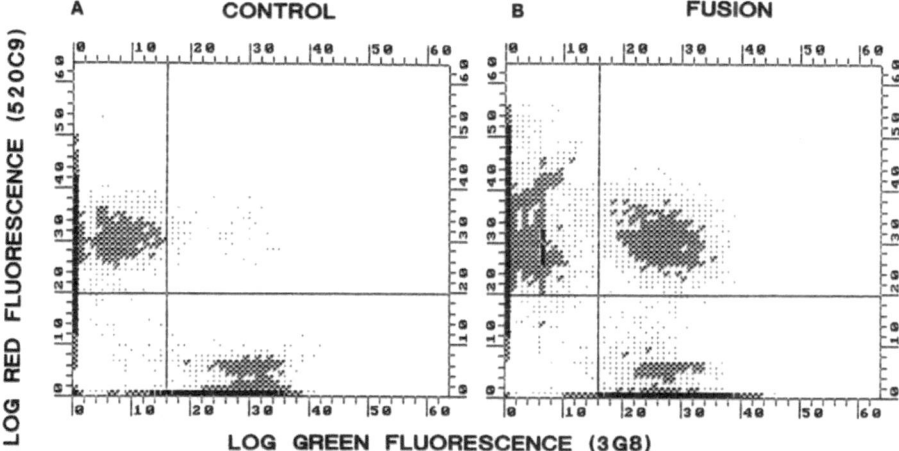

Fig. 1.   Sorting of dual fluorescent cells following PEG fusion. Hybridoma 520C9 was labeled with hydroethidine and hybridoma 3G8 with rhodamine 123. After PEG fusion, cells were sorted on the basis of dual fluorescence. Panel B shows production of doubly fluorescent hybrid hybridoma cells after fusion, while panel A shows cells that were mixed without fusing.

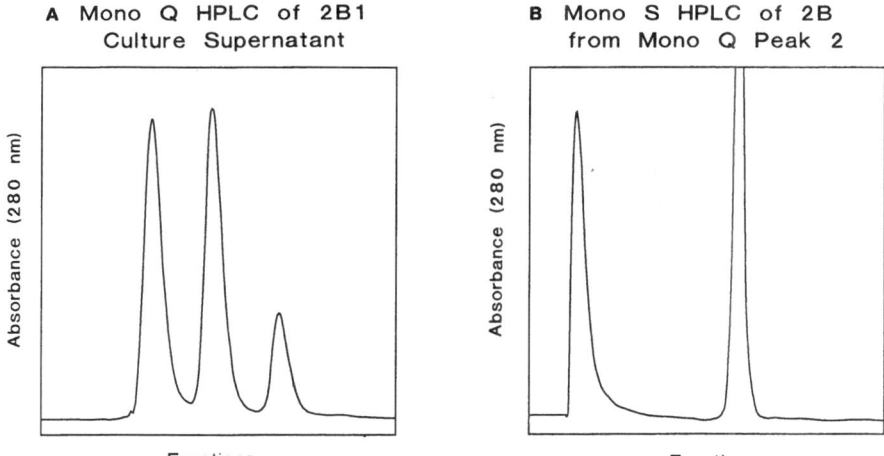

Fig. 2.   HPLC purification of 2B1 bispecific antibody. Concentrated serum-free culture supernatant of 2B1 hybrid hybridoma cells was applied to a mono Q column and eluted with a 40 ml linear gradient from 20 to 200 mM sodium phosphate, pH 7 (panel A). The pooled second peak fractions from mono Q chromatography were diafiltered, applied to a mono S column, and eluted with a 20 ml linear gradient from 0 to 450 mM NaCl in 20 mM sodium phosphate, pH 6.2.

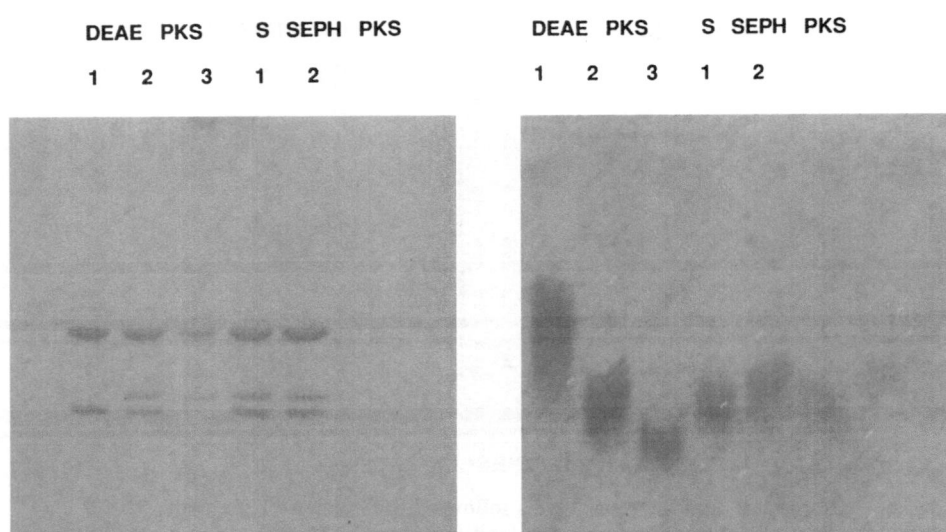

REDUCING SDS PAGE (20%)   ISOELECTRIC FOCUSING (pH 5-8)

Fig. 3.   Electrophoretic analysis of 2B1 purification fractions. 2B1 peaks from
           DEAE or S sepharose chromatography were pooled and samples were
           subjected to reducing SDS PAGE on a 20% acrylamide gel (panel A) or to
           isoelectric focusing on a pH 5-8 focusing gel (panel B). The unmarked lanes
           in panel B represent rechromatography of S sepharose peaks 1 and 2. In
           panel B, more acidic proteins are oriented toward the bottom of the gel.

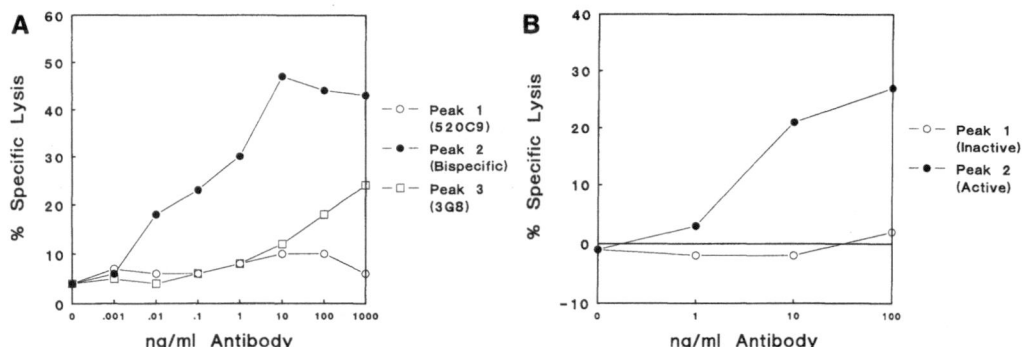

Fig. 4.   Targeted cytolysis by 2B1 purification fractions. Panel A, samples from
           DEAE sepharose chromatographic peaks were tested in a 3 hour ^{51}Cr release
           assay using human PBLs as effectors and SK-Br-3 cells as targets at a 27:1
           effector:target ratio in medium plus 5% FBS. Panel B, samples from S
           sepharose chromatographic peaks were tested in a 3 ^{51}Cr release assay using
           human PBLs as effectors and SK-Br-3 cells as targets at a 16:1 effector: target
           ratio in medium plus 10% autologous human serum.

Considering the possibility that the 520C9 binding site in the first 2B1 S sepharose peak might have lost binding activity due to variable region glycosylation, we labeled 2B1 cells by growth in tritiated galactose and glucosamine, and subsequently purified bispecific 2B1 by mono Q and mono S HPLC. In three experiments, 2B1 mono S peak 1 incorporated 50-70% more labeled sugars than mono S peak 2. However, when Fab fragments of the two sugar-labeled mono S peaks were run on nonreducing SDS PAGE and autoradiographed, no evidence was seen that labeled sugars had been incorporated into the Fab portions (data not shown). Further considering the possibility that the 520C9 binding site in 2B1 mono S peak 1 might be blocked by interaction with excess hinge region glycosylation, we tested the SK-Br-3 binding ability of mono S peak 1 and 520C9 Fab fragments. Figure 5 shows that mono S peak 1 Fab had almost exactly the same binding ability as 520C9 Fab, while intact mono S peak 1 antibody still displayed no SK-Br-3 binding.

All subsequent experiments were performed with 2B1 mono S or S sepharose peak 2 material. Figure 6 shows that 2B1 was able to promote specific lysis of SK-Br-3 cells in 3 hour ^{51}Cr release assays. A mixture of 520C9 and 3G8 parental antibodies did not cause SK-Br-3 lysis (panel A), but the lysis caused by 2B1 was comparable to that caused by a chemically linked 520C9-3G8 heteroconjugate (panel B). While the total lysis observed in 3 hour ^{51}Cr release assays was often 50% or less, extensive lysis was observed in longer assays. Figure 7 shows that 2B1 promoted complete lysis of SK-Br-3 target cells in a 3 day tritium release assay using effector cells (PBLs) from three human donors. (Lysis values over 100% occur when cytolysis releases more label than control lysis in 0.5% SDS.) In other ^{51}Cr and ^{3}H release assays (not shown), we have observed that 2B1 can mediate specific lysis of a wide range of cancer cell lines including breast cancer lines BT-20, BT-474, CaMa-1, MDA-MB-175 and MDA-MB-361, ovarian cancer line OVCAR3, and lung cancer line CaLu-3.

In targeted lysis by 2B1 acting with human PBLs, the predominant cytolytic effector cells are CD16-positive LGLs. We also observed that 2B1 could promote targeted lysis by cultured human monocytes, which become activated and express CD16 when maintained in vitro (Unkeless, 1989). Figure 8 shows that human monocytes cultured 3 days in 100 ng/ml M-CSF and supplied with 2B1 were able to target four human breast cancer and one human lung cancer cell lines, but not a normal human foreskin fibroblast used as a negative control.

If bispecific antibodies are used in vivo, they must function under conditions that may interfere with their cell complexing and targeting abilities. One concern is that the significant levels of human immunoglobulin found in serum may interfere with bispecific antibody binding to Fc receptors. Figure 9 shows that 2B1-targeted lysis of SK-Br-3 by PBLs from two human donors was not blocked by 100% autologous human serum.

CD16 or human Fcγ receptor III is found on three types of human blood cells -- LGLs, activated macrophages and PMNs (Fanger et al., 1989). We have shown that the first two cell types are active effector cells with 2B1, but experiments with 2B1 and PMNs have not shown significant effector activity (data not shown). Figure 10 shows that a 2:1 ratio of PMNs to PBLs (approximately the ratio expected in blood) significantly interfered with 2B1 targeting of SK-Br-3 cells, but did not completely eliminate targeting activity by cells from either of two human donors. An alternate way to investigate the effects of both human serum and PMNs is to carry out cytolytic assays in whole human blood. Figure 11 shows that 1 μg/ml 2B1 increased lysis of SK-Br-3 target cells by two fold in human blood and by 2-6 fold in blood supplemented with various cytokines.

Penetration of bispecific antibody to tumor sites in vivo is another concern, and suggests that early trials place antibody in situations where tumor access is relatively direct -- e.g. antigen-positive intraperitoneal tumor with i.p. administration of antibody. We have begun in vitro studies using effector and target cells derived from human ovarian cancer patients. Figure 12 shows that 2B1 promoted significant lysis of SK-Br-3 cells by total ovarian cancer ascites cells from six such patients.

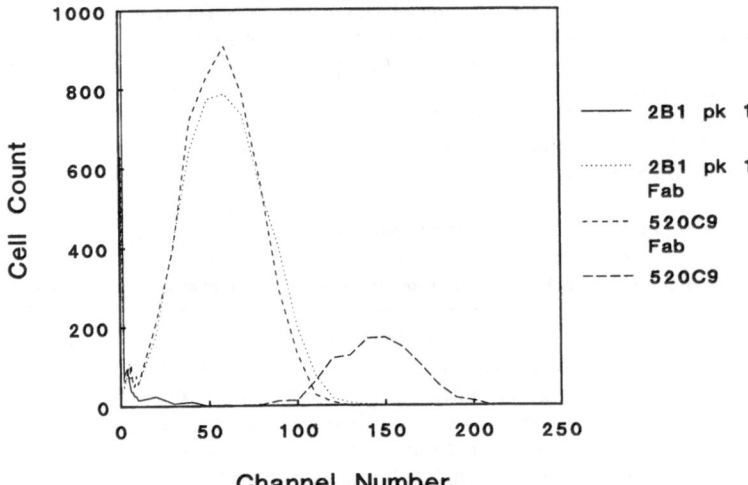

Fig. 5.   SK-Br-3 cell binding by Fab fragments of 2B1 S sepharose peak 1. SK-Br-3
          cell binding of parental antibody 520C9, 2B1 bispecific antibody from S
          sepharose peak 1, and their respective Fab fragments was compared by
          immunofluorescent flow cytometry.

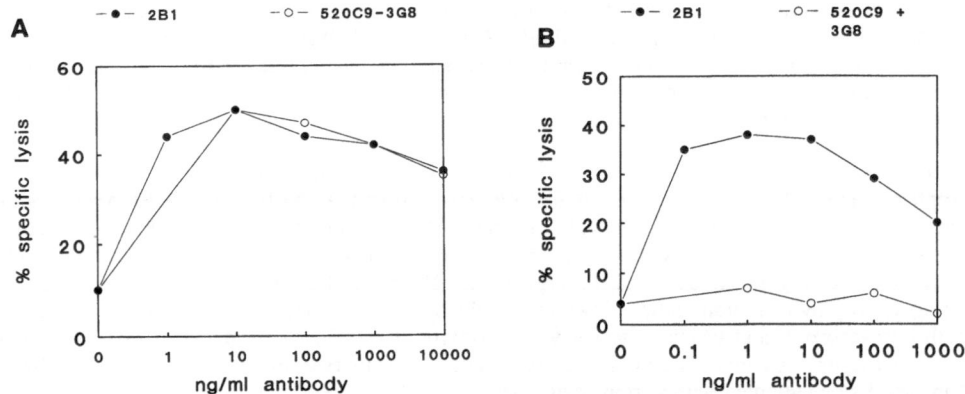

Fig. 6.   Targeted cytolysis of ^{51}Cr labeled SK-Br-3 cells by human PBLs with 2B1
          bispecific antibody or controls. Adherent SK-Br-3 cells were labeled with
          ^{51}Cr and exposed to various concentrations of 2B1 (both panels), mixed
          520C9 and 3G8 parental antibodies (panel A), or 520C9-3G8 heteroconjugate
          (panel B). Target cells were rinsed, and human PBLs were added at a 20:1
          effector:target ratio. Chromium release was measured after 3 hours.

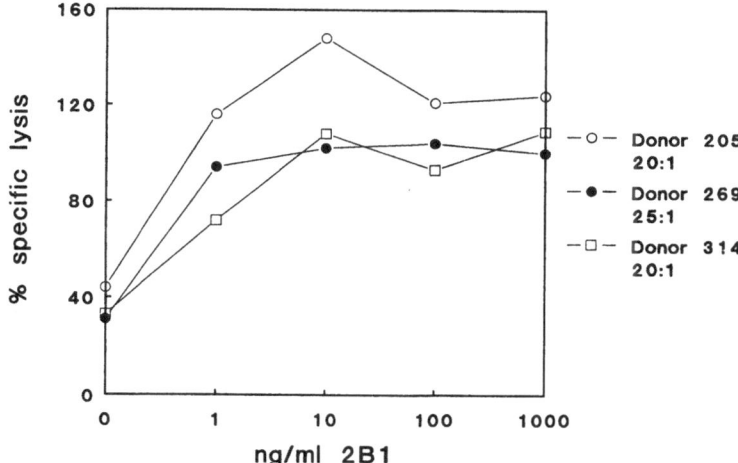

Fig. 7. Targeted cytolysis of ^{3}H labeled SK-Br-3 cells by human PBLs with 2B1 bispecific antibody. Adherent SK-Br-3 cells were labeled with tritiated thymidine and exposed to various concentrations of purified 2B1 bispecific antibody. Human PBLs from three donors (pre-incubated 3 days in 5 units/ml IL-2) were added at 20:1 or 25:1 effector:target ratios and incubated another 3 days in medium containing 10% autologous human serum before measuring tritium release.

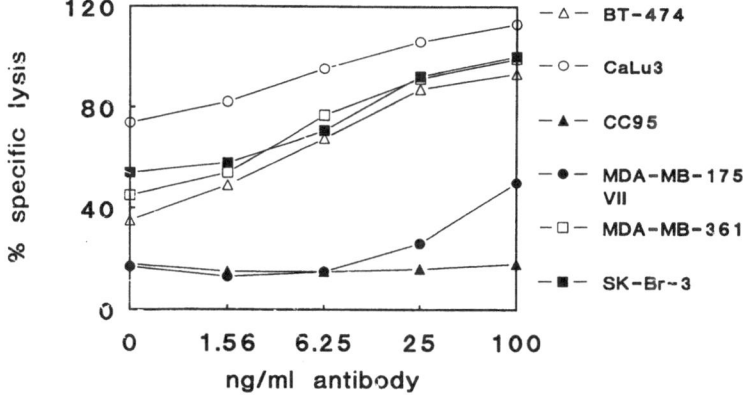

Fig. 8. Targeted cytolysis of six ^{3}H labeled human cancer cell lines by cultured human monocytes and 2B1 bispecific Fab'2 fragment. Five cancer cell lines and human foreskin fibroblast line CC95 (a negative control not expressing c-erbB-2) were labeled with tritiated thymidine and exposed to various concentrations of 2B1 bispecific antibody Fab'2 fragment. Human monocytes were cultured in 100 ng/ml of M-CSF, added to labeled target cells at effector:target ratios of 25 or 28:1 and incubated another 3 days in AIM V serum-free medium before measuring tritium release.

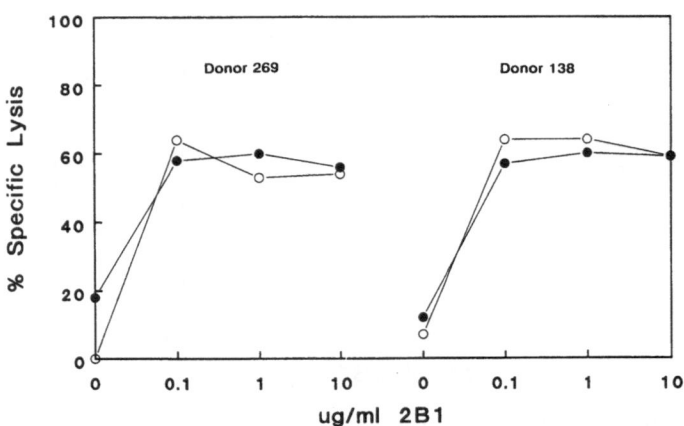

Fig. 9.   Targeted cytolysis of ^{51}Cr labeled SK-Br-3 cells by human PBLs with 2B1 bispecific antibody in 100% autologous human serum. SK-Br-3 target cells were exposed to various concentrations of 2B1 bispecific antibody, rinsed, and human PBLs from two donors were added at 40:1 effector:target ratio in either 10% or 100% autologous human serum for a 3 hour ^{51}Cr release assay.

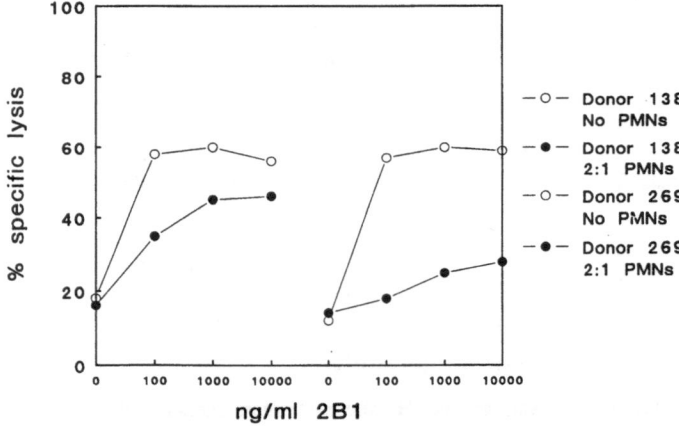

Fig. 10.   Targeted cytolysis of ^{51}Cr labeled SK-Br-3 cells by human PBLs with 2B1 bispecific antibody in the presence of human PMNs. SK-Br-3 target cells were exposed to various concentrations of 2B1 bispecific antibody, rinsed, and human PBLs from two donors were added at 40:1 effector:target ratio in the presence or absence of a 2:1 excess of autologous human PMNs for a 3 hour ^{51}Cr release assay.

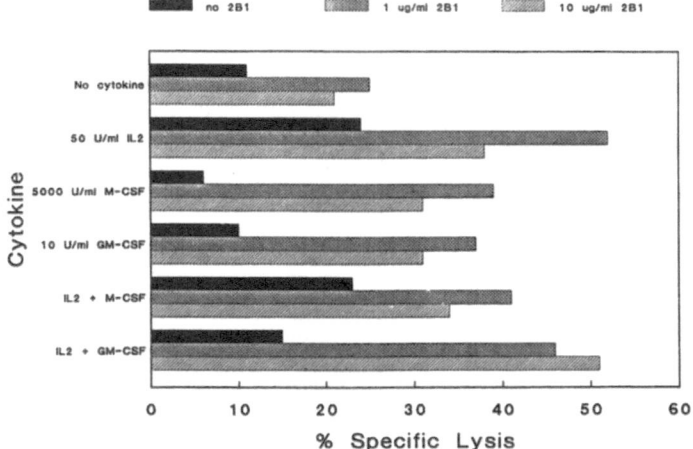

Fig. 11.   Targeted cytolysis of ^{3}H labeled SK-Br-3 cells by 2B1 in whole human
           blood. Twenty-five thousand ^{3}H thymidine-labeled SK-Br-3 cells were mixed
           with 0.5 ml fresh heparinized human blood, 1 or 10 μg/ml 2B1 bispecific
           antibody, and cytokines as indicated, and incubated 3 days at 37 °C in 2 ml
           microtiter wells before measuring tritium release.

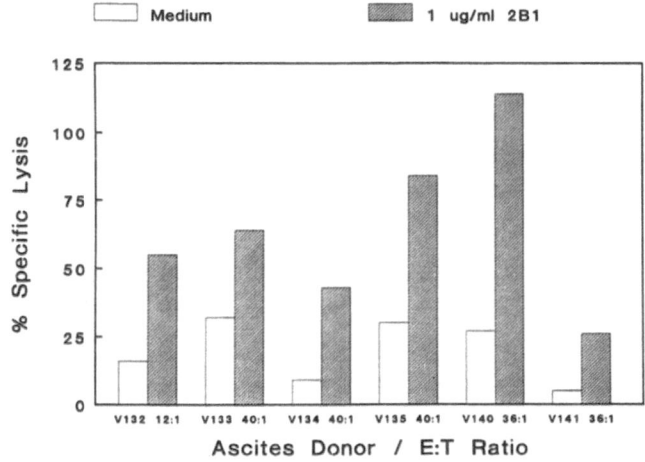

Fig. 12.   Targeted cytolysis of ^{3}H labeled SK-Br-3 cells by 2B1 with human ovarian
           cancer ascites cells. ^{3}H thymidine-labeled SK-Br-3 cells were mixed with 1
           μg/ml 2B1 bispecific antibody and human ovarian cancer ascites cells at
           effector:target ratios as indicated and incubated 3 days in AIM V serum-free
           medium before measuring tritium release.

DISCUSSION

A number of methods are available for making hybrid hybridomas. Some rely on placing different drug resistance markers in each parental cell line, and selecting fused cells for the ability to grow in the presence of both drugs (Reading, 1984; Milstein and Cuello, 1984; Martinis et al., 1985; Chervonsky et al., 1988). Alternatively, one cell may be marked for drug resistance and fused with another cell that was poisoned prior to fusion (Clark, Gilliland and Waldmann, 1988). Finally, cells may be labeled with different fluorescent dyes and sorted after fusion to select hybrids (Junker and Pedersen, 1981; Karawajew et al., 1987; Koolwijk et al., 1988). The method used to create our hybrid hybridomas falls into the last category. We have chosen the combination of hydroethidine and rhodamine 123 because these dyes stain living cells with relatively low toxicity. While rhodamine 123 is normally pumped out of cells via P glycoprotein (multiple drug resistance protein), this effect can be blocked and good staining can be achieved if verapamil is included in the medium to block P glycoprotein action.

Table 1 summarizes our results in fusing seven tumor cell reactive hybridomas with hybridoma 3G8. Six of seven pairings yielded clones that appeared to be hybrid hybridomas on the basis of combined evidence including (a) production of both parental antibodies by ELISA reactivity, HPLC, SDS PAGE and/or IEF; (b) double DNA content; or (c) production of cell complexing or cytolytic targeting activity not shown by parental antibodies or combinations. From the six pairings that yielded hybrid hybridomas, we were able to purify bispecific antibody that caused tumor cell : PMN complexing in three cases, and in two cases the bispecific antibody also promoted targeted cytotoxicity against antigen-positive tumor cells. As described above in Results, the bispecific antibody derived from 113F1 and 3G8 was able to complex tumor cells with CD16 positive cells, but was not able to mediate targeted cytotoxicity.

Table 1. Summary of Hybrid Hybridoma Fusions

Hybridomas	Antigenic Target	Fusions	Outcome
520C9 / 3G8	200 K glycoprotein (c-erbB-2)	2	obtained active bispecific clone 2B1
15D3 / 3G8	P-glycoprotein	1	obtained active bispecific clone 1A7
113F1 / 3G8	40/60/100/200 K breast cancer glycoprotein complex	11	obtained clones making bispecific antibody that bound FcR III but not tumor antigen; complexed cells but did not target cytolysis
17F9 / 3G8	P-glycoprotein	1	obtained bispecific clones; little activity observed; could not purify
34F2 / 3G8	42 K cancer glycoprotein	3	obtained bispecific clones; little activity observed; could not purify
OVB3 / 3G8	ovarian cancer	3	obtained bispecific but observed no cytolytic activity
317G5 / 3G8	42 K cancer glycoprotein	7	no active clones obtained

There are several reasons that fused hybridoma cells may not become viable hybrid hybridomas, or may not produce significant amounts of active bispecific antibody. If the nuclei of two cells that fuse are not in mitotic synchrony, condensation of the leading nucleus at the beginning of mitosis triggers premature condensation of the lagging nucleus, leading to loss of all or most of one set of chromosomes instead of successful integration (Johnson and Rao, 1970). We have analysed a number of "hybrid hybridoma" clones several weeks after fusion and sorting for double fluorescence, and found that many produce only one parental antibody and have a single hybridoma DNA content. Our assumption is that these clones lost one parental nucleus at an early point by the mechanism suggested above. It is also possible for hybrid hybridomas to lose individual chromosomes over time, and this can result in the loss of one or more antibody heavy or light chains just as observed for hybridomas cultured over long periods without subcloning. Hybrid hybridoma 2B1 has been karyotyped to have a chromosome count in the hyperhexaploid range (data not shown), as opposed to the theoretically expected octoploid count.

When a stable hybrid hybridoma making both light and both heavy antibody chains is obtained, the yield of bispecific antibody containing one binding site from each parental antibody will depend on which heterologous light / heavy and heavy / heavy chain combinations assemble successfully. If heterologous heavy chain assembly does not occur, no bispecific antibody will form. If heterologous light / heavy chain assemble occurs freely and equal amounts of all chains are made, then only 2/16 or 12.5% of the total antibody made will have one binding site of each parental specificity, and the desired bispecific antibody may be difficult to purify from other hybrid species. Our fusions of 17F9 and 34F2 with 3G8 may represent such cases. Only if heterologous heavy / heavy chain pairing is allowed but heterologous heavy / light chain pairing is disfavored will a large fraction (up to 50%) of the assembled antibody be bispecific. In fact, that situation seems to occur with reasonable frequency (at least 3 of 7 cases in our experience). There is a probably a prejudice in favor of homologous versus heterologous light / heavy chain pairing; since B cell clones are stimulated and expand only if they assemble functional antibody binding sites, homologous light / heavy pairing is guaranteed to be allowed, while the heterologous pairing opportunities that occur in hybrid hybridomas have no such guarantee.

When significant amounts of a bispecific antibody are made and successfully purified, its properties may or may not prove suitable for an intended application such as targeted cytotoxicity. In the case of 2B1, we observed two forms of apparently bispecific antibody that could be resolved on mono S or S sepharose chromatography. One form actively targeted cytotoxicity and the other was inactive, apparently due to lack of a functional 520C9 binding site.

In attempting to explain the inactive bispecific form of 2B1, we considered three hypotheses: (1) that the 520C9 binding site was inactivated by variable region glycosylation (Margni and Binaghi, 1988; Wallick, Kabat and Morrison, 1988); (2) that the 520C9 binding site was inactivated by interaction with excess hinge region glycosylation; and (3) that the molecule was not bispecific, but resulted from some other arrangement of parental antibody heavy and light chains. Sugar labeling experiments suggested that the inactive form of 2B1 was indeed more heavily glycosylated than the active form, but the sugar labels we employed were not incorporated into the Fab fragments of either form, arguing against the hypothesis of variable region glycosylation. We could also reject the third hypothesis, since the inactive form of 2B1 showed CD16 binding ability and its Fab fragments regained the ability to bind c-erbB-2, proving that the molecule did contain both 3G8 and 520C9 binding sites. Thus, we favor the hypothesis that heavier hinge region glycosylation in the inactive form of 2B1 occludes the activity of the 520C9 binding site through a direct or indirect steric interaction.

Using the mono S peak 2 form of purified 2B1, we have demonstrated targeted lysis of c-erbB-2 positive breast, lung and ovarian cancer cells using either human PBLs or cultured human monocytes as effectors. Addition of autologous human serum at concentrations up to 100% caused no perceptible interference with either cell complexing or cytolytic targeting by 2B1. On the other hand, addition of human PMNs (which bind 2B1 but do not act as cytolytic effectors against tumor cells) reduced 2B1 targeted lysis. This could occur both because PMNs compete with LGLs to bind the available supply of 2B1, and also because 2B1 decorated PMNs bind to tumor cells and block access by LGLs. The former effect could in theory be avoided by using a higher concentration of 2B1, while the latter may be more difficult to eliminate. Despite the blocking effects of PMNs, we were able to show that 2B1 targets lysis of tumor cells by either whole human blood or total ascites cells from human ovarian cancer patients. We look forward to future evaluation of 2B1 and other bispecific antibodies in animal models and human clinical trials.

REFERENCES

Brennan, M., Davison, P.F. and Paulus, H. (1985) Preparation of bispecific antibodies by chemical recombination of monoclonal immunoglobulin G1 fragments. Sci. 229:81.

Bucana, C., Saiki, I. and Nayar, R. (1986) Uptake and accumulation of the vital dye hydroethidine in neoplastic cells. J. Histochem. Cytochem. 34:1109.

Clark, M., Gilliland L. and Waldmann, H. (1988) Hybrid antibodies for therapy. Prog. in Allergy 45:31.

Chervonsky, A.V., Faerman, A.I., Evdonina, L.V., Jazova, A.K., Kazarov, A.R. and Gussev, A.I. (1988) A simple metabolic system for selection of hybrid hybridomas (tetradomas) producing bispecific monoclonal antibodies. Molec. Immunol. 25:913.

Fanger, M.W., Shen, L., Grazianno, R.F. and Guyre, P.M. (1989) Cytotoxicity mediated by human Fc receptors for IgG. Immunol. Today 10:92-99.

Frankel, A.E., Ring, D.B., Tringale, F. and Hsieh-Ma, S.T. (1985) Tissue distribution of breast cancer associated antigens defined by monoclonal antibodies. J. Biol. Resp. Modif. 4:273-286.

Johnson, R.T. and Rao, P.N. (1970) Mammalian cell fusion: induction of premature chromosome condensation in interphase nuclei. Nature 226:717.

Johnson, L.V., Walsh, M.L., Bockus, B.J. and Chen, L.B. (1981) Monitoring of relative mitochondrial membrane potential in living cells by fluorescence microscopy. J. Cell Biol. 88:526.

Junker, S. and Pedersen, S. (1981) A universally applicable method of isolating somatic cell hybrids by two-colour flow sorting. Biochem. Biophy. Res. Comm. 102:977.

Karawajew, J., Micheel, B., Behrsing, O. and Gaestel, M. (1987) Bispecific antibody-producing hybrid hybridomas selected by a fluorescence activated cell sorter. J. Immunol. Meth. 96:263.

Karpovsky, B., Titus, J.A., Stephany, D.A. and Segal, D.M. (1984) Production of target specific effector cells using hetero-crosslinked aggregates containing anti-target cell and anti-Fcγ receptor antibodies. J. Exp. Med. 160:1686.

Koolwijk, P., Rozenmuller, E., Stak, R.K., De Lau, W.B.M. and Bast, B.J.E.G. (1988) Enrichment and selection of hybrid hybridomas by Percoll density gradient centrifugation and fluorescent-activated cell sorting. Hybridoma 7:217.

Margni, R. and Binaghi, R. (1988) Nonprecipitating asymmetric antibodies. Ann. Rev. Immunol. 6:535.

Martinis, J., Kull, J.F., Franz, G. and Bartholemew, M. (1985) Monoclonal antibodies with dual specificity. In: H. Peters, (Ed.) Proceedings of the 30th Colloquium on Protides of the Biological Fluids. Pergamon Press, New York, p. 311.

Milstein, C. and Cuello, A.C. (1984) Hybrid hybridomas and the production of bi-specific monoclonal antibodies. Immunol. Today 5:299.

Neyfakh, A.A., Dmitrevskaya, T.V. and Serpinskaya, A.S. (1988) The membrane transport system responsible for multidrug resistance is operating in nonresistant cells. Exp. Cell Res. 178:513.

Reading, C.L. (1984) Procedures for in vitro immunization and monoclonal antibody production. In: B. H. Tom and J.P. Allison (Eds), Hybridomas and Cellular Immortality. Plenum Press, New York, p. 235.

Ring, D.B., Kassel, J.A., Hsieh-Ma, S.T., Bjorn, M.J., Tringale, F., Eaton, A.M., Reid, S.A., Frankel, A.E. and Nadji, M. (1989) Distribution and physical properties of BCA200, a 200,000 dalton glycoprotein selectively associated with human breast cancer. Cancer Res. 49:3070.

Unkeless, J.C. (1979) Characterization of a monoclonal antibody directed against mouse macrophage and lymphocyte Fc receptors. J. Exp. Med. 150:580.

Unkeless, J.C. (1989) Function and heterogeneity of human Fc receptors for immunoglobulin G. J. Clin. Invest. 83:355.

Wallick, S.C., Kabat, E.A. and Morrison, S.L. (1988) Glycosylation of a V_H residue of a monoclonal antibody against α(1->6) dextran increases its affinity for antigen. J. Exp. Med. 168:1099-1109.

Willingham, M.C., Fitzgerald, D.J. and Pastan, I. (1987) Pseudomonas exotoxin coupled to a monoclonal antibody against ovarian cancer inhibits the growth of human ovarian cancer cells in a mouse model. PNAS USA 84:2474.

MECHANISMS CONTROLLING STEROID RECEPTOR BINDING TO SPECIFIC DNA SEQUENCES

Dean P. Edwards, Patricia A. Estes, Sergio Oñate,
Candace A. Beck, Angelo M. DeMarzo, and Steven K. Nordeen

University of Colorado Health Sciences Center, Department of
Pathology (B216), 4200 E. 9th Ave., Denver, CO 80262

INTRODUCTION

The biological actions of steroid hormones are mediated by specific nuclear receptors to regulate an increase or decrease in transcription of specific genes and thereby exert their influence on development and growth of normal and malignant tissues of the reproductive tract. In hormone-dependent breast cancer, the primary hormone responsible for stimulating cellular proliferation is estrogen. A number of gene products under estrogen control have been identified and characterized,[1-4] and several of these estrogen regulated proteins are under investigation for their value as clinical prognostic markers.[5,6] The physiological effects of progesterone in breast cancer and the gene products under progesterone control have been studied less extensively.[7] In general, progestins have been observed to be antiproliferative in experimental studies with human breast cancer cells[8-10] and progestin therapy is receiving renewed attention in treatment of hormone dependent breast cancer.[11,12]

Steroid receptors are hormone-dependent transcriptional activators that bind to specific cis-acting DNA sequences of hormonally regulated genes. These sequences, termed hormone response elements (HREs), have been shown by mutagenesis studies to be responsible for mediating induction of gene transcription by the steroid, implying that receptor binding directly to HREs is an important step in regulation of gene transcription.[13,14] Receptors for the sex steroids require a hormone-dependent "activation" step in order to gain affinity for DNA. The mechanism, however, for the process of receptor activation and subsequent binding to HREs remains as yet poorly understood. The present paper is a review of studies from our laboratory on the mechanisms that control the binding of steroid receptors to specific DNA sites. Studies have been performed with progesterone receptors (PR) in T47D human breast cancer cells utilizing the progesterone response elements (PRE) of the mouse mammary tumor virus (MMTV) as the target gene.[15,16]

EXPERIMENTAL

Cell Cultures and Monoclonal Antibodies: T47D human breast cancer cells were cultured as previously described.[17] Progesterone receptors (PR) are expressed in T47D cells as two different sized proteins of 120 kDa and 94 kDa, termed PR-B and PR-A, respectively.[18] PR-A has been shown to be a truncated form of PR-B, missing N-terminal sequences, but otherwise containing DNA and steroid binding domains identical to PR-B.[19] Although the mechanism whereby PR-A originates is in question,[20,21] it appears to be a true cellular product. The functional role of PR-A and PR-B are not completely understood, but they have been reported to exhibit different target gene specificities.[22] Monoclonal antibodies (MAbs) designated AB-52 and B-30 (mouse IgGs) were prepared against purified PR from T47D cells as previously described.[17] AB-52 recognizes both the A and B forms of PR while B-30 recognizes PR-B only.

Immune-Isolation of Progesterone Receptors: Buffers used for cellular extraction of PR contained 10 mM Tris-OH, pH 7.4, 1 mM dithiothreitol, 1 mM EDTA, and 10% glycerol plus a mixture of protease inhibitors.[17] Cytosolic PR from non-hormone treated cells were prepared as a 105,000 × g supernatant of cellular homogenates. To stabilize the unactivated 8-10S cytosol receptor complex, 20 mM sodium molybdate was included in homogenization buffers. To prepare activated PR in cell-free cytosol, sodium molybdate was omitted and cytosols were incubated with the synthetic progestin, R5020 for 4 h at 4°C followed by exposure to 0.5 M NaCl for 1 h at 4°C. Nuclear PR, activated by addition of R5020 to intact cells for 1 h at 37°C, were prepared as previously described as a 0.5 M NaCl extract of isolated nuclei.[23] Samples of cytosol or nuclear PR were dialyzed against TEDG buffer to reduce the salt concentration prior to immune isolation and DNA binding assays. Protein A Sepharose was pre-coated noncovalently with receptor specific MAbs (AB-52 or B-30) for use as an immunoabsorbent. Aliquots of cytosol or nuclear PR were incubated with MAb coated beads on an end-over-end rotator for 4 h at 4°C. Resins were then washed three times in TEG containing 0.3 M NaCl.[24]

SDS-Gel Electrophoresis and Western Immunoblotting: Immobilized PR complexes were eluted directly from protein A Sepharose with SDS sample buffer. Electrophoresis and Western immunoblotting of extracted PR was carried out as previously described using ^{35}S-protein A and autoradiography of dried nitrocellulose for detection.[25]

DNA Binding Assay. Gel-mobility shift assays were performed as previously described.[25] Approximately 0.06 pmole (based on steroid binding assay) of PR in cellular extracts were incubated for 20 min at room temperature with 0.1 ng of a [^{32}P]end-labeled synthetic oligonucleotide (5'-gatc-TTTGGTTACAAACTGTTCTTAAAACGAG-3: 3'-AAACCAATGTTTGACAAGAATTTTGCTCctag-5') corresponding to the distal most (-189 to -162 from the transcription start site) progesterone/glucocorticoid response element (PRE/GRE) of mouse mammary tumor virus (MMTV).[15,16] This will be referred to as PRE. The DNA binding reactants were submitted to electrophoresis and autoradiography as described.[25]

RESULTS

Hormone-Dependence of Progesterone Receptor Binding to Specific DNA Sites

Receptors for sex steroid hormones are bound weakly to DNA/chromatin in nuclei of hormone-free cells and are recovered largely in the cytosol fraction as 8-10S oligomeric complexes composed of the steroid binding polypeptide in physical association with another cellular protein, identified as the 90 kDa heat shock protein (hsp 90). Receptors require a hormone-dependent "activation" step in order to gain affinity for DNA/chromatin in nuclei of intact cells. It appears that the receptor-associated hsp 90 functions to repress DNA binding of receptors in the absence of hormone. In support of this, hormone-dependent activation either *in vitro* or in intact cells is found to result in dissociation of hsp 90 from receptors and their conversion from the 8-10S non-DNA binding form to the 4S DNA binding form. Therefore, at least one step in the receptor activation process is thought to be hormone-induced release of hsp 90.[26,27]

A fundamental issue with respect to receptor activation that remains unclear is whether hormone in fact promotes binding of all classes of steroid receptors to their specific DNA sites. The activity of a gene regulatory protein can be controlled either at the level of DNA binding or through modification of the protein already bound to specific DNA sites. Several studies have suggested that either mode of regulation may occur with steroid receptors.[28-31] These discrepancies may reflect differences in methodology or more likely may be the result of receptor classes or receptors from different tissues or species displaying different behavior with respect to interaction with hsp 90. To examine the question of the hormone-dependence of human PR binding to specific DNA sites, we have compared different forms of native human PR extracted from T47D cells for their DNA binding activity in a gel-mobility shift assay. These included: 1) unoccupied 8-10S cytosol PR prepared from non-hormone treated T47D cells; 2) cytosol 4S PR activated by cell-free binding of R5020 followed by treatment with 0.3 M salt; 3) and 4S nuclear PR activated by binding of R5020 in intact cells prior to their extraction from nuclei with 0.5 M salt. When a [32P]labeled oligonucleotide corresponding to the progesterone response element (PRE) of MMTV was incubated with cytosol or nuclear extracts and the resultant PR-DNA complexes separated from free DNA by non-denaturing polyacrylamide gel electrophoresis, the unoccupied 8-10S cytosol PR failed to show any binding to the PRE-oligonucleotide (Fig. 1, lane 1). Activated cytosol PR-R5020 complexes, bound with low but detectable activity resulting in reduced mobility of some [32P]DNA (Fig. 1, lane 2). Activated nuclear PR-R5020 complexes, however, displayed a substantially higher level of DNA binding (Fig. 1, lane 3). From multiple experiments activated nuclear PR was found to exhibit a range of 5-7 fold higher DNA binding activity than activated cytosol PR. The specificity of the receptor-DNA complexes is shown in the other lanes of Fig. 1 (lanes 4-10). The presence of PR in the upshifted [32P]DNA complex is shown by the further mobility shift produced by the addition of anti-PR MAbs (AB-52 and B-30). Specificity of PR for the PRE-DNA over that of non-specific DNA is shown by the effective competition with excess unlabeled PRE DNA and lack of competition with excess unrelated oligonucleotides (Fig. 1, lanes 7-9). Based on these results, we conclude that human PR binding to specific gene sequences is highly dependent on hormone, thus supporting a model for receptor activation in which hormone functions to increase the affinity of receptors for HREs. The substantially higher DNA binding activity of activated nuclear PR, over that of the activated cytosol PR-hormone complex, was of great interest to us because hsp 90 has apparently been dissociated from both receptor forms. This suggests that hsp 90 dissociation from receptors is necessary but not sufficient for maximal activation of receptor DNA binding function. We have investigated the role of receptor interaction with nuclear protein factor(s), and PR dimerization as other possible factors, that might account for the higher DNA-binding activity of nuclear PR activated by hormone binding in intact cells.

Nuclear Protein(s) Enhance PR Binding to DNA: We reasoned that an activated receptor in cytosol may bind less well to DNA than an activated receptor in nuclear extracts if there are factor(s) in nuclei that facilitate PR-DNA interactions. To test this possibility, we have performed mixing experiments by adding nuclear extracts lacking PR to an activated cytosol PR preparation. Addition of increasing amounts of nuclear protein resulted in a dose-dependent increase in the amount of [32P]PRE-DNA bound by the cytosol PR (Fig. 2). Enhanced DNA binding occurred with a constant amount of DNA and cytosol PR, indicating that a factor(s) is present that acts in a cooperative manner with receptors to enhance DNA binding. At

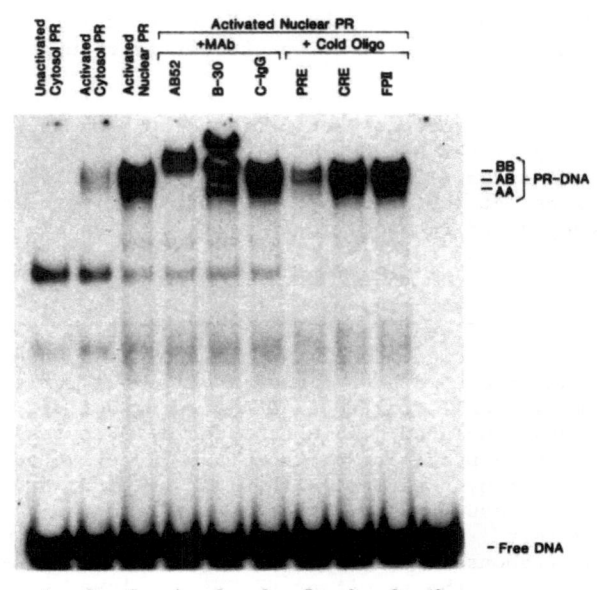

Fig. 1. Binding of different receptor forms to an end-labeled PRE-oligonucleotide by gel-mobility shift assay (lanes 1-3). Arrows indicate positions of free [^{32}P]DNA and PR-DNA complexes. Addition of receptor specific MAbs (AB-52, B-30) and an unrelated MAb, C-IgG (lanes 4-6). Competition with excess unlabeled PRE DNA and unrelated oligo-nucleotides (lanes 7-9) (reproduced from Ref. 24).

best, however, a 3 fold increase in DNA binding of cytosol PR was obtained by the addition of nuclear proteins. Thus, other cellular factors present in nuclei appear to account only in part for the higher DNA binding activity of nuclear PR over that of cytosol PR. It is of interest to note that addition of nuclear proteins produced an increase in the amount of DNA binding without a change in mobility of the PR-DNA complex. This suggests that the factor(s) responsible for enhanced PR-DNA binding are either present in low amounts in cytosol, or that the nuclear factor enhances PR binding to DNA without itself entering into the PR-DNA complex. We have begun to characterize the nuclear factor(s) responsible for enhanced DNA binding of receptors. The activity was found in non-hormone responsive (MDA-231), as well as hormone responsive cells (T47D), suggesting it may be ubiquitous. The activity is both heat and trypsin sensitive and fractionates by differential dialysis in the size range of 50-75,000 molecular weight, suggesting it is associated with a protein(s).

Progesterone Receptor Binds to DNA as a Dimer: A commonly employed technique to demonstrate whether a protein binds to DNA as a dimer is to perform gel mobility shift assays with protein size variants of two different lengths. If the protein binds as a dimer three distinct mobility protein-DNA complexes are expected, the outer two representing homodimers of the long and short version of the protein and the intermediate mobility complex a heterodimer composed of one long and one short subunit.[28] We have been able to exploit the fact that PR are naturally expressed as two different sized proteins (PR-A and PR-B) to demonstrate by gel-mobility shift that PR binds to its cognate HRE as a dimer. As illustrated in Fig. 1, three distinct PR-DNA complexes are obtained by gel-mobility shift performed with cell extracts containing both truncated PR-A and full length PR-B. This is consistent with PR complexing to PRE-DNA as dimers composed of the possible subunit combinations between PR-A and PR-B (AA dimers, AB heterodimers and BB dimers). In further support of this, addition of the A-B specific MAb (AB-52) further shifted the mobility of all three DNA complexes indicating the presence of both PR-A and PR-B (Fig. 1, lane 4). Addition of the B specific MAb, B-30, produced a further shift of only the two upper complexes indicating that no PR-B is present in the faster mobility complex (Fig. 1, lane 5). Also, gel-mobility shift assays performed with a nuclear

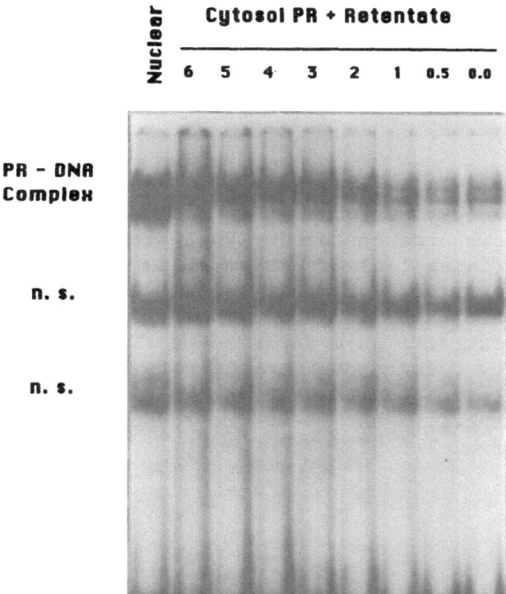

PR - DNA
Complex

n. s.

n. s.

free

Fig. 2. Effects of nuclear proteins on binding of cytosol PR to DNA. Gel mobility shift assay of nuclear PR activated *in vivo* (far left lane), cytosol PR activated *in vitro* (far right lane) and activated cytosol PR after addition of nuclear proteins (middle lanes). T47D nuclear extracts added to cytosol PR were depleted of PR with a specific MAb column, dialyzed (retentate) and then added in increasing amounts to activated cytosol PR (adapted from Ref. 23).

extract that contained PR-A alone (by depletion of PR-B with the B-30 MAb), produced only the fastest mobility DNA complex indicating that it in fact contains PR-A exclusively (Figure 3).

Receptor Dimerization in the Absence of DNA

Although the gel-mobility shift experiments illustrate that PR are complexed to DNA as dimers, they do not address the question of whether PR dimerization occurs in solution in the absence of DNA or only upon contact with DNA. To explore this question, we have again taken advantage of the fact that PR are expressed as two different sized proteins. If receptor dimerization between the A and B forms of PR were to occur in the absence of DNA, then immune isolation of crude cell extracts with the PR-B specific MAb (B-30) should result in co-isolation of PR-A. Immune isolations, therefore, were performed with the B-30 MAb and the isolated receptor complexes were than analyzed by Western blot with the AB-52 MAb (A and B specific) for the presence of both PR-A and PR-B. This assay was applied to the same three native PR forms analyzed in Fig. 1 for DNA binding activity. Little or no PR-A was found to associate with PR-B from the unoccupied 8S cytosol PR complex. However, significant amounts of PR-A co-isolated with activated 4S cytosol and with activated 4S nuclear PR complexes (Fig. 4). These experiments, therefore, provide direct evidence that activated PR are capable of forming stable PR-A•PR-B oligomers in the absence of DNA. Results of the co-immune isolation assays also revealed a positive correlation between the ability of different PR forms to oligomerize in solution and their ability to bind to DNA (compare Figs. 1 and 4). Unoccupied 8S cytosol PR failed to bind to PRE DNA and to oligomerize in solution, activated cytosol PR displayed some capability of forming A-B oligomers and exhibited low DNA

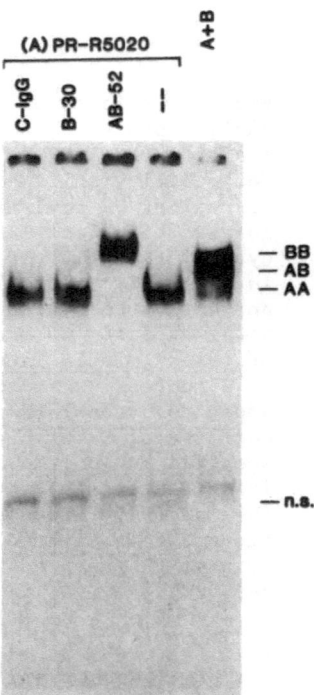

Fig. 3. Gel-mobility shift assays performed with nuclear extracts containing PR-A alone (lanes 1-4) or with nuclear extracts containing both PR-A and PR-B (lane 5). Nuclear extracts were depleted of PR-B by passage over an affinity resin coupled with the PR-B specific MAb, B-30 (reproduced from Ref. 25).

binding activity, while activated nuclear PR exhibited an even higher extent of A-B oligomerization and the highest amount of DNA binding. This supports the hypothesis that dimerization in solution may be an important regulatory step required for receptor recognition of specific DNA sites.

Factors Controlling Dimerization of Progesterone Receptor: Using this convenient co-immune isolation assay to detect PR oligomerization, we have begun to explore the factors that may control PR dimerization in the absence of DNA. By preparing cytosol receptors that contained varying amounts of associated hsp 90, we found an inverse relationship between the amount of hsp 90 association with PR and the extent of PR-A•PR-B oligomerization that occurred in solution (Fig. 5). This suggested to us that hsp 90 may repress DNA binding function of receptors indirectly by blocking dimerization. Despite the apparent role of hsp 90 in negatively modulating PR dimerization, activated nuclear PR were found to be capable of more efficient oligomerization than activated cytosol PR, yet both PR forms have essentially released all hsp 90 (Fig. 5, right panel). This indicates that additional factors may control dimerization.

Progesterone receptors are phosphoproteins and have been shown to exhibit basal phosphorylation in the absence of hormone and to undergo increased phosphorylation upon binding hormone in intact cells.[29] We have begun to explore the possible involvement of hormone-dependent phosphorylation in modulating dimerization. A distinctive feature of hormone-dependent phosphorylation of PR in T47D cells is an apparent upshift in molecular weight of both PR-A and PR-B proteins on SDS-gels.[29] We have observed a positive correlation between hormone-induced phosphorylation of PR and the ability of PR to oligomerize in solution. Only nuclear PR activated by hormone-binding in intact cells have undergone hormone-dependent phosphorylation. Activated cytosol PR-hormone complexes, which show a lower

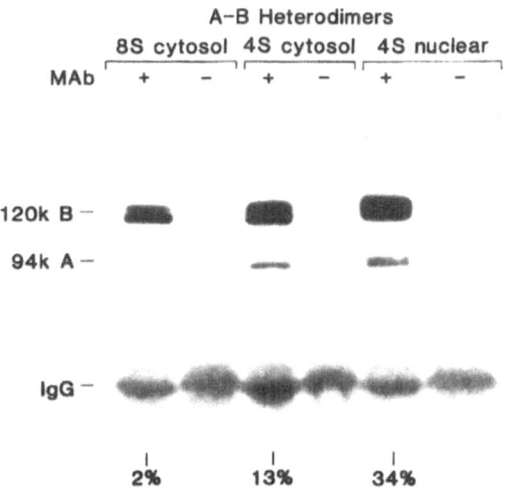

Fig. 4. Co-immune isolation assay to detect PR-A•PR-B oligomers
in the absence of DNA. The native forms of PR indicated were
immune-isolated (by use of protein A Sepharose as an immunoabsor-
bant) with the PR-B specific MAb, B-30 (+) or a control unrelated
MAb (-). Immobilized PR complexes were then extracted from
protein A Sepharose and analyzed for the presence of both PR-A and
PR-B by Western blot with AB-52. The ratio of PR-A to PR-B in
each isolated complex was estimated by direct scanning of the
Western blot and expressed as the amount of PR-A as a percent
relative to PR-B (adapted from Ref. 24).

ability to oligomerize, are not post-translationally modified in this manner
as evidenced by their failure to exhibit an upward shift in apparent molecular weight on SDS-PAGE (Fig. 6). We have also observed that the higher
phosphorylated form of PR-A (designated PR-A$_2$ in Fig. 6), that predominates
in the presence of hormone, forms an association with PR-B that is more
stable to salt dissociation than that formed with the underphosphorylated
form of PR-A (designated PR-A$_1$) that predominates in the absence of
hormone (not shown, see ref. 24). These results suggest that hormone-
dependent phosphorylation may contribute to the stabilization of PR dimers
in solution.

DISCUSSION

     The process of activation of steroid receptors and their subsequent
binding to specific gene sequences is emerging as a more complex mechanism
than perhaps anticipated. Ligand-promoted release of the inhibitory molecule, hsp 90, is thought to be an important step required for activation of
receptor DNA binding function. However, substantial differences were detected in DNA binding activity of two different forms of activated PR (cytosol
and nuclear) despite the fact that hsp 90 has been dissociated from both
(Fig. 1 and 5). This suggests that hsp 90 dissociation from receptors is
necessary, but not sufficient for maximal DNA binding function. This also
suggests that hormone binding in intact cells (activated nuclear PR) induces
modifications in receptor required for maximal activation that are not

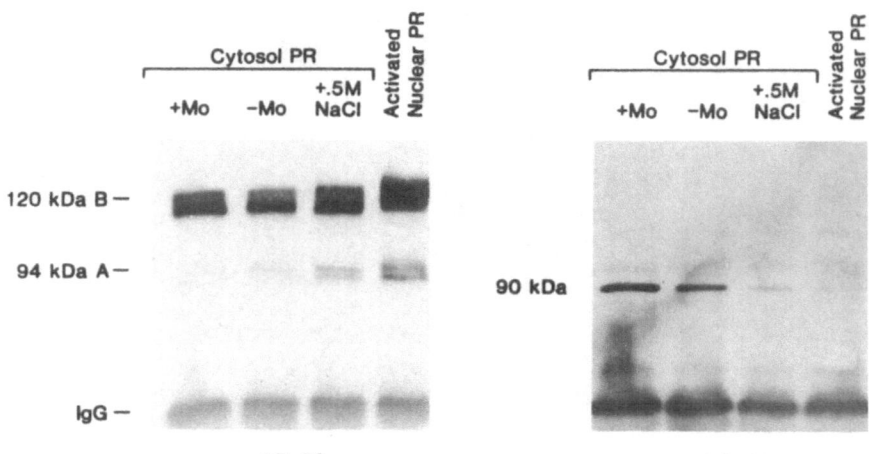

Fig. 5. Dissociation of hsp 90 from receptors is correlated with formation of PR oligomers in the absence of DNA. Different native forms of PR indicated were immune isolated with the B-30 MAb as in Fig. 4 and the isolated PR complexes were then analyzed by Western blotting with either AB-52 for detection of associated PR-A or with a MAb (AC-88) specific for hsp 90 for detection of co-associated hsp 90 (reproduced from Ref. 24).

mimicked *in vitro* (cytosol PR). Similar results and conclusions regarding hsp 90 have been obtained with chick oviduct PR[30] and with studies from another laboratory working with human PR.[31]

The present studies suggest two additional regulatory steps that may be involved in activation of receptors; 1) receptor interaction with other nuclear proteins, and 2) receptor dimerization. Studies from several laboratories have described the presence of nuclear protein(s) that enhance the binding of steroid receptors to specific DNA sequences. This was first described for estrogen receptor (ER) binding to the estrogen response element (ERE) of the vitellogenin gene. Two non-histone chromosomal proteins (NHPI and NHPII) have been identified and purified that themselves bind to EREs and also enhance receptor binding to EREs.[32] Two studies with *in vitro* synthesized thyroid receptor (TR) have also shown that TR binding to thyroid response elements is highly dependent upon other nuclear protein(s).[33,34] Whether the nuclear protein(s) responsible for the enhanced binding of cytosol PR to DNA are the same as those described with estrogen or thyroid receptors is not known. Accessory proteins of the nature described in these studies may play an important physiological role, both in facilitating receptor DNA binding, and in mediating receptor-dependent gene transcription.

Several studies have now demonstrated that steroid receptors both dimerize in solution in the absence of DNA and exist as dimers when complexed to their cognate DNA sites. This has been shown for estrogen receptors,[35,36] glucocorticoid receptors,[37,38] and progesterone receptors.[24] Whether receptor dimerization represents an intermediate regulatory step in the activation process, required for recognition of specific DNA sites, has not been fully explored for all classes of steroid receptors. Nor have the factors that control dimerization been examined to any great extent.

A recent report by Fawell et al.[36] has provided one of the only studies as yet to directly address the question of whether steroid receptor

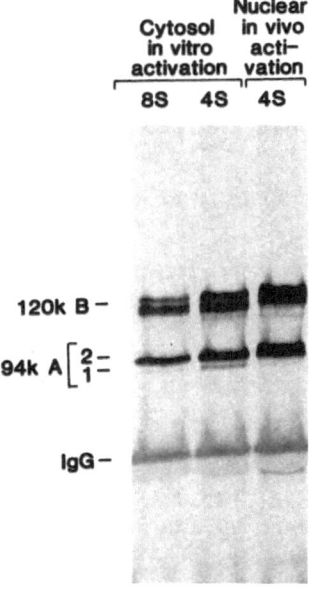

Cytosol in vitro activation | Nuclear in vivo activation
8S 4S | 4S

120k B —
94k A [2=
[1=

IgG —

Fig. 6. The different native forms of PR indicated were immune-isolated from cytosol or nuclear extracts with the A-B specific MAb, AB-52. Immobilized PR complexes were extracted from protein A Sepharose and analyzed by Western immunoblotting with AB-52 (adapted from Ref. 24).

dimerization in solution is a requirement for recognition of specific DNA sites. By deletion and site directed mutagenesis they have identified a conserved region of heptad repeat hydrophobic residues in the steroid binding domain, that is responsible for mediating dimerization of the estrogen receptor in the absence of DNA.[36] They also reported a direct correlation between the ability of ER mutants to form dimers in solution and their ability to bind to estrogen response elements. These results support the hypothesis that ER dimerization largely controls DNA binding function. Our biochemical analysis in the present study, revealing a positive correlation between the ability of different PR forms to oligomerize in solution and their ability to bind to specific DNA sites (Figs. 1, 4 and 5), are also consistent with a model in which dimerization is a critical factor controlling receptor recognition of dyad symmetrical DNA. With respect to factors that control dimerization, the present findings that dissociation of hsp 90 from receptors correlated with the ability of PR to oligomerize in solution, suggests that hsp 90 may negatively regulate receptor dimerization (Fig. 5). This raises the possibility that hsp 90 may inhibit receptor DNA binding indirectly by blocking dimerization, instead of masking the DNA binding domain as generally presumed.[26] The present studies also suggest the possibility that hormone-dependent phosphorylation may impart stability to PR dimers in solution. There is precedent for phosphorylation regulating dimerization of a gene regulatory protein. This has been shown for the cAMP dependent response element binding protein, CREB.[40]

An interesting aspect of proteins that utilize dimerization to control DNA binding function is the possibility to form heterodimers as a mechanism to expand diversity of functional activities with a limited number of regulatory proteins. This phenomenon has now been described for several gene regulatory proteins including studies suggesting that thyroid receptors and retinoic acid receptors may be capable of forming functional heterodimers.[39] The present findings that PR-A and PR-B are capable of oligomerizing raises the possibility that PR may be capable of forming both homodimers (AA or BB) and heterodimers (AB) as a mechanism to create novel receptor molecules, each with different functional activities.

113

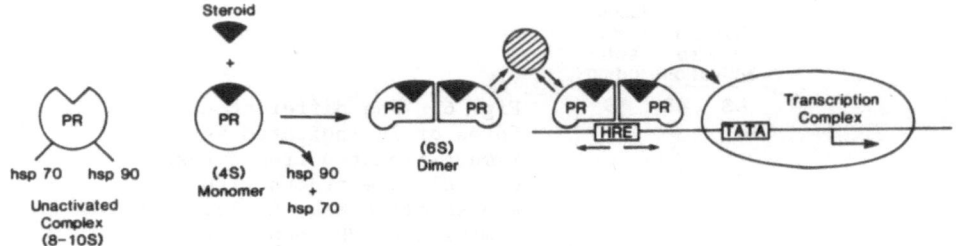

Fig. 7. Multistep model for the process of receptor activation and binding to specific DNA sequences.

Our biochemical studies on PR binding to DNA are consistent with the hypothesis that hormonal activation of receptors to the DNA binding form is a multistep process. A working hypothesis is depicted in Fig. 7. In the absence of hormone, receptors are bound with hsp 90 and perhaps with other proteins to form a large 8-10S unactivated complex. Hsp 90 represses DNA binding activity, rendering receptors unavailable for gene transcription until needed. Hormone-binding is predicted to induce release of hsp 90, and this is followed by oligomerization of PR polypeptide chains to form a 6S dimeric molecule as an intermediate step prior to DNA binding. Release of hsp 90 from receptors, though necessary, is not sufficient to promote maximal activation of DNA binding function. Hormone-dependent phosphorylation is proposed to regulate dimer stability in the absence of DNA and thus indirectly modulate DNA binding. Other nuclear protein(s) (depicted as shaded circle) can also facilitate binding of PR to their cognate HREs.

REFERENCES

1. K. B. Horwitz, M. B. Mockus, and B. A. Lessey, Variant T47D human breast cancer cells with high progesterone-receptor levels despite estrogen and antiestrogen resistance, Cell 28:633 (1982).
2. A. M. C. Brown, J.-M. Jeltsch, M. Roberts, and P. Chambon, Activation of *pS2* gene transcription is a primary response to estrogein in the human breast cancer cell line MCF-7, Proc. Natl. Acad. Sci. USA 81:6344 (1984).
3. S. A. W. Fuqua, M. Blum-Salingaros, and W. L. McGuire, Induction of the estrogen-regulated "24K" protein by heat shock, Cancer Res. 49:4126 (1989).
4. V. Cavailles, P. Augereau, M. Garcia, and H. Rochefort, Estrogens and growth factors induce the mRNA of the 52K-pro-cathepsin-D secreted by breast cancer cells, Nucleic Acids Res. 16:1903 (1988).
5. M. C. Rio, J. P. Bellocq, B. Gairard, U. B. Rasmussen, A. Krust, C. Koehl, H. Calderoli, V. Schiff, R. Renaud, and P. Chambon, Specific expression of the pS2 gene in subclasses of breast cancers in comparison with expression of the estrogen and progesterone receptors and the oncogene *ERBB2*, Proc. Natl. Acad. Sci. USA 84:9243 (1987).
6. A. K. Tandon, G. M. Clark, G. C. Chamness, J. M. Chirgwin, and W. L. McGuire, Cathepsin D and prognosis in breast cancer, New Eng. J. Med. 322:297 (1990).
7. C. Joyeux, H. Rochefort, and D. Chalbos, Progestin increases gene transcription and messenger ribonucleic acid stability of fatty acid synthetase in breast cancer cells, Mol. Endocrinol. 4:681 (1989).

8. P. G. Gill, F. Vignon, S. Bardon, D. Derocq, and H. Rochefort, Difference between R5020 and the antiprogestin RU486 in antiproliferative effects on human breast cancer cells, <u>Breast Canc. Res. Treat</u>. 10:37 (1987).

9. R. Poulin, J.-M. Dufour, and F. Labrie, Progestin inhibition of estrogen-dependent proliferation in ZR-75-1 human breast cancer cells: Antagonism by insulin, <u>Breast Canc. Res. Treat</u>. 13:265 (1989).

10. R. L. Sutherland, R. E. Hall, G. Y. N. Pang, E. A. Musgrove, and C. L. Clarke, Effect of medroxyprogesterone acetate on proliferation and cell cycle kinetics of human mammary carcinoma cells, <u>Cancer Res</u>. 48:5084 (1988).

11. S. M. Sedlacek, An overview of megestrol acetate for the treatment of advanced breast cancer, <u>Seminars in Oncology</u> 15:3 (1988).

12. G. C. Chamness, Progestin action and progesterone receptors in breast cancer, <u>Cancer Res</u>. 49:7176 (1989).

13. R. M. Evans, The steroid and thyroid hormone receptor superfamily, <u>Science</u> 240:889 (1988).

14. M. Beato, Gene regulation by steroid hormones, <u>Cell</u> 56:335 (1989).

15. A. C. B. Cato, R. Miksicek, G. Schütz, J. Arnemann, and M. Beato, The hormone regulatory element of mouse mammary tumour virus mediates progesterone induction, <u>The EMBO J</u>. 5:2237 (1986).

16. S. K. Nordeen, B. Kühnel, J. Lawler-Heavner, D. A. Barber, and D. P. Edwards, A quantitative comparison of dual control of a hormone response element by progestins and glucocorticoids in the same cell line, <u>Mol. Endocrinol</u>. 3:1270 (1989).

17. P. A. Estes, E. J. Suba, J. Lawler-Heavner, D. Elashry-Stowers, L. L. Wei, W. P. Sullivan, D. O. Toft, K. B. Horwitz, and D. P. Edwards, Immunological analysis of progesterone receptors in human breast cancer I: Purification of transformed receptors and production of monoclonal antibodies, <u>Biochemistry</u> 26:6250 (1987).

18. K. B. Horwitz, M. D. Francis, and L. L. Wei, Hormone-dependent covalent modification and processing of human progesterone receptors in the nucleus, <u>DNA</u> 4:451 (1985).

19. H. Gronemeyer, B. Turcotte, C. Quirin-Stricker, M. T. Bocquel, M. E. Meyer, Z. Krozowski, J. M. Jeltsch, T. Lerouge, J. M. Garnier, and P. Chambon, The chicken progesterone receptor: sequence, expression and functional analysis, <u>EMBO J</u>. 6:3985 (1987).

20. O. M. Conneely, B. L. Maxwell, D. O. Toft, W. T. Schrader, and B. W. O'Malley, The A and B forms of the chicken progesterone receptor arise by alternate initiation of translation of a unique mRNA, <u>Biochem. Biophys. Res. Commun</u>. 149:493 (1987).

21. P. Kastner, A. Krust, B. Turcotte, U. Stropp, L. Tora, H. Gronemeyer, and P. Chambon, Two distinct estrogen-regulated promoters generate transcripts encoding the two functionally different human progesterone receptor forms A and B, <u>EMBO J</u>. 9:1603 (1990).

22. L. Tora, H. Gronemeyer, B. Turcotte, M.-P. Gaub, and P. Chambon, The N-terminal region of the chicken progesterone receptor specifies target gene activation, <u>Nature</u> 333:185 (1988).

23. D. P. Edwards, B. Kühnel, P. A. Estes, and S. K. Nordeen, Human progesterone receptor binding to a mouse mammary tumor virus DNA: Dependence on hormone and non-receptor nuclear factor(s), <u>Mol. Endocrinol</u>. 3:381 (1989).

24. A. M. DeMarzo, C. A. Beck, S. A. Oñate, and D. P. Edwards, Evidence for dimerization of mammalian progesterone receptors in the absence of DNA: Possible modulation of dimerization by the 90 kDa heat shock protein, <u>Proc. Natl. Acad. Sci. USA</u>, in press (1990).

25. D. El-Ashry, S. Oñate, S. K. Nordeen, and D. P. Edwards, Human progesterone receptor complexed with the antagonist RU 486 binds to a hormone response element in a structurally altered form, <u>Mol. Endocrinol</u>. 3:1545 (1989).

26. W. B. Pratt, E. R. Sanchez, E. H., Bresnick, S. Meshinchi, L. C. Scherrer, F. C. Dalman, and M. J. Welsh, Interaction of the glucocorticoid receptor with the M_r 90,000 heat shock protein: An evolving model of ligand-mediated receptor transformation and translocation. <u>Cancer Res</u>. 49:2222s (1989).

27. S. L. Kost, D. F. Smith, W. P. Sullivan, W. J. Welch, and D. O. Toft, Binding of heat shock proteins to the avian progesterone receptor, <u>Mol. Cell Biol</u>. 9:3829 (1989).

28. I. A. Hope, and K. Struhl, GCN4, A eukaryotic transcription activator protein, binds as a dimer to target DNA, <u>EMBO J</u>. 6:2781 (1987).

29. P. L. Sheridan, M. D. Francis, and K. B. Horwitz, Synthesis of human progesterone receptors in T47D cells. Nascent A- and B-receptors are active without a phosphorylation-dependent post-translational maturation step, <u>J. Biol. Chem</u>. 264:7054 (1989).

30. L. A. Denner, N. L. Weigel, W. T. Schrader, and B. W. O'Malley, Hormone-dependent regulation of chicken progesterone receptor deoxyribonucleic acid binding and phosphorylation, <u>Endocrinology</u> 125:3051 (1989).

31. M. K. Bagchi, S. Y. Tsai, M.-J. Tsai, and B. W. O'Malley, Identification of a functional intermediate in receptor activation in progesterone-dependent cell-free transcription, <u>Nature</u> 345:547 (1990).

32. I. M. Feavers, J. Jiricny, B. Moncharmont, H. P. Saluz, and J. P. Jost, Interaction of two nonhistone proteins with the estradiol response element of the avian vitellogenin gene modulates the binding of estradiol-receptor complex, <u>Proc. Natl. Acad. Sci. USA</u> 84:7453 (1987).

33. J. Burnside, D. S. Darling, and W. W. Chin, A nuclear factor that enhances binding of thyroid hormone receptors to thyroid hormone response elements, <u>J. Biol. Chem</u>. 265:2500 (1990).

34. M. B. Murray, and H. C. Towle, Identification of nuclear factors that enhance binding of the thyroid hormone receptor to a thyroid hormone response element, <u>Mol. Endocrinol</u>. 3:1434 (1989).

35. V. Kumar, and P. Chambon, The estrogen receptor binds tightly to its responsive element as a ligand-induced homodimer, <u>Cell</u> 55:145 (1988).

36. S. E. Fawell, J. A. Lees, R. White and M. G. Parker, Characterization and colocalization of steroid binding and dimerization activities in the mouse estrogen receptor, <u>Cell</u> 60:953 (1990).

37. S. Y. Tsai, J. Carlstedt-Duke, N. L. Weigel, K. Dahlman, J.-Å Gustafsson, M.-J., and B. W. O'Malley, Molecular interactions of steroid hormone receptor with its enhancer element: Evidence for receptor dimer formation, <u>Cell</u> 55:361 (1988).

38. Ö. Wrange, P. Eriksson, and T. Perlmann, The purified activated glucocorticoid receptor is a homodimer, <u>J. Biol. Chem</u>. 264:5253 (1989).

39. C. K. Glass, S. M. Lipkin, O. V. Devary, and M. G. Rosenfeld, Positive and negative regulation of a gene transcription by a retinoic-acid-thyroid hormone receptor heterodimer, <u>Cell</u> 59:697 (1989).

40. K. K. Yamamoto, G. A. Gonzalez, W. H. Biggs III, and M. R. Montminy, Phosphorylation-induced binding and transcriptional efficacy of nuclear factor CREB, <u>Nature</u> 328:175 (1987).

ORIGIN AND NATURE OF THE MILK LIPID GLOBULE MEMBRANE

Thomas W. Keenan and Daniel P. Dylewski[*]

Department of Biochemistry, Virginia Polytechnic
Institute, Blacksburg, VA, and [*]Department of Botany
and Microbiology, Auburn University, Auburn, AL

INTRODUCTION

Membranous material which surrounds and stabilizes the triacylglycerol-rich milk lipid globules is commonly referred to as the milk-fat or milk-lipid globule membrane (MLGM hereafter). This membranous material originates from two distinct sites within mammary epithelial cells, from specialized regions of apical plasma membrane, and from regions of endoplasmic reticulum (ER). That portion of the MLGM derived from apical plasma membrane, termed the primary membrane, appears in electron micrographs as a typical bilayer membrane with an electron-dense material on the inner membrane face. Regions of apical plasma membrane differentiated by the presence of a coat material on the cytoplasmic face envelop globules during their extrusion from the cell. That component derived from ER lacks bilayer membrane structure. This material, primarily composed of proteins and polar lipids, covers the surface of the lipid droplets within the cell. Constituents of this coat material mediate intracellular fusions through which droplets grow in volume and also may be involved in interaction of droplets with plasma membrane. The origin and nature of these components of the MLGM is reviewed briefly herein. This is not a comprehensive review. In the interest of brevity, citations in many cases are to recent publications or to reviews.

ENDOPLASMIC RETICULUM CONTRIBUTION TO MLGM

Earliest intracellular precursors of milk lipid globules originate by blebbing or budding from ER. Triacylglycerols appear to accumulate at focal points on or in the ER membrane (1). Whether this accumulation of triacylglycerols is due to localized synthesis or accretion is unknown. It has been suggested that triacylglycerols accumulate between the halves of the bilayer membrane and are released from ER into the cytoplasm as droplets coated with the outer or cytoplasmic half of the ER membrane (2, 3). Some morphological evidence supporting this suggestion has been obtained (4, 5), but information which would prove or disprove this hypothesis is lacking.

By whatever mechanism they originate, milk lipid globule precursors first appear in the cytoplasm as small (diameters < 0.5 μm) droplets which have a triacylglycerol-rich core surrounded by a coat material which lacks unit-like or bilayer membrane structure (1, 6). Small lipid droplets, termed microlipid droplets, appear to grow in volume by fusions with each other. Fusions give rise to larger droplets, termed cytoplasmic lipid droplets; operationally defined as those droplets with diameters > 1 μm. In electron micrographs of tissue specimens fixed in osmium, ferrocyanide, or permanganate, the electron-dense material on the periphery of both small and large droplets appeared to be granular (1). Over much of the droplet surface this material was between 6 and 10 nm in thickness, but in localized regions this material appeared to be thickened considerably and to be composed of two or more layers of tripartite lamellar structure (Fig. 1a-c). Some small, droplet-like structures in the immediate vicinity of ER in basal cell regions had ribosomes associated with this surface coat, but ribosomes were not seen on larger droplets, or on small droplets in apical cell regions (1, 6).

In addition to observations made by electron microscopic examination of fixed and sectioned material, evidence for surface coat material on intracellular lipid droplets has been obtained by their isolation and compositional analysis (1, 6). Analyses have been made with droplets isolated from cow and rat mammary glands. Droplets were isolated by density gradient centrifugation, taking advantage of the fact that they have lower densities than do organelles and vesicles derived from components of the endomembrane system. Droplets had densities ranging from < 1 to 1.12 g/cc, and as expected density was inversely related to volume. Smaller, more dense droplets also may be present in homogenates, but these would overlap with membrane vesicles. For many of our analyses droplets of < 1 g/cc (cytoplasmic lipid droplets) and droplets of 1 to 1.12 g/cc (microlipid droplets) were collected as separate fractions. Droplets of different density classes from cow mammary gland had lipid to protein ratios ranging from about 1.5:1 to 40:1. ER fractions had lipid to protein ratios of 0.45:1. On a protein basis, droplets were enriched greatly in triacylglycerols, and to a lesser extent in cholesterol, relative to ER. ER and various density classes of droplets had similar phospholipid to protein ratios. Surface coat material of droplets and ER membranes had the same five major phospholipid classes; sphingomyelin and the phosphoglycerides of choline, ethanolamine, inositol and serine. Droplets differed from ER in that sphingomyelin accounted for a higher proportion of the total phospholipid (9 to 15% versus 4% for ER) and phosphatidyl choline acounted for a correspondingly lower proportion of the total phospholipid. Cytoplasmic lipid droplets also had monohexosyl- and dihexosylceramides and gangliosides in their surface coat material. These glycosphingolipids were but trace consitiuents of ER. Microlipid droplets also have gangliosides in their surface coat material; on a protein basis gangliosides were lower in amount in microlipid droplets than in cytoplasmic lipid droplets (7).

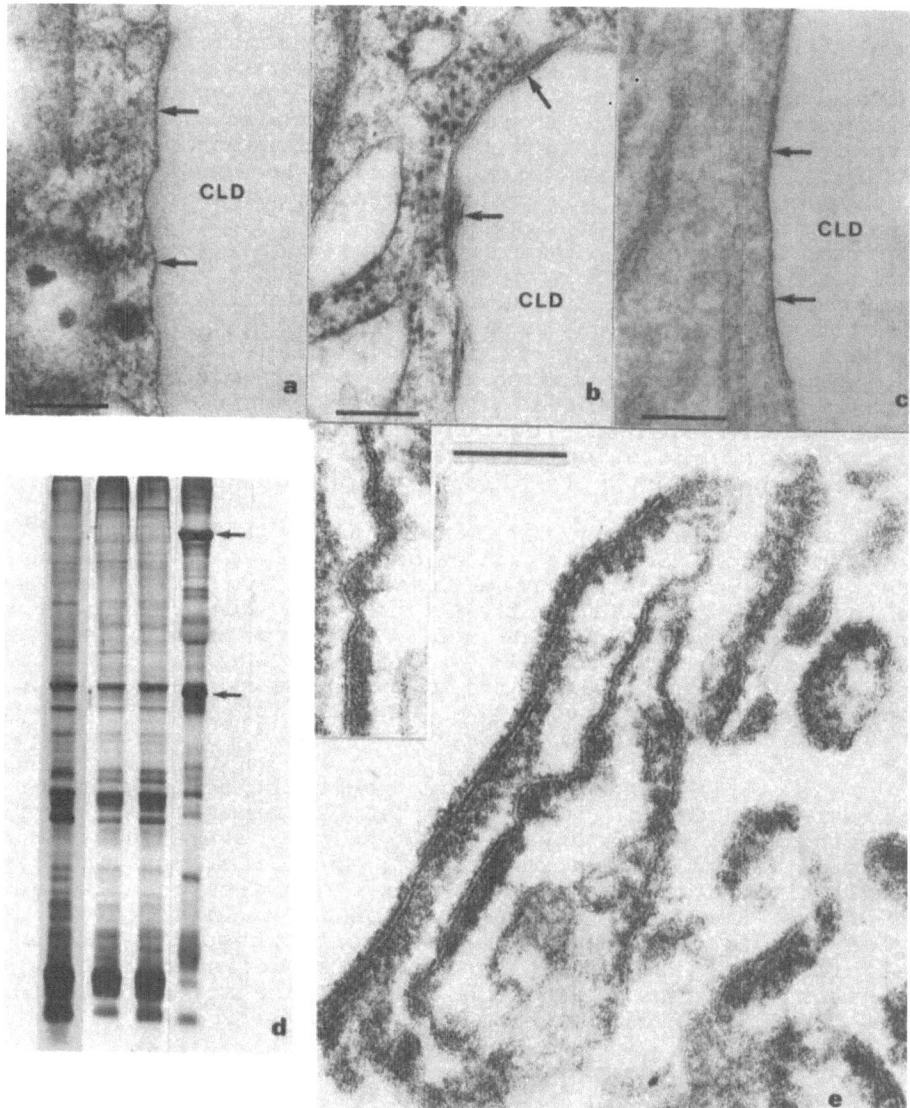

Fig. 1. (a, b, c) Surface coat appearance on cytoplasmic lipid
droplets (CLD) of cow in specimens fixed simultaneously with
glutaraldehyde and osmium tetroxide (a, b), or with potassium
permanganate (c). Arrows denote surface material separating
lipid droplets from cytoplasm. Localized thickened regions of
this material displayed tripartite lamellar structure (arrows
in b). Bars, 0.2 μm. (d) Polypeptide patterns of, from left
to right, microlipid droplets, cytoplasmic lipid droplets, ER
membranes and MLGM from cow. Arrows denote xanthine oxidase
(M_r 155,000) and butyrophilin (M_r 66,000). (e) Isolated MLGM
from cow largely consists of sheets of membranes with the inner
(originally cytoplasmic) face coated with a dense staining
material of variable thickness. This coat material is
amorphous, but occassionally regularly arranged globules or
ridges were observed (insert). Bar, 0.1 μm.

When separated in SDS-polyacrylamide gels, micro- and cytoplasmic lipid droplets had virtually identical polypeptide patterns, and these patterns qualitatively were very similar to those of ER membrane (1, 6). Nearly all polypeptides in the surface coats of lipid droplets corresponded in mobility to polypeptides of ER membranes (Fig. 1d). ER membranes contained additional polypeptides which did not correspond in mobility to polypeptides of lipid droplet surface coat material. Polyclonal antibodies raised against protein fractions from MLGM and from cytoplasmic lipid droplets recognized common polypeptides in microlipid droplets and ER membrane (6). Using immunogold conjugate electron microscopic detection, antibodies raised against surface coat proteins of cytoplasmic lipid droplets were found on surfaces of microlipid droplets and on ER in situ. Antibody binding did not occur uniformly over the reticular network. Instead, immunogold particles appeared in clusters over localized regions of ER cisterna. This corresponds with morphological observations of budding or blebbing of lipid droplets from localized regions of ER.

Intracellular lipid droplets of all density classes examined had little activity in several enzymes which serve as markers for various membrane fractions (1). Droplets were distinguished from ER membranes in that the former lacked detectable activity in NADH- and NADPH-cytochrome c reductase activities. In addition, lipid droplets had very low activities in acyltransferases utilizing glycerol-3-phosphate or diacylglycerol as acceptors (8). In contrast, acyltransferase activities were enriched in ER fractions.

In a series of experiments, radiolabeled precursors of lipids and proteins were injected intraperitoneally into lactating rats. At intervals from 3 min to several hours after injection, animals were sacrificed, and ER, micro- and cytoplasmic lipid droplets, and milk lipid globules were isolated. Kinetics of incorporation of glycerol and palmitate into triacylglycerols and phospholipids (6), orthophosphate into phospholipids, and leucine into proteins (unpublished results) in all cases were consistent with the interpretation that microlipid droplets originate from ER.

In summary, morphological observations and biochemical data are consistent with an ER origin of intracellular lipid droplet precursors of milk lipid globules. That the surface coat of droplets differs from ER membrane in morphology and does not faithfully reproduce ER membrane in composition can be interpreted as evidence that the coat represents a selective part of the ER membrane. Whether this part is derived by structural and compositional rearrangement of that portion of the ER membrane which coats droplets, or whether this coat is the outer half of the bilayer membrane of ER cannot be ascertained from information available. The hypothesis that this coat is the outer monolayer of the ER membrane is attractive, but this interpretation is inconsistent with the polar lipid composition of the coat. In a variety of mammalian cells, phosphatidyl choline, sphingomyelin, and glycosphingolipids largely are restricted to the exoplasmic leaflet (9). The exoplasmic leaflet of the plasma membrane corresponds to the lumenal leaflet of intracellular endomembranes. Given this, if the coat on lipid droplets represented the cytoplasmic leaflet of ER membrane, one would predict that this material would be low in sphingomyelin and phosphatidyl choline; just the opposite situation was found. In addition to the obvious interpretation that this data excludes origin of the coat material from the cytoplasmic leaflet of ER membrane, alternative possibilities must be considered. One possibility is that assymetric lipid distribution in membranes of cell types in which it has been studied differs from that of membranes of differentiated mammary epithelial cells, which have not been studied with respect to lipid asymmetry. Second, it may be that ER membrane becomes rearranged in those localized regions where lipid droplets are formed and released, or that specific translocation of certain lipids across the bilayer (flip-flop) occurs in these regions (9, 10). Another possibility is that sphingomyelin and phosphatidylcholine are incorporated at the surface of the droplet during stages of its formation. During the process of budding from ER those phospholipids preexisting at the droplet surface may add to those polar lipids derived from the cytoplasmic leaflet of ER. These questions beg experimental resolution. The presence of glycosphingolipids in lipid droplet coat material causes an additional difficulty in interpretation. Because of their localization in the exoplasmic leaflet of plasma membrane (9), their presence in the cytoplasmic leaflet of endomembranes would not be predicted. In addition, glycosphingolipids are formed through glycosylation of ceramide by enzymes concentrated in Golgi apparatus (9, 11). There exists data suggesting that glycosphingolipids can be transported from Golgi apparatus not only to plasma membrane, but also to ER (12). The presence of gangliosides on lipid droplet surfaces has been confirmed by immunogold electron microscopic demonstration of binding of monoclonal antibody to the major ganglioside of mammary gland to the surfaces of lipid droplets in situ and in isolated fractions (7).

ROLE OF INTRACELLULAR LIPID DROPLET COAT MATERIAL

Undoubtedly the surfactant coat material is required to stabilize the triacylglycerol-rich core of lipid droplets and prevent their coalescence in the cytoplasm. Beyond this stabilization role, the coat material appears to participate also in droplet fusions, and in droplet-plasma membrane interactions. If cytoskeletal elements function in guiding lipid droplets from their sites of origin to their sites of secretion from the cell, coat constituents would be expected to participate in interaction with elements of the cytoskeleton. Unfortunately, mechanisms responsible for unidirectional transit of lipid droplets through the cytoplasm to apical cell regions are not understood. Evidence that microtubules or microfilaments may be involved in this process has been obtained, but the evidence in some cases is weak and contradictory (discussed in 13). To date, no definitive information has been obtained on what is responsible for this unidirectional transfer mechanism. Resolution of this question awaits development of cell-free systems in which transit can be reconstituted, or development of cell lines which secrete milk lipid globules.

Milk lipid globules range in size from under 0.2 to over 10 μm in diameter. Globule size distribution has been studied most thoroughly with cow's milk, but data available suggest a similar size range of globules in milks of humans and some other species (14). While globules in the 1 to 8 μm diameter range account for 90% or more of the volume of milk lipid, globules with diameters under 1 μm account for 80% of the total number of lipid globules. Within the cell, one mechanism for growth of lipid droplets appears to be fusions of microlipid droplets with each other to form larger droplets. Microlipid droplets also can fuse with larger, cytoplasmic lipid droplets, providing triacylglycerols for further volume increases in larger droplets (1, 6). Morphometric evidence suggests that much of the growth of larger droplets occurs in apical regions of cells, just prior to or during the secretion process (16, 17). While images interpreted as microlipid droplet-microlipid droplet and microlipid droplet-cytoplasmic lipid droplet fusions are seen commonly in electron micrographs, several investigators have failed to find morphological evidence for fusions between larger, cytoplasmic lipid droplets (1, 15, 16, 17).

From research to date the size range of lipid globules in milk can be accounted for by the fusion process. Smaller milk lipid globules arise most probably from secretion of microlipid droplets which have undergone no or but a few fusions. Larger droplets originate by continued fusions with microlipid droplets. Morphological and kinetic evidence in support of this interpretation have been obtained. However, this evidence is insufficient to allow the interpretation that fusion of droplets is the sole or major mechanism for droplet growth. Other possible mechanisms for this growth, for example carrier proteins which convey triacylglycerols from their site of synthesis to growing lipid droplets, cannot be excluded (18).

The process of microlipid droplet fusion has been reconstituted in a cell-free system (7). Fusion appears to involve constituents of the surface coat of lipid droplets. As droplets grow, excess coat material is lost from the surface, as would be expected from geometric consideration of surface area (area = $4\pi r^2$) to volume (volume = $4/3\pi r^3$) ratios of what are in essence spherical particles. The fate of coat material shed from droplet surfaces during fusion within cells is unknown. In the cell-free system, fusion was promoted by calcium, and by a protein fraction of cytosol. Gangliosides of the surface coat of lipid droplets also may be involved in some manner in fusion. Enrichment of droplets with exogenously supplied gangliosides promoted fusion. Addition to the cell-free system of a monoclonal antibody against the major ganglioside of the surface coat of lipid droplets abolished fusion. In agreement with morphological observations of cells, in the cell-free system microlipid droplet-microlipid droplet and microlipid droplet-cytoplasmic lipid droplet fusions occurred, but cytoplasmic lipid droplet-cytoplasmic lipid droplet fusions did not. The reasons why cytoplasmic lipid droplets do not fuse with each other is not apparent. Within the scope of compositional analyses performed to date, coat materials on micro- and cytoplasmic lipid droplets largely are indistinguishable, except for the increased level of gangliosides per unit protein in cytoplasmic lipid droplets. Resolution of this question awaits more detailed study.

The process of secretion of lipid droplets has been described repeatedly since Bargmann and Knoop (19) observed lipid droplets to be budded directly from apical cell regions; during budding droplets became completely surrounded by a layer of apical plasma membrane (4, 13, 14). Wooding (15) suggested an alternate mechanism, one in which fat droplets contacting the apical plasma membrane also become surrounded with secretory vesicles which fuse with each other and the plasma membrane. This resulted in formation of intracellular vacuoles containing membrane-coated lipid droplets. Release of droplets, surrounded partially in apical plasma membrane, was envisioned to occur via exocytosis of the vesicle contents. Morphological evidence for which of these alternative processes occurs or predominates is equivocal. However, the bulk of biochemical evidence favors the interpretation that the major mechanism for secretion of milk lipid globules involves envelopment of droplets directly and completely in plasma membrane. A direct, minor contribution from Golgi apparatus derived secretory vesicle membrane cannot be excluded (13, 14).

Regions of plasma membrane which associate with lipid droplets are characterized in electron micrographs by the appearance of an electron-dense material on the cytoplasmic face of the membrane (15, 20, 21). Droplets contact not the plasma membrane directly, but rather this material. What constituents of this electron-dense material and of the coat of lipid droplets participate in recognition between droplet and differentiated plasma membrane regions remains to be elucidated. Based on immunomicroscopic studies of differentiated mammary epithelial cells (22, 23), and extensive biochemical studies with MLGM (13, 21, 24), it appears probable that butyrophilin and xanthine oxidase, two prominent proteins associated with the MLGM, are major constituents of the electron dense material on the cytoplasmic face of that portion of apical plasma membrane which participates in envelopment of lipid droplets. By immunomicroscopy butyrophilin, a hydrophobic transmembrane glycoprotein, appears to be restricted to the apical surface of milk secreting cells (22, 24). Xanthine oxidase is distributed throughout the cytoplasm, but appears to be concentrated at the apical cell surface (23). Butyrophilin, which has been characterized extensively, is acylated and binds phospholipids tightly (21, 25). Because of its apical localization and properties, it has been believed that butyrophilin must be involved in mediating interaction between lipid droplets and apical plasma membrane. Recently, the gene for butyrophilin was cloned and sequenced (26). From the inferred primary amino acid sequence, it is not apparent how butyrophilin could act as a 'receptor' for lipid droplets. Since butyrophilin has an exoplasmic N-terminus and a single membrane-spanning domain, interaction with lipid droplets must occur with the 257 residue C-terminal domain, if in fact this protein does interact with lipid droplets. The C-terminal domain has no obvious hydrophobic regions, so one must conclude that this interaction must be with proteins (26). This could be with proteins of the lipid droplet surface directly, or through complexes formed with cytoplasmic proteins. Since butyrophilin and xanthine oxidase show a propensity to associate with each other (H. W. Heid, personal communication), a butyrophilin-xanthine oxidase complex could be involved in lipid droplet interaction. Other possibilities for participation of butyrophilin in lipid globule secretion have been discussed by Jack and Mather (26). The function which xanthine oxidase may play in the recognition or envelopment process remains obscure. An understanding of the mechanisms responsible for recognition of lipid droplets and their envelopment in plasma membrane will require identification and characterization of those molecules which mediate these processes.

MILK LIPID GLOBULE MEMBRANE

The membrane surrounding lipid globules in milk resembles closely plasma membrane in morphology in that it has a typical bilayer appearance, with the space between bilayers being comparable to plasma membrane, and has an externally disposed glycocalyx (21, 22, 27). This membrane is differentiated from plasma membrane, other than differentiated regions of apical plasma membrane, by the appearance of the electron-dense material associated with the inner face of the membrane (21, 22, 28). Some of the plasma membrane which initially surrounds globules may be lost following secretion, within alveolar lumina or in expressed milk (29), but estimates of the extent of this loss vary widely (13). In regions of globules where the bilayer membrane has been lost, a granular material remains on the surface. This material may be equivalent to the coat material on intracellular droplets.

Membranes can be released from milk lipid globule suspensions by several processes, including freezing and thawing, vigorous agitation (churning), or suspension of droplets in polar, aprotic solvents (30) (Fig. 1e). After release, membranous material normally is collected by centrifugation at g forces of 100,000 or more. Milk lipid globules and membranes derived therefrom are identical or nearly so in phospholipid composition. But, upon SDS-polyacrylamide gel separation, major quantitative differences are evident in polypeptide composition of intact globules and isolated MLGM (14, 31). In particular, part of the xanthine oxidase and most of a glycoprotein of M_r about 44,000 associated with intact globules are not recovered with the membrane. In addition, many minor polypeptides of globules also are not recovered entirely with the MLGM. Dependent upon the method and temperature of globule disruption, these proteins remain associated with the congealed lipid (butter) or are in the aqueous phase. When suspensions of washed lipid globules are churned at low temperatures, a considerable amount of proteinaceous material remains associated with the surfaces of congealed lipid droplets (6) (Fig. 2). Several of the polypeptides in this material have electrophoretic mobilities identical to polypeptides of the surface coat of intracellular lipid droplets. However, polypeptide patterns of this material do not faithfully reproduce the pattern of polypeptides from micro- or cytoplasmic lipid droplets. Protein fractions recovered from the lipid phase after churning have much larger relative amounts of polypeptides migrating with butyrophilin and xanthine oxidase than do protein fractions from intracellular droplets. In the latter, xanthine oxidase and butyrophilin are minor constituents. This observation implies that some intermixing of constituents

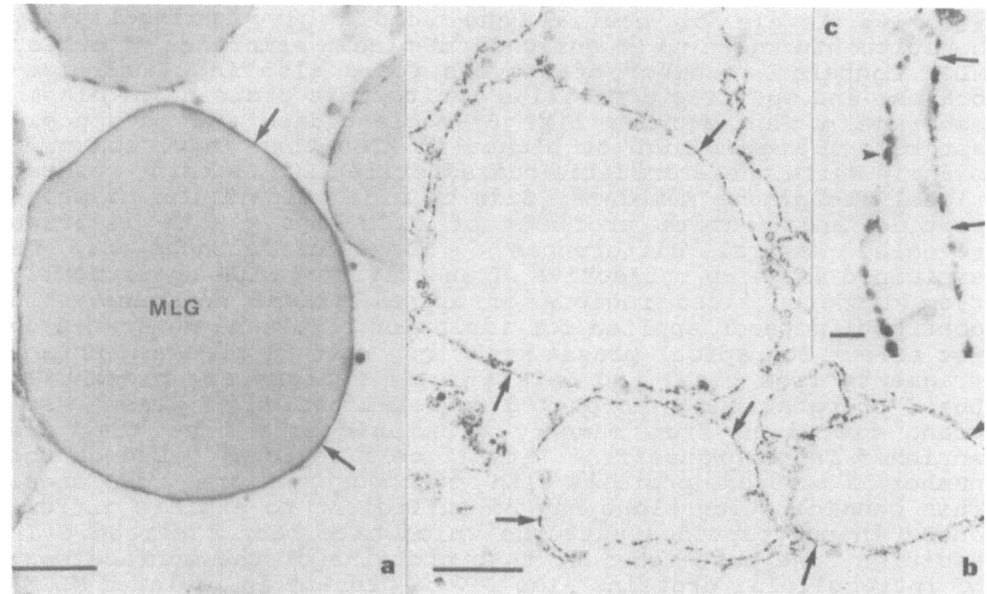

Fig. 2. (a) Cow milk lipid globule completely surrounded by a unit-like membrane (arrows) with dense-staining coat material on the inner membrane face. Bar, 1.0μm. (b) Portiions of the dense staining coat material remained associated with surfaces of lipid globules after MLGM was removed by churning at low temperature. Bar, 1.0 μm. (c) Higher magnification of (b) showing localized thickened regions of this coat material. This surface material lacked unit-like membrane structure. Bar, 0.1 μm.

occurs between material coating the droplet surface on the cell and the apical plasma membrane-derived component of MLGM. Whether this mixing or rearrangement occurs during or after secretion, or whether it is induced by the method used to disrupt globules, remains to be determined. That butyrophilin and xanthine oxidase remain associated with droplet surfaces upon disruption of globules suggests the possibility of a strong interaction between either or both of these proteins and constituents present on the surface of intracellular droplets.

MLGM from cows has been characterized extensively, particularly with respect to lipid and enzymic composition, and polypeptide patterns (reviews, 4, 13, 14, 32, 33, 34). MLGM from several other species also have been characterized, but to a more limited extent. Several authors have made comparisons of MLGM with plasma membrane-rich fractions from lactating mammary gland (35, 36, 37, 38, 39). Results of such comparisons show extensive similarity of MLGM and plasma membrane in terms of phospholipid composition. Invariably, MLGM appears to be enriched in total lipids, and especially in triacylglycerols, relative to plasma membrane. Undoubtedly this is a consequence of acquisition of triacylglycerols by the membrane from the core lipid; possibly this occurs during dissociation of MLGM from globules. MLGM also contains high relative levels of neutral and acidic glycosphingolipids, constituents known to be enriched in plasma membranes of cells. MLGM contains a number of enzymes found also in plasma membranes, and specific activities (units/unit protein) of plasma membrane marker enzymes like 5'-nucleotidase and phosphodiesterase I are as high or higher in MLGM than in plasma membrane. While these criteria characterize MLGM as being related closely to plasma membrane, side by side comparisons of polypeptide and protein profiles of MLGM and plasma membrane revealed several differences. These differences can be explained as being reflective of an origin of MLGM specifically from differentiated regions of apical plasma membrane. In contrast, methods applied for isolation of plasma membranes do not select for apical plasma membrane, rather, plasma membrane fragments from basal and lateral cell regions are present in these preparations. Morphological examination of plasma membrane fractions from mammary gland suggests that they are enriched in membranes from lateral cell regions, based on the number of membrane profiles with attached junctional complexes. This consideration alone may be sufficient to explain differences in polypeptides patterns which have been observed. In addition, one must consider that proteins of the surface coat of intracellular proteins also may be present in isolated MLGM; the extent of their presence in the isolated membrane probably is related to the method used for isolation of MLGM. Based on high specific activities of plasma membrane marker enzymes in MLGM preparations, one can argue that surface coat proteins of droplets dilute plasma membrane-derived MLGM proteins but little. However, alternative explanations are not excluded. Preparative methods may well dissociate peripheral proteins from MLGM to a greater extent than from plasma membrane. Also possible is that, as is apparently the case for butyrophilin, certain plasma membrane enzymes become concentrated in differentiated regions of plasma membrane participating in envelopment of lipid droplets. Questions such as this could be addressed if a method for isolation specifically of apical plasma membrane from mammary epithelial cells were available. While such a method has not yet been described, the known concentration of butyrophilin in apical membrane and its transmembrane orientation suggest that antibodies to epitopes in the N-terminal portion of butyrophilin might profitably be applied for immunoprecipitation of apical plasma membranes.

In general terms, much is known about the nature of MLGM and its sites of intracellular origin. Much of what we believe about the mechanisms of origin and secretion of lipid globules and formation of lipid globule membrane is based on interpretation of static images provided by morphological studies. While many biochemical/compositional studies have been made which have provided useful information, interpretation of the findings has been handicapped by lack of suitable reference cell membrane fractions. In molecular terms, we know remarkably little about how droplets are formed, how they grow prior to secretion, and how they acquire those components which form the MLGM. Progress in this area has been handicapped greatly by the lack of availability of differentiated mammary epithelial cell lines which secrete milk lipid globules, and by slow progress in development of cell-free systems in which steps in formation and secretion of lipid globules can be reconstructed.

ACKNOWLEDGEMENTS

Much of our own research discussed herein was supported by U. S. Public Health Service Grant GM31244 from the National Institutes of Health. We thank many colleagues for useful discussion of several issues discussed herein. Especially we thank Drs. Ian H. Mather and Hans W. Heid for many stimulating, detailed discussions on milk lipid globule formation and secretion, and Dr. W. W. Franke for provision of the micrograph in Fig. 1e. Figures 1a, b, c and d are from reference #1, 1e is from Cytobiologie 14:259 (1977), and 2 is from reference # 6. These are reproduced with permission of the publisher.

REFERENCES

1. D. P. Dylewski, C. H. Dapper, H. M. Valivullah, J. T. Deeney and T. W. Keenan, Morphological and biochemical characterization of possible intracellular precursors of milk lipid globules, Eur. J. Cell Biol. 35:99 (1984).

2. C. A. Long and S. Patton, Formation of intracellular fat droplets: Interrelation of newly synthesized phosphatidylcholine and triglycerides in milk, J. Dairy Sci. 61:1392 (1978).

3. R. O. Scow, E. J. Blanchette-Mackie and L. C. Smith, Transport of lipid across capillary endothelium, Feder. Proc. 39:2610 (1980).

4. S. Patton and T. W. Keenan, The milk fat globule membrane, Biochim. Biophys. Acta 415:273 (1975).

5. M. Zaczek and T. W. Keenan, Morphological evidence for an endoplasmic reticulum origin of milk lipid globules using lipid-selective staining procedures, Protoplasma in press (1990).

6. J. T. Deeney, H. M. Valivullah, C. H. Dapper, D. P. Dylewski and T. W. Keenan, Microlipid droplets in milk secreting mammary epithelial cells: evidence that they originate from endoplasmic reticulum and are precursors of milk lipid globules, Eur. J. Cell Biol. 38:16 (1985).

7. H. M. Valivullah, D. R. Bevan, A. Peat and T. W. Keenan, Milk lipid globules: Control of their size distribution, Proc. Nat. Acad. Sci. USA 85:8775 (1988).

8. H. M. Valivullah, D. P. Dylewski and T. W. Keenan, Distribution of terminal transferases of acylglycerol synthesis in cell fractions from lactating mammary gland, Int. J. Biochem. 18:799 (1986).

9. G. van Meer, Lipid traffic in animal cells, Annu. Rev. Cell Biol. 5:247 (1989).

10. W. R. Bishop and R. M. Bell, Assembly of phospholipids into cellular membranes: Biosynthesis, transmembrane movement and intracellular translocation, Annu. Rev. Cell Biol. 4:579 (1988).

11. T. W. Keenan, Membranes of mammary gland IX. Concentration of glycosphingolipid galactosyl and sialyltransferases in Golgi apparatus from bovine mammary gland, J. Dairy Sci. 57:187 (1974).

12. G. R. Matyas and D. J. Morre', Subcellular distribution and biosynthesis of rat liver gangliosides. <u>Biochim. Biophys. Acta</u> 921:599 (1987).

13. I. H. Mather and T. W. Keenan, Function of endomembranes and the cell surface in the secretion of organic milk constituents, <u>in</u>: "Biochemistry of Lactation," T. B. Mepham, ed., Elsevier, Amsterdam (1983).

14. T. W. Keenan, I. H. Mather and D. P. Dylewski, Physical equilibria: Lipid phase, <u>in</u>: "Fundamentals of Dairy Chemistry," 3rd ed., N. P. Wong, R. Jenness, M. Keeney and E. H. Marth, eds., Van Nostrand Reinhold, New York (1988).

15. F. B. P. Wooding, The mechanism of secretion of the milk fat globule, <u>J. Cell Sci</u>. 9:805 (1971).

16. B. H. Stemberger, R. M. Walsh and S. Patton, Morphometric evaluation of lipid droplet associations with secretory vesicles, mitochondria and other components in the lactating cell, <u>Cell Tissue Res</u>. 236:471 (1984).

17. B. H. Stemberger and S. Patton, Relationships of size, intracellular location, and time required for secretion of milk fat droplets, <u>J. Dairy Sci</u>. 64:422 (1981).

18. S. Patton, Origin of the milk fat globule, <u>J. Amer. Oil Chem. Soc</u>. 50:178 (1973).

19. W. Bargmann and A. Knoop, Ueber die morphologie der Milchsekretion. Licht- und electronenmikroskopische Studien an der Milchdruese der Ratte, <u>Z. Zellforsch</u>. 49:344 (1959).

20. F. B. P. Wooding, Comparative mammary fine structure, <u>in</u> "Comparative Aspects of Lactation," M. Peaker, ed., Academic Press, New York (1977).

21. C. Freudenstein, T. W. Keenan, W. N. Eigel, M. Sasaki, J. Stadler and W. W. Franke, Preparation and characterization of the inner coat material associated with fat globule membranes from bovine and human milk. <u>Exp. Cell Res</u>. 118:277 (1979).

22. W. W. Franke, H. W. Heid, C. Grund, S. Winter, C. Freudentein, E. Schmid, E-D. Jarasch and T. W. Keenan, Antibodies to the major insoluble milk fat globule membrane associated protein: specific location in apical regions of lactating epithelial cells. <u>J. Cell Biol</u>. 89:4-85 (1981).

23. E-D. Jarasch, C. Grund, G. Bruder, H. W. Heid, T. W. Keenan and W. W. Franke, Localization of xanthine oxidase in mammary gland epithelium and capillary endothelium, <u>Cell</u> 25:67 (1981).

24. L. M. Neira and I. H. Mather, Biochemical and immunological comparison of bovine butyrophilin with a butyrophilin-like glycoprotein in guinea pig milk fat globule membrane: Evidence that the guinea pig protein is developmentally regulated and specifically expressed in lactating mammary tissue, <u>Protoplasma</u> in press (1990).

25. T. W. Keenan, H. W. Heid, J. Stadler, E-D. Jarasch and W. W. Franke, Tight attachment of fatty acids to proteins associated with milk lipid globule membrane. <u>Eur. J. Cell Biol</u>. 26:270 (1982).

26. L. J. W. Jack and I. H. Mather, Cloning and analysis of cDNA encoding bovine butyrophilin, an apical glycoprotein expressed in mammary tissue and secreted in association with milk fat globule membrane during lactation, <u>J. Biol. Chem</u>. 265:14,481 (1990).

27. M. Sasaki and T. W. Keenan, Ultrastructural character-
    ization of carbohydrate distribution on milk lipid
    globule membrane, <u>Cell Biol. Int. Reports</u> 3:67 (1979).
28. F. B. P. Wooding, The structure of the milk fat globule
    membrane, <u>J. Ultrastruct. Res</u>. 37:388 (1971).
29. F. B. P. Wooding, Milk fat globule membrane material in
    skim milk, <u>J. Dairy Res</u>. 41:331 (1974).
30. C. H. Dapper, H. M. Valivullah and T. W. Keenan, Use of
    polar aprotic solvents to release membrane from milk
    lipid globules, <u>J. Dairy Sci</u>. 70:760 (1987).
31. I. H. Mather and T. W. Keenan, Studies on the structure of
    milk fat globule membrane, <u>J. Membrane Biol</u>. 21:65
    (1975).
32. C. Kanno, Secretory membranes of the lactating mammary
    gland, <u>Protoplasma</u> in press (1990).
33. T. W. Keenan, D. P. Dylewski, T. A. Woodford and R. H.
    Ford, Origin of milk fat globules and the nature of
    the milk fat globule membrane, <u>in</u>: "Developments in
    Dairy Chemistry-2," P. F. Fox, ed., Applied Science,
    London (1983).
34. A. V. McPherson and B. J. Kitchen, Reviews of the progress
    of dairy science: The bovine milk fat globule membrane
    - its formation, composition, structure and behaviour
    in milk and dairy products, <u>J. Dairy Res</u>. 50:107
    (1983).
35. C. Kanno, H. Hattori and K. Yamauchi, Characterization of
    plasma membrane proteins from lactating mammary gland,
    <u>Agric. Biol. Chem</u>. 51:1325 (1987).
36. C. Kanno, H. Hattori and K. Yamauchi, Lipid composition of
    plasma membranes isolated from lactating bovine mammary
    gland, <u>Agric. Biol. Chem</u>. 51:2995 (1987).
37. J. W. Huggins, T. P. Trenbeath, R. W. Chesnut, C. A. C.
    Carraway and K. L. Carraway, Purification of plasma
    membranes of mammary gland: Comparisons of subfractions
    with rat milk fat globule membrane, <u>Exp. Cell Res</u>.
    126:279 (1980).
38. T. W. Keenan, D. J. Morre', D. E. Olson, W. N. Yunghans
    and S. Patton, Biochemical and morphological comparison
    of plasma membrane and milk fat globule membrane from
    bovine mammary gland, <u>J. Cell Biol</u>. 44:80 (1970).
39. T. W. Keenan, H. M. Valivullah and J. T. Dunlevy, Isolation
    of plasma membranes from mammary gland by two phase
    polymer partitioning, <u>Anal. Biochem</u>. 177:194 (1989).

N-LINKED OLIGOSACCHARIDE SYNTHESIS AND CELLULAR SOCIOLOGY

Armando J. Parodi

Instituto de Investigaciones Bioquimicas "Fundacion Campomar"
Antonio Machado 151
1405 Buenos Aires, Argentina

From a simple enzymatic reaction to interactions among cells, there are practically no biological events that do not have recognition of shapes as the key factor determining the uniqueness of the event. In this presentation I would like to introduce protein-linked oligosaccharides as elements carrying enormous potential information, essential for the specificity of biological events.

In two other types of molecules, peptides and oligonucleotides, the specificity of the information is determined solely by the number and sequence of the monomers, whereas in oligosaccharides the information may be also carried by the position and the anomeric configuration (α or β) of the bonds. Thus, a disaccharide formed by the same hexose (for instance two mannoses) may form 8 different disaccharides, whereas there could be a single dipeptide or a single dinucleotide formed by the same amino acid or nucleotide. The difference is more impressive as the complexity of the polymers increases. For instance, four different hexoses may form 35,600 different tetrasaccharides but four different amino acids or nucleotides may only form 24 different tetrameric molecules. It should be mentioned that the number of possible tetrasaccharides is actually much higher than 35,600 as several substituents (sulfate, phosphate) may be linked to the hydroxyl groups.

Is this diversity in the informational potential that makes protein-linked oligosaccharides substances worth being studied in cancer-related problems.

Initial steps in the synthesis of N-linked oligosaccharides

N-glycosylation is initiated by the transfer of an oligosaccharide containing 3 glucose, 9 mannose and 2 N-acetylglucosamine residues from a dolichol-P-P derivative to asparagine units in nascent polypeptide chains[1]. Dolichol is a polyprenol lipid containing from 11 isoprene units in certain protozoa to 20-21 in mammalian cells[2,3]. The oligosaccharide transferred has a well conserved structure in evolution as the same oligosaccharide is transferred in mammals, plants and fungi (Fig. 1).

Breast Epithelial Antigens, Edited by R.L. Ceriani
Plenum Press, New York, 1991

As soon as the oligosaccharide is transferred, its processing is initiated by the removal of the glucose units catalyzed by two specific glucosidases located in the lumen of the endoplasmic reticulum. Glucosidase I is an α (1,2) glucosidase that removes the more external glucose unit, whereas glucosidase II, an α (1,3) glucosidase, excises the two remaining glucose units[1]. There is at least one (and probably two) α (1,2) mannosidase in the endoplasmic reticulum that may remove certain peripheral mannose units[4,5]. The glucose-free, protein-linked oligosaccharides are then transiently re-glucosylated directly from UDP-Glc and deglucosylated again by glucosidase II. The role of this apparently futile process is unknown. It occurs in mammalian, plant, fungal and protozoan cells and it is apparently related to the adoption by glycoproteins of their final tertiary structure, a process that precedes their passage to the cis cisternae of the Golgi apparatus[6].

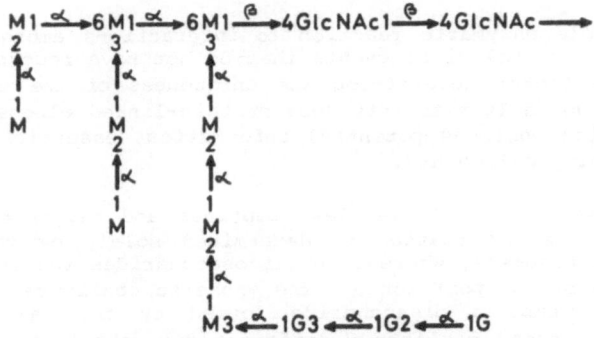

Fig. 1.- Structure of the oligosaccharide transferred in protein N-glycosylation. M stands for mannose and G for glucose.

The processing occurring in the endoplasmic reticulum is schematically represented in Fig. 2.

Further processing reactions occurring in the Golgi apparatus of mammalian, plant, and fungal cells differ significantly. In this presentation I will only mention those occurring in the first ones, but for the sake of comparison, processing reactions occurring in Saccharomyces cerevisiae are depicted in Fig. 3.

Processing of oligosaccharides in the Golgi apparatus. Sorting of lysosomal enzymes

Processing of the oligosaccharides in mammalian cells continues by further trimming of α (1,2) mannose units by Golgi mannosidase I, an enzyme located in the cis cisternae (Fig. 4).

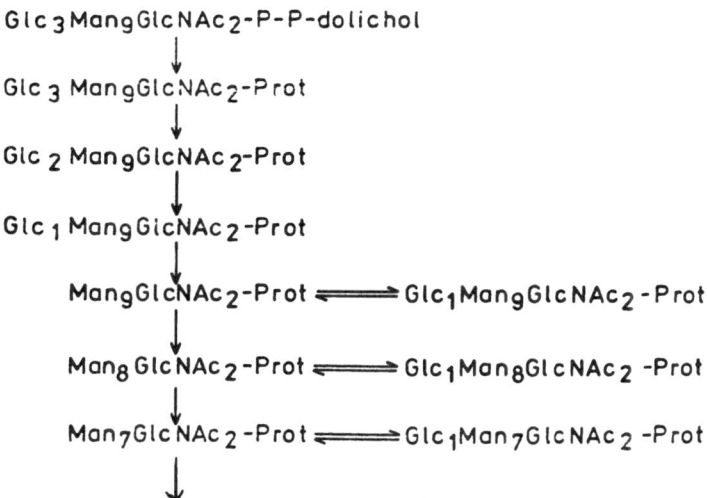

Fig. 2.- <u>Processing</u> <u>of</u> <u>N-linked</u> <u>oligosaccharides</u> <u>occurring</u> <u>in</u> <u>the</u> <u>endoplasmic</u> <u>reticulum.</u>

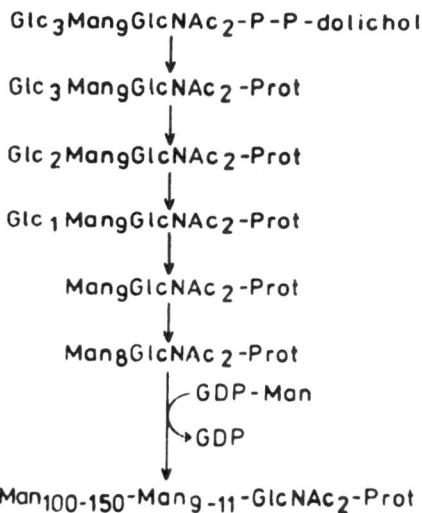

Fig. 3.- <u>Processing</u> <u>of</u> <u>N-linked</u> <u>oligosaccharides</u> <u>occurring</u> <u>in</u> <u>the</u> <u>yeast</u> <u>Saccharomyces</u> <u>cerevisiae.</u>

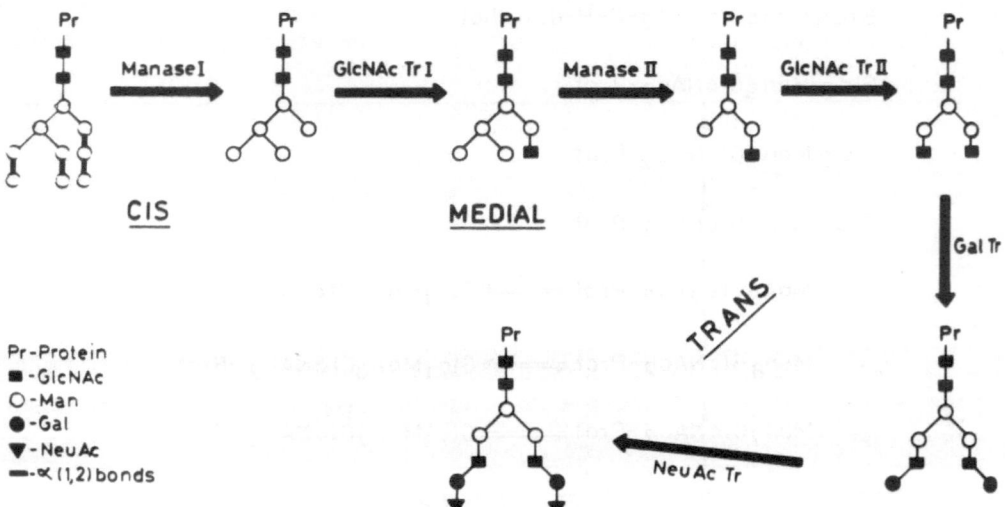

Fig. 4.- Schematic representation of the processing of N-linked oligo-saccharides occurring in the Golgi apparatus.

We will now see one of the best studied examples of recognition phe-nomena involving protein-linked oligosaccharides, that is, the routing of lysosomal enzymes to their final destination[8,9]. Lysosomal enzymes are glycoproteins having asparagine-linked oligosaccharides. In the cis cisternae of the Golgi apparatus, there is a membrane-bound enzyme that may transfer a N-acetylglucosamine-phosphate unit from UDP-GlcNAc to position 6 of mannose residues in the N-linked oligosaccharides (Fig. 5).

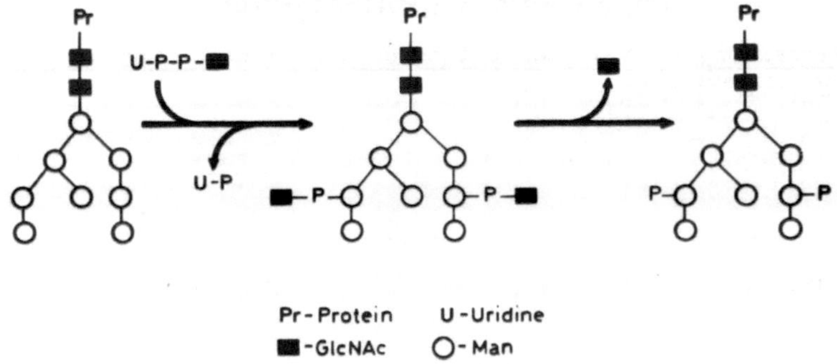

Fig. 5.- Reactions leading to the formation of Man-6-P groups in lyso-somal enzymes.

A second enzyme (N-acetylglucosamine-1-phosphodiester N-acetyl-glucosaminidase, also present in the cis cisternae of the Golgi apparatus) removes then the GlcNAc residue thus yielding phosphate units linked to position 6 of the mannose residues.

It has been shown, in cell-free assays, that the GlcNAc-1-P trans-ferase phosphorylates lysosomal enzymes at least two orders of magnitude better than other non-lysosomal glycoproteins[10]. It must be noted that, at this stage of processing, both lysosomal enzymes and non-lysosomal glycoproteins share the same compartments within the Golgi apparatus. The basis for this discrimination has been found not to be the presence of an amino acid sequence common to all lysosomal enzymes, but apparently to certain lysine units scattered at both sides of the oligosaccharide but that become proximal once the lysosomal enzymes are properly folded. Two receptors specific for the Man-6-P units have been detected, isolated and sequenced[11]. They have been located by immunocytochemistry in all the Golgi cisternae, that is, the cis, medial and trans[12]. This fact, together with the presence in lysosomal enzymes not only of phosphory-lated oligosaccharides composed of mannose and N-acetylglucosamine units, but also of complex oligosaccharides containing galactose and sialic acid residues suggests that the ligand-receptor complex migrates from the cis to the medial, and then to the trans Golgi cisternae from where it exits in coated vesicles and is delivered to a prelysosomal compartment. As will be seen below, galactose and sialic acid units are added to gly-coproteins in the trans cisternae.

The acidic environment of the prelysosomal compartment induces a dissociation of the ligand-receptor complex. The receptor then recycles back to the Golgi apparatus to pick up another ligand molecule. A schema-tic representation of the mechanism of routing of lysosomal enzymes is represented in Figs. 5 and 6.

A small proportion of lysosomal enzymes (5-20%) are normally secreted to the external medium but may be recognized by receptors present in the plasma membrane, internalized and delivered to lysosomes.

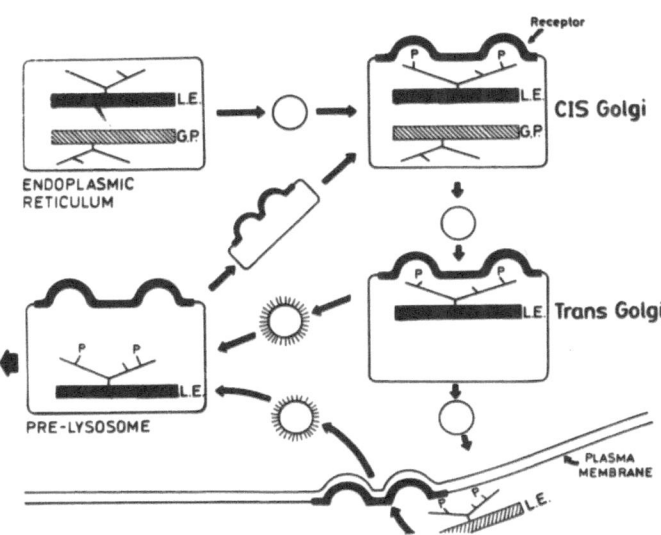

Fig. 6.- Schematic representation of the sorting mechanism of lysosomal enzymes. L.E.: lysosomal enzyme; G.P.: glycoprotein.

As mentioned above, two different receptors have been characterized. One of them is independent from divalent cations for binding activity and has a 44 amino acids amino terminal sequence, a 2269 residue extracytoplasmic domain, a single 23 residue transmembrane region and a 163 residue carboxyl terminus cytoplasmic domain. The extracytoplasmic domain has 19 potential glycosylation sites, of which at least 2 are occupied. The extracytoplasmic domain has a highly repetitive structure composed of 15 contiguous units of 147 amino acids. The homologies between units are of about 16-38%. The spacings are composed of cysteine rich regions. The second receptor is dependent on divalent cations for ligand binding, has a 28 residue amino terminal signal sequence, a 159 residue extracytoplasmic domain, a single 25 residue transmembrane region and a 67 carboxyl terminal cytoplasmic domain. The entire extracytoplasmic domain of the cation-dependent receptor is similar to each of the repeating units from the cation independent receptor with a sequence homology of 28%. This suggests that both receptors evolved from one ancestral gene. Cells lacking the large cation independent receptor secrete 70% of newly synthesized lysosomal enzymes. The residual sorting appears to be mediated by the cation-dependent receptor. Both receptors are also in the plasma membrane, but only the cation independent one is capable of internalizing lysosomal enzymes.

A schematic representation of both receptors is shown in Fig. 7.

There are more than 30 separate lysosomal enzyme storage diseases, the majority of which result from the absence of one particular enzyme. There are other diseases, such a I cell disease (mucolipidosis II) and pseudo Hurler polydistrophy (mucolipidosis III) that have a deficiency (or a very low amount) of all lysosomal enzymes. Fibroblasts from

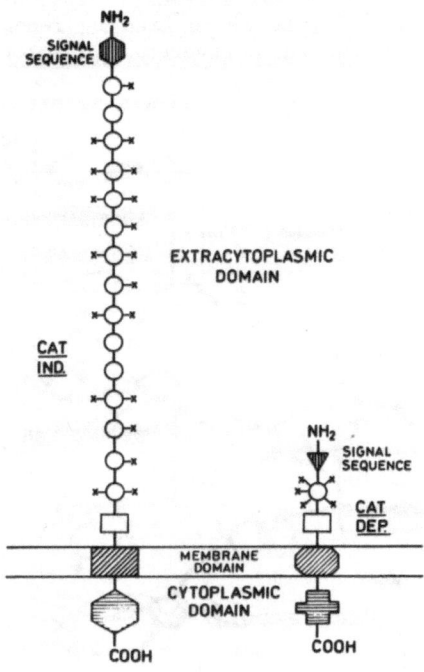

Fig. 7.- Schematic representation of the structure of cation-independent and dependent receptors. Crosses indicate potential glycosylation sites.

patients suffering from those diseases are deficient in the N-acetylglucosamine phosphotransferase (the first enzyme of the sorting pathway) and secrete the lysosomal enzymes to the external milieu.

Processing of oligosaccharides in the medial Golgi apparatus. Recognition events related to reactions occurring therein

I will continue now describing the processing of non-lysosomal enzyme glycoproteins. Some of the oligosaccharides, those belonging to the high mannose-type, have the composition $Man_{5-9}GlcNAc_2$ even in mature, fully processed glycoproteins. Further processing of the oligosaccharides is stopped probably because the oligosaccharides are not accessible to the glycosidases and glycosyltransferases of the exit route of glycoproteins[13]. Alternatively, the already mentioned mannosidase I, located in the cis cisternae of the Golgi apparatus may fully degrade the oligosaccharides to $Man_5GlcNAc_2$, by removing all $\alpha(1,2)$ linked units.

A N-acetylglucosaminyltransferase (N-acetylglucosaminyltransferase I) may then transfer a GlcNAc unit to yield $GlcNAc_1Man_5GlcNAc_2$. Mannosidase II removes then two additional mannose units, thus producing $GlcNAc_1Man_3GlcNAc_2$ (Fig. 4)[14].

It is worth mentioning that the presence of the GlcNAc is an absolute requirement for the action of mannosidase II. Both N-acetylglucosaminyltransferase I and mannosidase II are located in the medial Golgi cisternae.

There are six different N-acetylglucosaminyltransferases[15]. Their specificities are indicated in Fig. 8.

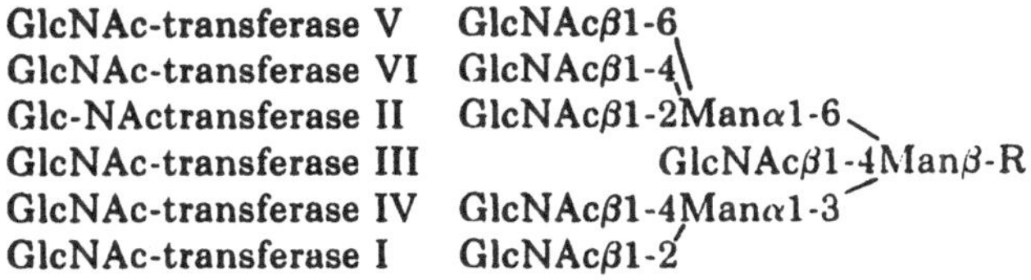

GlcNAc-transferase V GlcNAcβ1-6

GlcNAc-transferase VI GlcNAcβ1-4

Glc-NActransferase II GlcNAcβ1-2Manα1-6

         GlcNAcβ1-4Manβ-R

GlcNAc-transferase III

GlcNAc-transferase IV GlcNAcβ1-4Manα1-3

GlcNAc-transferase I GlcNAcβ1-2

Fig. 8.- Reactions catalyzed by the medial Golgi N-acetylglucosaminyltransferases.

Galactosyl and sialyltransferases, located in the trans Golgi cisternae may then add galactosyl and sialyl residues thus forming the so called complex-type oligosaccharides (Fig. 4). Each stretch of sialyl-galactosyl-N-acetylglucosaminyl residues is called an antenna and complex-type oligosaccharides may have up to five antennae.

I will describe now how interferring with the action of mannosidase II (that is, blocking the formation of complex-type oligosaccharides) may affect the metastatic characteristics of certain tumor cells, and furthermore how in some cases the metastatic potential of tumor cells is directly linked to the action of N-acetylglucosaminyltransferase V. Mannosidase II is inhibited by a plant alkaloid, swainsonine. Taking advantage of this fact, murine melanoma cells (B16-F10, highly metastatic) were grown for approximately two generations in the presence of the alkaloid and then injected into the lateral vein of C57BL/6 mice[16]. A dramatic swainsonine-dose dependent decrease of pulmonary colonization 14 days after the injection was observed. Treatment of the murine melanoma cells with swainsonine, however, did not modify the tumorigenicity as judged by the fact that mice did not change the rate the appearance of palpable tumors, which killed the animals at the same rate as those produced by untreated cells. Additionally, treated and untreated cells grew equally well in soft agar. Moreover, the same number of cells treated and untreated with swainsonine (both labeled with [^{125}I]-iododeoxyuridine) were initially (2 min) retained by lung cells. Both sets of cells were cleared from the lungs at negative exponential rates, but the swainsonine-treated ones were more rapidly lost than the untreated cells. This suggests that the mechanism by which swainsonine inhibits metastatic colonization is by interferring with the retention of cells by the target organ.

I will discuss now an example of how the $\beta(1,6)$ branching catalyzed by N-acetylglucosaminyltransferase V is directly associated with metastasis. The reaction catalyzed by this enzyme is depicted in Fig. 8. In general, tumor cells tend to have glycoproteins with larger, more branched N-linked oligosaccharides and to have a higher sialic acid content. This last fact obviously correlates with the higher branching as the antennae generally end up with sialic acid units.

The highly metastatic murine tumor cells MDAY-D2 were grown in the presence of leuco phytohemagglutinin (L-PHA), a lectin recognizing specifically the $\beta(1,6)$-linked lactosamine antenna (the disaccharide Gal-GlcNAc is called lactosamine), whose synthesis is initiated by transferase V. Two lectin-resistant clones were isolated[17]. They were highly tumorigenic but poorly metastatic when injected subcutaneously or intravenously to syngeneic or nude mice. A main glycoprotein of the external surface (gp 130) contained sialic acid in both wild type and resistant cells, but did bind L-PHA only in the wild type ones, thus indicating that the defect was specifically related to the absence of $\beta(1,6)$ branches in the resistant variants. The transferase V content of these cells was only 20% of that of the wild type ones.

The reverse experiment was also attempted, that is, to convert tumorigenic non-metastatic cells into metastatic ones. Cells of the murine mammary adenocarcinoma, non-metastatic SP1 line, were transfected with pSVneo (neomycin-resistant), alone or with pSVneo linked to activated T24H-ras oncogene or non-activated cH-ras. Neomycin-resistant clones were isolated and tested for the expression of transfected genes, metastasis in syngeneic mice and expression of L-PHA binding in gp 130. The highest metastatic cells were those transfected with pSVneo-T24H and they had the highest L-PHA binding capacity. Most significantly, cells transfected with the two other constructs had lower metastatic capacity and lower L-PHA binding. However, when clones were isolated from the small number of metastasis produced by them, they all had high L-PHA binding capacity. That is, the presence of the $\beta(1,6)$-linked lactosamine antenna correlated with the metastatic potential. On the other hand, the level of gp 130 was similar in all cells. The only difference was in the β·(1,6) branching.

Processing of oligosaccharides in the trans Golgi cisternae. Related recognition events

As mentioned above, galactosyl and sialyl units may be added to the distal N-acetylglucosaminyl units in the trans cisternae of the Golgi apparatus, thus completing the formation of the complex-type oligosaccharides[1]. The glycoproteins thus processed may be then transported to the plasma membrane, to intracellular locations or secreted. I will mention now a last example of a recognition event involving N-linked oligosaccharides, in this case specifically sialic acid units. Influenza virus is an enveloped virus containing a lipid membrane that is obtained during maturation by budding from the plasma membrane of infected cells[18,19]. Two membrane glycoproteins coded by the virus are hemagglutinin (HA) and neuraminidase. HA is synthesized as a single polypeptide that is cleaved proteolytically by removal of Arg 329 into two chains HA_1 (36,336 D) and HA_2 (25,750 D). These chains are linked covalently by a disulfide bond between amino acid 14 (HA_1) and 137 (HA_2). The two chain monomers are associated noncovalently to form trimers on the surface of the membrane. HA binds to sialic acid residues in glycoproteins or glycolipids in the surface of target cells. This binding leads to a fusion of membranes and to infection. As a result of enzymatic desialylation of target cells, no infection occurs. Moreover, enzymatic re-sialylation of target cells or addition of sialylated glycolipids (that become integrated into the membranes) to them restores the binding of HA and the capacity of cells of being infected by the virus.

The carbohydrate binding site in HA forms a pocket that protrudes from the membrane and is composed of amino acids that are largely conserved in the numerous strains of the virus. As an example of specificity, it may be mentioned that when the primary sequence of wild type HA of human virus was compared with that of mutants having decreased affinity for NeuAc α(2,6)Gal and marked increased affinity for NeuAc α(2,3)Gal, it was found that a Leu at position 226 in the parental strain had been changed for a Glu in the mutants. Similar studies in avian wild type isolates and their variants showed the reverse change (from α(2,3) to α(2,6)-linked sialic acid in specificity in wild type and mutants, respectively), and a change from Glu to Leu.

In my presentation, I have only mentioned a few of the increasingly numerous examples of recognition events in which protein-linked oligosaccharides are involved. I am sure that the wealth of information coming from carbohydrate biochemistry and tumor biology will show that they are not distant fields of interest, but closely related ones.

REFERENCES

1. R. Kornfeld & S. Kornfeld. Assembly of asparagine-linked oligosaccharides. Annu. Rev. Biochem. 54: 631 (1985).
2. A.J. Parodi & L.A. Quesada-Allue. Protein glycosylation in Trypanosoma cruzi. I. Characterization of dolichol-bound monosaccharides and oligosaccharides synthesized "in vivo". J. Biol. Chem. 257: 7637 (1982).
3. L.A. Quesada-Allue & A.J. Parodi. Novel mannose carrier in the trypanosomatid Crithidia fasciculata behaving as a short -saturated polyprenyl phosphate. Biochem. J. 212: 123 (1983).
4. J. Bischoff & R. Kornfeld. Evidence for an α-mannosidase in the

endoplasmic reticulum of rat liver. J. Biol. Chem. 258: 7907 (1983).

5. L.J. Rizzolo & R. Kornfeld. Post-translational protein modification in the endoplasmic reticulum. Demonstration of fatty acylase and deoxymannojirimycin-sensitive α-mannosidase activities. J. Biol. Chem. 263: 9520 (1988).

6. S. Trombetta, M. Bosch & A.J. Parodi. Glucosylation of glycoproteins by mammalian, plant, fungal & trypanosomatid protozoa microsomal membranes. Biochemistry 28: 8108 (1989).

7. A.J. Parodi. The mechanism of synthesis of the polysaccharide part of mannan in Saccharomyces cerevisiae. Arch. Biochem. Biophys. 210: 372 (1981).

8. S. Kornfeld. Trafficking of lysosomal enzymes in normal and disease states. J. Clin. Invest. 77: 1 (1986).

9. K. von Figura & A. Hasilik. Lysosomal enzymes and their receptors. Annu. Rev. Biochem. 55: 167 (1986).

10. L. Lang, M.L. Reitman, J. Tang, R.M. Roberts & S. Kornfeld. Lysosomal enzyme phosphorylation. Recognition of a protein-dependent determinant allows specific phosphorylation of oligosaccharides present on lysosomal enzymes. J. Biol. Chem. 259: 14663 (1984).

11. N.M. Dahms, P. Lobel & S. Kornfeld. Mannose 6-phosphate receptors and lysosomal enzyme targeting. J. Biol. Chem. 264: 12115 (1989).

12. H.J. Geuze, J.W. Slot, G.J.A.M. Strous, A. Hasilik & K. von Figura. Ultrastructural localization of the mannose 6-phosphate receptor in rat liver. J. Cell. Biol. 98: 2047 (1984).

13. P. Hsieh, M. Rich-Rosner & P.W. Robbins. Selective cleavage by endo-β-N-acetylglucosaminidase H at individual glycosylation sites of Sindbis virion envelope glycoprotein. J. Biol. Chem. 258: 2555 (1983).

14. I. Tabas & S. Kornfeld. The synthesis of complex-type oligosaccharides. III. Identification of an α-D-mannosidase activity involved in a late stage of processing of complex-type oligosaccharides. J. Biol. Chem. 253: 7779 (1978).

15. I. Brockhausen, E. Hull, O. Hindsgaul, H. Schachter, R.W. Shah, S.W. Michnick & J. Carver. Control of glycoprotein synthesis. Detection and characterization of a novel branching enzyme from hen oviduct, UDP-N-acetylglucosamine:GlcNAc β1-6 (GlcNAc β1-2) Man -R (GlcNAc to Man) β-4-N-acetylglucosaminyltransferase VI. J. Biol. Chem. 264: 11211 (1989).

16. M.J. Humphries, K. Matsumoto, S.L. White & K. Olden. Oligosaccharide modification by swainsonine treatment inhibits pulmonary colonization by B16-F10 murine melanoma cells. Proc. Natl. Acad. Sci. U.S.A. 83: 1752 (1986).

17. J.W. Dennis, S. Laferte, C. Waghorne, M.L. Breitman & R.S. Kerbel. β1-6 branching of Asn-linked oligosaccharides is directly associated with metastasis. Science 236: 582 (1987).

18. D.C. Wiley & J.J. Skehel. The structure and function of the hemagglutinin membrane glycoprotein of influenza virus. Annu. Rev. Biochem. 56: 365 (1987).

19. W. Weis, J.H. Brown, S. Cusack, J.C. Paulson, J.J. Skehel & D.C. Wiley. Nature 333: 426 (1988).

DEVELOPMENT OF ENZYME IMMUNOASSAYS FOR BREAST CARCINOMA-ASSOCIATED

MUCIN ANTIGENS

Joseph P. Brown[1], Rose Beer[1], Sidney
Hallam[1], Patricia Stewart[1], Kristi
Stob[1] and Ted A. W. Splinter[2]

[1]Genetic Systems, Seattle,
Washington, [2]University Hospital,
Rotterdam

*Hybridomas producing monoclonal antibodies directed against epitopes of
breast carcinoma-associated mucin were obtained by immunizing mice with
mucin preparations and fusing their spleen cells with mouse myeloma
cells. Fourteen antibodies were selected for biochemical and
serological characterization. Four of the antibodies (M15, M22, M23,
and M27) bound to deglycosylated mucin and to a synthetic peptide
containing the mucin core repeated sequence, and one antibody (M26)
bound to a carbohydrate epitope, sialyl-Le^x. The epitopes recognized by
the other antibodies were not identified. Prototype "sandwich"
immunoassays were constructed by using pairs of antibodies, one to
capture antigen from the test sample, and the other, labeled with
peroxidase, to detect bound antigen. The assays were compared by
testing panels of sera from breast cancer patients and healthy donors as
controls. Stepwise discriminant analysis showed that the antibodies
differed significantly in their ability to distinguish between the two
groups. Antibody M29 provided the best discrimination, and antibody M26
gave additional information. These two antibodies, together with
antibody M38, which was used as the peroxidase conjugate in both assays,
were used to develop two enzyme immunoassays, referred to as CA M29 and
CA M26. These assays were compared with CA 15.3 and CEA by testing
panels of sera from breast cancer patients and controls and further
evaluated by testing serial samples from breast cancer patients
receiving chemotherapy.*

Mucins are heavily glycosylated, high molecular weight glycoproteins,
which are synthesized by epithelial tissues and released into the
circulation in large amounts by most carcinomas, including breast
carcinoma (1-4). The mucins associated with breast carcinoma are
structurally and antigenically complex and are expressed in both normal
and tumor tissues (5-7). However, mucins from normal breast epithelial
cells and tumor cells differ in their relative epitope densities (8, 9).

Monoclonal antibodies obtained by immunizing mice with human milk fat
globule membranes or purified mucin have been used to develop
immunoassays to detect mucins in the sera of breast carcinoma patients
(10-14). None of these assays, however, is entirely specific for breast

carcinoma, since the antigens are found in detectable quantities in the sera of healthy individuals, and in larger amounts in sera of patients with non-malignant disease and with other types of carcinoma.

The aim of the work summarized in this chapter was to identify epitopes that would be more specific for the mucin produced by breast carcinoma, in order to develop new diagnostic tests. Accordingly, we generated fourteen new monoclonal antibodies against breast carcinoma-associated mucin. The generation and characterization of these antibodies has been described previously (15).

Biochemical and immunological characterization of the antibodies showed that while they all appeared to bind to the same mucin, they recognized epitopes differing in density and degree of tumor specificity. The antibodies that were most suitable for use in immunoassays were selected by testing panels of sera from breast cancer patients and healthy donors as controls. In this manuscript, we describe the selection of the most suitable antibodies and the development and evaluation of two assays, referred to as CA M26 and CA M29.

MONOCLONAL ANTIBODIES TO BREAST CARCINOMA-ASSOCIATED MUCINS

Hybridomas producing antibodies to breast carcinoma-associated mucin were generated by immunizing mice with mucin and fusing their spleen cells with mouse myeloma cells. Fourteen antibodies were selected and characterized biochemically and serologically (15). Four antibodies (M15, M22, M23, M27) bound to a synthetic peptide containing the mucin core repeated sequence (16) as shown in Fig. 1. One of the antibodies (M26) bound to a carbohydrate epitope, sialyl-Le^x. The epitopes recognized by the other antibodies were not identified.

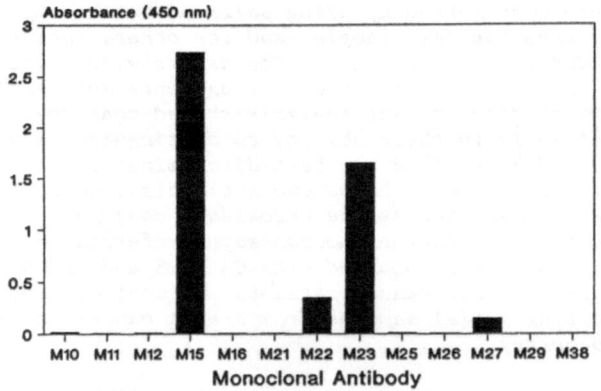

Figure 1. *Binding of monoclonal antibodies to synthetic peptide. Microtiter wells were coated with peptide SAPDTRPAPGSTAPPAHGVT and incubated with antibody-peroxidase conjugates, followed by chromogenic substrate.*

To determine which antibodies were most suitable for use in immunoassays, prototype assays were set up, in which one antibody was used to capture antigen from the test sample, and peroxidase-conjugated antibody was used to detect bound antigen. In some cases the same antibody was used as both the solid phase and the conjugate. In other cases antibody M38, which recognizes an abundant epitope, was used as the conjugate.

The assays were used to test panels of sera from breast carcinoma patients and controls, and the results were analyzed by stepwise discriminant analysis, a statistical technique that allows one to identify the assay or combination of assays that best discriminates between two sample populations (Fig. 2).

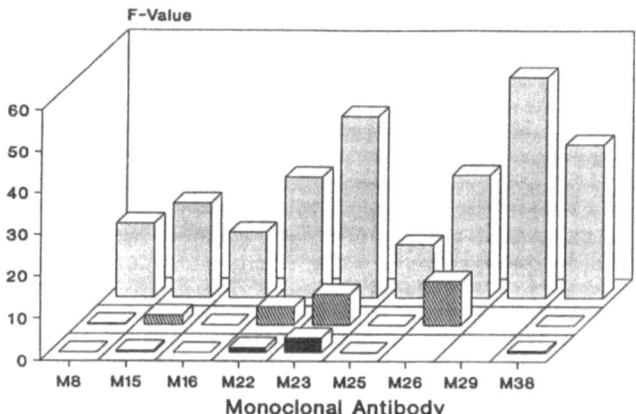

Figure 2. Stepwise discriminant analysis to compare nine prototype assays. Antibodies adsorbed to microtiter wells were incubated with diluted sera from forty patients with metastatic breast carcinoma and forty healthy donors. Bound mucin was detected by incubation with peroxidase-conjugated antibody, followed by incubation with chromogenic substrate. The same antibody was used as solid phase and conjugate, except for antibodies M26 and M29, for which antibody M38 was used as conjugate. F-values are a measure of the ability of each assay to distinguish between the two sample populations. The results of the three steps of the analysis are shown from back to front respectively.

The first step of the analysis (back row) showed that antibody M29, which has the highest F-value, discriminates between the two populations of test samples better than any of the other antibodies. The second step of the analysis (middle row) showed that antibody M26 provided the most additional information. In contrast, antibodies M8, M16, M25, and M38 did not provide any significant additional information, while antibody M23 and, to a lesser extent, antibodies M22 and M15, provided some information. In the third step of the analysis, antibody M23 had a modest, though not statistically significant F-value. Antibodies M26 and M29 were selected for further assay development.

ASSAY DEVELOPMENT

Assay procedure. The CA M26 and CA M29 EIAs were designed as quantitative "sandwich" immunoassays. The solid phase consists of microwells to which purified antibodies, M26 and M29 respectively, are adsorbed. The antibody-coated microtiter plates are dried and stored at room temperature. Full details of the procedure may be obtained from the manufacturer, Genetic Systems Corporation.

Test samples are diluted 1:11, and 0.05 ml of the diluted sample is added to the antibody-coated wells and incubated for one hour at room temperature. During this time, antigen present in the sample binds to the antibody-coated wells. At the end of the incubation, unbound material is removed by washing with buffer. Positive and negative controls and calibration standards are also tested in each run.

Peroxidase-conjugated antibody M38 (0.05 ml) is then added. After incubation for one hour at room temperature the wells are washed to remove unbound conjugate, and 0.1 ml buffered substrate/chromogen reagent (hydrogen peroxide and 3,3',5,5' tetramethylbenzidine) is added to each well. If antigen is present, blue color develops in a reaction with bound conjugate present in the well during the final one-hour incubation period.

The reaction is stopped by addition of sulfuric acid, which results in a color change from blue to yellow. The absorbance values for positive control, standards, and test samples are determined spectrophoto-metrically at 450 nm by using a microplate reader. Standard curves are constructed for each test by plotting absorbance value versus concentration for each standard. The antigen concentrations of diluted samples and diluted positive control are read from the standard curves, in units per milliliter, based upon reference standards maintained at Genetic Systems.

Reproducibility: Reproducibility was measured by testing four samples in duplicate on three occasions with three different manufactured lots of CA M26 and CA M29. The data were evaluated by analysis of variance, and the results are expressed as the coefficients of variation (CV) both between run and between lot. The results are summarized in Table 1. Coefficients of variation were in the range 3% to 10% between run and 7% to 11% between lot, except for the lowest CA M26 concentration tested.

Table 1. *Reproducibility of CA M26 and CA M29 EIA.*

Test	Sample	Mean (u/ml)	Between Run (% CV)	Between Lot (% CV)
CA M26	A	59.9	10.2	25.5
CA M26	B	117.8	8.7	11.1
CA M26	C	230.3	4.1	8.3
CA M26	D	730.9	3.2	7.7
CA M29	E	12.3	6.6	7.0
CA M29	F	28.8	7.2	6.7
CA M29	G	37.3	8.6	9.2
CA M29	H	67.4	5.5	9.9

Analytical Specificity: Lipemic samples, and samples spiked with hemoglobin (up to 52 mg/dl) and bilirubin (up to 7.7 mg/dl) did not interfere with the recovery of CA M26 or CA M29.

Analytical Sensitivity: Analytical sensitivity, a measure of the minimum amount of antigen that can be distinguished from a negative control, was approximately 0.50 units/ml for CA M26 and 0.02 units/ml for CA M29.

Linearity: Linearity of the CA M26 EIA was assessed by testing five dilutions (1:11, 1:16.5, 1:22, 1:44, 1:88) of eight serum samples. The results were analyzed by linear regression. Correlation coefficients ranged from 0.995 to 0.999 for CA M26 and from 0.997 to 0.999 for CA M29.

CLINICAL PERFORMANCE

The CA M26 EIA was evaluated on serum samples from breast cancer patients and controls. The results are summarized in Table 2.

Table 2. *CA M26 and CA M29 in breast cancer patients and controls. Results are expressed as the percent of specimens above the reference values of 100 units/ml and 20 units/ml for CA M26 and CA M29 respectively.*

Serum Sample	Number Tested	Percent Elevated	
		CA M26	CA M29
Breast cancer			
Present	245	50.2	67.0
No evidence of disease	302	3.0	18.5
Other cancer			
Present	388	11.6	40.3
No evidence of disease	265	1.5	13.9
Controls			
Benign disease	134	1.5	10.4
Healthy	600	2.2	0.3

Cancer patients: 547 serum samples from 208 patients with a previous diagnosis of breast cancer were tested. CA M26 levels were at least 100 u/ml in 50.2% of the 245 samples from patients with clinically evident breast cancer, and 3.0% of the 302 samples from patients who were clinically disease-free. CA M29 levels were at least 20 u/ml in 67.0% of the samples from patients with clinically evident breast cancer, and 18.5% of the 302 samples from patients who were clinically disease-free. 653 samples from patients with other malignancies (primarily carcinomas) were also tested. CA M26 levels were elevated in 11.6% of the 388 samples from patients with clinically evident disease and 1.5% of the 265 samples from patients who were clinically disease-free. CA M29 levels were elevated in 40.3% and 13.9% of the same serum samples.

Non-malignant disease: Serum samples from 134 patients with non-malignant disease were tested; 98.5% contained CA M26 levels less than 100 u/ml, and all contained less than 200 u/ml. Of the same sera, 89.6% contained CA M29 levels less than 20 u/ml, and 98.5% contained less than 40 u/ml.

Healthy donors: Serum samples from 600 healthy blood donors were tested for CA M26; 97.8% contained less that 100 u/ml, and 99.7% contained less than 200 u/ml. For CA M29, 99.7% contained less that 20 u/ml, and all contained less than 40 u/ml.

Comparison with CA 15.3 and CEA: Sera from 78 patients with recurrent breast cancer and 25 controls were tested for CA M26, CA M29, CA 15.3

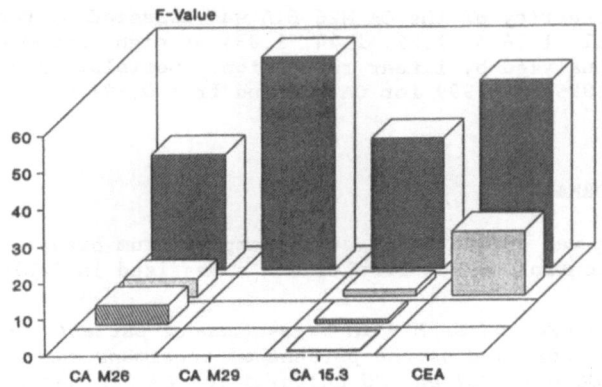

Figure 3. *Stepwise discriminant analysis to compare four immunoassays. Panels of sera from breast cancer patients and controls were tested for CA M26, CA M29, CA 15.3 (Centocor) and CEA (Hybritech). F-values are a measure of the ability of each antibody to discriminate between the two populations of samples. In each step of the analysis, the test with the highest F-value is selected. The results of the three steps of the analysis are shown from back to front respectively.*

(Centocor) and CEA (Hybritech), and the data were analyzed by stepwise discriminant analysis. The results of the analysis are shown in Fig. 3. In the first step of the analysis (back row), CA M29 had the highest F-value, followed by CEA. The second step shows that CEA provided the most additional information, with some provided by CA M26. In the third step, CA M26 provided additional information.

Monitoring breast cancer patients during therapy: The CA M26 and CA M29 tests were used in a retrospective analysis of serum samples obtained from breast cancer patients during therapy. All samples were tested coded. Results from three of the patients are shown here. A manuscript describing the completed study of 96 patients is in preparation.

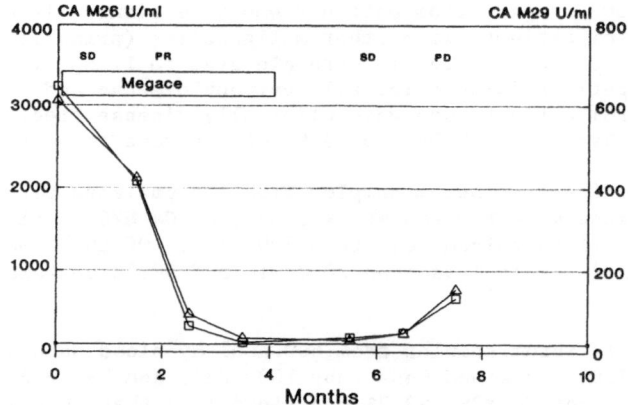

Figure 4. *Serum samples from a patient with recurrent breast carcinoma tested for CA M26 and CA M29 over an eight-month period. SD, stable disease; PR, partial response; PD, progressive disease.*

The patient shown in Fig. 4 was treated with megace for four months, during which time a partial response was observed. Levels of CA M26 (squares) and CA M29 (triangles) fell progressively, but did not reach the normal range (CA M26, 100 u/ml; CA M29, 20 u/ml). Levels of both markers started to rise after three months, and progressive disease was noted.

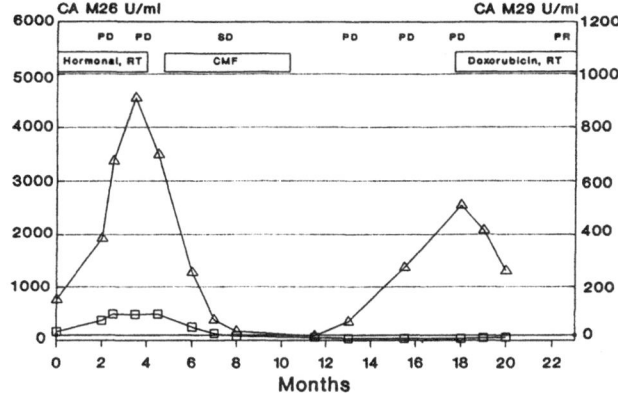

Figure 5. *Serum samples from a patient with recurrent breast carcinoma tested for CA M26 and CA M29 over an 20-month period.*

In the patient shown in Fig. 5, levels of CA M29 (triangles) rose as disease progressed (first four months and months 12 to 18) and fell during therapy with cytoxan/methotrexate/fluorouracil (CMF), which stabilized the disease, and during therapy with doxorubicin and radiation, which led to a partial response. However, levels of CA M26 (squares) fell to within the normal range during CMF-treatment, and did not rise subsequently.

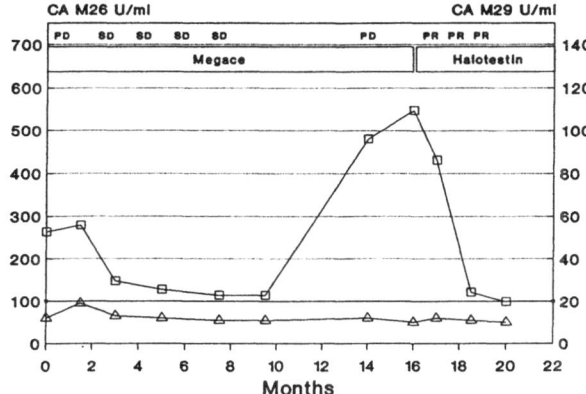

Figure 6. *Serum samples from a patient with recurrent breast carcinoma tested for CA M26 and CA M29 over an 20-month period.*

In the patient shown in Fig. 6, levels of CA M29 (triangles) remained within the normal range throughout the period of observation. However, levels of CA M26 (squares) initially fell in response to treatment with megace and then rose with progressive disease, only to fall again when a partial response was obtained with halotestin.

DISCUSSION

The goal of the work described here was the development of better assays for monitoring breast carcinoma patients. In our initial studies we used a novel screening assay to test a panel of antibodies for their ability to recognize antigens that were elevated in the sera of breast carcinoma patients. We found that the three positive antibodies all recognized mucin epitopes (14).

In subsequent experiments we found that monoclonal antibodies to epitopes of breast carcinoma-associated mucin could be readily generated by immunizing mice with purified mucin. These antibodies proved to recognize many different epitopes. Four antibodies were shown to bind a synthetic peptide containing the mucin core repeat sequence, and one (M26) bound sialyl-Le^x, which had previously been identified as a tumor-associated antigen (17, 18). For most of the antibodies, however, including M29 and M38, the epitopes have yet to be identified. Virtually all the antibodies were distinguishable serologically, and the amount of antibody binding, both to cells and to mucin in serum, varied greatly, presumably due to differences in avidity and epitope density.

The antigenic complexity of the breast carcinoma-associated mucin was felt to be an opportunity, in that certain combinations of antibodies might be much more suitable for assay development than others. The problem was how to identify the optimal combination of antibodies, since almost three hundred different assays could in principle be constructed with the fourteen antibodies described here.

We determined empirically that the best approach was to use an antibody to a low-abundance epitope for antigen capture, and an antibody to a high-density epitope in the enzyme conjugate. The specificity of the resulting assay was determined primarily by the first antibody, while its analytical sensitivity was determined by the second antibody. This can be explained by the first antibody binding a subpopulation of the mucin from the test sample.

After some initial studies with pools of sera from breast cancer patients and healthy donors, we found that the most effective and objective method was to test panels of sera from breast cancer patients and controls and use stepwise discriminant analysis to compare the results. This method showed that the single most informative antibody was M29 and that antibody M26 provided additional information.

The CA M26 and CA M29 assays were developed as microtiter "sandwich" enzyme immunoassays. We were initially concerned about the precision and reproducibility of this assay format. However, we found that the assays performed well in this regard, and the microtiter format required only small amounts of sample, while allowing up to two hundred specimens to be run in parallel.

Initial evaluation of the clinical performance of the tests was encouraging. When CA M26 and CA M29 were compared with CA 15.3 and CEA by stepwise discriminant analysis, it was found that CA M29 gave the best discrimination, followed by CEA, with CA M26 providing additional information.

The assays were evaluated further by testing serial serum specimens from breast carcinoma patients during therapy. In general a good correlation was seen between clinical course and changing antigen levels, although in a significant proportion of patients with active disease both markers remained within the normal range. A report of the completed study of 96 patients is in preparation.

The value of the additional information provided by the CA M26 assay was highlighted by these studies. In some patients (Fig. 4) the two markers behave very similarly. However, in others (Figs. 5 and 6) one of the markers was elevated, while the other was in the normal range. Fig. 6, for example, shows a patient in whom CA M29 levels were in the normal range, but progressively rising CA M26 values coincided with the change from stable disease to progressive disease, only to fall again during the partial response achieved with chemotherapy.

In conclusion, it is clear from these and other studies that the antigenic complexity of mucins can be dissected by monoclonal antibodies to provide additional information about the disease status of breast cancer patients. In these studies we used two-site enzyme immunoassays, which measure the total amount of mucin expressing both of two different epitopes. It is interesting to speculate that if one were able to analyze in more detail the antigenic composition of the mucins in a serum sample, just as population of lymphocytes can be analyzed by the fluorescence-activated cell sorter, one might be better able to resolve minor populations of carcinoma-derived mucin from the larger amounts of mucin derived from normal tissues and obtain much greater clinical sensitivity and specificity.

ACKNOWLEDGEMENTS

We wish to thank Drs. Gary Goodman, Donald Mercer, Joyce Taylor-Papadimitriou and Morton Schwartz for their advice, encouragement and collaboration.

REFERENCES

1. Rittenhouse, H.G., Manderino, G.L., and Hass, G.M. (1985) Mucin-type glycoproteins as tumor markers. Laboratory Medicine 16:556-560.
2. Arklie, J., Taylor-Papadimitriou, J., Bodmer, W., Egan, M., and Millis, R. (1981) Differentiation antigens expressed by epithelial cells in lactating breast are also detectable in breast cancers. Int. J. Cancer 28:23-29.
3. Ceriani, R.L., Peterson, J.A., Lee, J.Y., Moncada, R., and Blank, E.W. (1983) Characterization of cell surface antigens of human mammary epithelial cells with monoclonal antibodies prepared against human milk fat globule. Somatic Cell Genetics 9:415-427.
4. Hilkens, J., Buijs, F., Hilgers, J., Hageman, P., Calafat, J., Sonnenberg, A., and Vander Valk, M. (1984) Monoclonal antibodies against human milk-fat globule membranes detecting differentiation antigens of the mammary gland and its tumors. Int. J. Cancer 34:197-206.
5. Shimizu, M., and Yamauchi, K. (1982) Isolation and characterization of mucin glycoproteins in human milk fat globule membrane. J. Biochem., Tokyo, 91:515-524.
6. Omerod, M.G., McIlhinney, J., Steele, K., and Shimizu, M., Glycoprotein PAS-O from the milk fat globule membrane carries antigenic determinants for epithelial membrane antigen. Mol. Immunology 22:265-269 (1985).

7. Sekine, H., Ohno, T. and Kufe, D.W. Purification and characterization of a high molecular weight glycoprotein detectable in human milk and breast carcinomas. J. Immunology 135:3610-3615 (1985).

8. Burchell, J.,Durbin, H., and Taylor-Papadimitriou, J. (1983) Complexity of antigenic expression of antigenic determinants recognized by monoclonal antibodies HMFG-1 and HMFG-2 in normal and malignant human mammary epithelial cells. J. Immunology 131:508-513.

9. Kufe, D., Inghirami, G., Abe, M., Hayes, D., Justi-Wheeler, J.H., and Schlom, J. (1984) Differential reactivity of a novel monoclonal antibody (DF3) with human malignant versus benign breast tumors. Hybridoma 3:223-232.

10. Hilkens, J., Kroezen, V., Bonfrer, J.M.G., De Jong-Bakker, M., and Bruning, P.F. (1986) MAM-6 antigen, a new serum marker for breast cancer monitoring. Cancer Research 26:2582-2587.

11. Hayes, D.F., Sekine, H., Ohno, T., Abe, M., Keefe, K., and Kufe, D.W. (1985) Use of a murine monoclonal antibody for detection of circulating plasma DF3 antigen levels in breast cancer patients. J. Clin. Invest. 75:1671-1678.

12. Burchell, J., Wang, D., and Taylor-Papadimitriou, J. (1984) Detection of the tumor-associated antigens recognized by the monoclonal antibodies HMFG-1 and 2 in sera from patients with breast cancer. Int. J. Cancer 34:763-768.

13. Papsidero, L.D., Nemoto, T., Croghan, G.A., and Chu, T.M. (1984) Expression of ductal carcinoma antigen in breast cancer sera as defined using monoclonal antibody F36/22. Cancer Research 44:4653-4657.

14. Linsley, P.S., Ochs, V., Laska, S., Horn, D., Ring, D.B., Frankel, A.E., and Brown, J.P. (1986) Elevated levels of a high molecular weight antigen detected by antibody W1 in sera from breast cancer patients. Cancer Research 46:5444-5450.

15. Linsley, P.S., Brown, J.P., Magnani, J.L., and Horn, D. (1988) Monoclonal antibodies reactive with mucin glycoproteins found in sera from breast carcinoma patients. Cancer Research 48:2138-2148.

16. Gendler, S.J., Burchell, J.M., Duhig, T., Lamport, D., White, R., Parker, M. and Taylor-Papadimitriou, J. (1987) Cloning of a partial cDNA encoding differentiation and tumor-associated mucin glycoproteins expressed by human mammary epithelium. Proc. Natl. Acad. Sci. USA 84:6060-6064.

17. Fukushima, K. Hirota, M. Terasaki, P.I., Wakisaka, A. Togashi, H. Chia, D., Suyama, N. Fukushi, Y. Nudelman, E. and Hakomori, S.I. (1984) Characterization of sialylated Lewis-x as a new tumor-associated antigen. Cancer Research 44:5279-5285.

18. Chia, D., Terasaki, P.I., Suyama, N., Galton, J., Hirota, M. and Katz, D. (1985) Use of monoclonal antibodies to sialylated Lewis-x and sialylated Lewis-a for serological tests of cancer. Cancer Research 45:435-437.

CLINICAL STUDIES AND NEW DEVELOPMENTS WITH Hybri-BREScan (CA-549), A MONOCLONAL ASSAY FOR BREAST CANCER-ASSOCIATED ANTIGEN

Kurtis R. Bray, Isaac A. Mizrahi, Marie-Jeanne Yerna

Hybritech Incorporated
San Diego, California and Liege, Belgium

ABSTRACT

Hybri-BREScan antigen (formerly CA-549) is a circulating tumor associated antigen that is often elevated in the sera of breast cancer patients. Preliminary studies of this antigen have shown it to have utility in the monitoring of breast cancer disease status. Biochemical analyses have shown that this antigen is a member of the heterogeneous family of high molecular weight, mucin-like glycoproteins found in human milk fat globule membranes.

Clinical trials at four sites using Hybri-BREScan immuno-radiometric assay to measure antigen in serum showed good assay performance: interassay precision and intra-assay precision were 6.2% and 5.2% cv, respectively, with a minimum dètectable concentration of 0.12 U/ml. The normal range was determined to be 0-12 U/ml. Using a cut-off of 5.0 ng/ml for CEA retrospective studies of sera from patients with various benign and malignant diseases revealed:

Status	n	Hybri-BR.+	CEA+	%Positive
Healthy	514	30	ND	5.8
Benign Breast	230	15	–	6.5
	155	–	9	5.8
Metastatic	159	88	–	55
Breast Cancer	136	–	77	56
Ovarian Cancer	76	35	–	46
	76	–	11	14

Ongoing prospective monitoring studies of breast cancer patients has revealed patients in which CEA was not elevated but in which the Hybri-BREScan assay is proving to be a useful monitor of disease status.

We have also recently developed an enzyme immunometric assay for Hybri-BREScan antigen which avoids the safety and shelf-life problems of radiometric assays. This assay also has good assay performance in measuring the antigen in the sera of breast cancer patients and excellent correlation to the immuno-radiometric assay: n=62, y=1.05x-0.87, r=.986.

INTRODUCTION

Monitoring the post-treatment status of breast cancer by means of certain immunoassays has proven to be a valuable tool in detecting recurrences and for selecting or modifying thera-peutic regimens for subsequent treatment. Carcinoembryonic antigen (CEA) has long been recognized and used as a useful monitor in those breast cancer cases where it is elevated in the patients' sera (1). More recently, antigens found in human milk fat globule antigen membranes (MFGM) have also proven use-ful for breast cancer monitoring (2-5).

We have previously described an immunoassay originally called CA-549 and now known by the trade name Hybri-BREScan (6). This monoclonal antibody-based immunoradiometric assay measures a circulating antigen that is part of the heteroge-neous MFGM family of antigens, and preliminary studies have suggested that it may be useful for breast cancer monitoring (7,8). To confirm these observations, formal clinical trials were undertaken to determine both the analytical performance of the Hybri-BREScan assay as well as its clinical utility. This study reports the results of retrospective and ongoing prospec-tive clinical trials of Hybri-BREScan. In addition, we report the development of an enzyme-based immunometric Hybri-BREScan assay intended to provide excellent assay performance but with-out the problems associated with the in-vitro use of radioac-tive reagents.

MATERIALS AND METHODS

Clinical Specimens

For clinical studies, sera were obtained either retrospec-tively from frozen serum banks at the clinical trial sites or prospectively from freshly drawn specimens from patients en-rolled in the ongoing program at the sites. Benign and malig-nant diagnoses, stagings, and assessments of disease status were made by the examining physician without prior knowledge of Hybri-BREScan values.

Measurement of Antigens

Hybri-BREScan antigen was measured by an immunoradiometric assay (IRMA) as previously described (6). Briefly, antibody-coated polystyrene beads were incubated with 20 ul serum and 300 ul assay buffer for 2 hours at room temperature with con-tinuous shaking on a horizontal rotator; the beads were then washed twice with 2 ml of wash buffer. Two hundred ul of ra-dioiodinated tracer antibody were then added and incubated for an additional 2 hours at room temperature with shaking. The beads were then washed a second time and counted on a gamma counter.

152

Enzyme immunometric assays for Hybri-BREScan antigen were performed by reacting antibody-coated beads with 20 ul of serum and 300 ul of assay buffer for 1 hour at room temperature with shaking. The beads were then washed twice with 2 ml of wash buffer and to each was added 200 ul of antibody conjugated to alkaline phosphatase. They were again incubated for 1 hour at room temperature and again washed twice with wash buffer. To each bead was then added 200 ul of substrate which was incubated for 30 minutes at room temperature without shaking, then quenched and read spectrophotmetrically at 405 nm.

CEA was measured by commercial kits from Hybritech Inc. (San Diego, CA) which were run according to protocols in the kit inserts. CA 15-3 was measured by commercial kits from Centocor (Malvern, PA) and was run according to the kit insert.

RESULTS

Analytical Performance

In order to determine the normal range of Hybri-BREScan antigen, 168 male and 346 female apparently healthy individuals were tested for serum antigen levels. The upper limit of the normal range was defined as the mean value plus 2 standard deviations. The mean for males was 5.89 U/ml SD = 3.12 and for females was 5.72 U/ml SD = 3.17. Thus, there were no significant differences observed in normal male and female levels. The mean level of Hybri-BREScan antigen for all normals was 5.78 U/ml SD = 3.17. The rounded-off upper limit of the normal range was thus calculated to be 12.0 U/ml.

The assay sensitivity was determined by measuring the minimal amount of antigen that could be reliably detected as defined by the amount of antigen corresponding to cpm of the mean of the 0 calibrator plus 2 SD of the mean. Across the clinical trial sites, the average minimal detectable concentration was shown to be 0.12 U/ml.

To assess assay precision, a performance panel of 7 serum pools ranging in antigen level from 5.5 U/ml to 95 U/ml were prepared and tested. For 20 replicates of each pool with a single assay, the intra-assay precision was determined to be 5.2% CV. For replicate runs across 20 separate assays spanning at least 10 days, the interassay precision was determined to be 6.2% CV.

Retrospective Clinical Performance

Serum samples collected from patients with various benign and malignant conditions and stored frozen at the clinical sites were tested for Hybri-BREScan antigen levels. CEA levels were often measured or already known for many but not all of these patients. Using 12.0 U/ml as the positive/negative cutoff for Hybri-BREScan, the numbers and percentages of positives were determined. Table I presents a summary of the serum Hybri-BREScan data. For comparison, Table II presents a similar summary table for CEA using 5.0 ng/ml as the cut-off.

TABLE I

Elevated Hybri-BREScan Levels Among Malignant Patients

Malignant Disease	Hybri-BR.+	Total	%Positive
Breast Cancer			
Stage 1	3	60	5.0
Stage 2	10	64	15.6
Stage 3	20	59	33.9
Stage 4	68	100	68.0
Ovarian Cancer	35	76	46.0
Hepatoma	45	93	48.4
Lung Cancer	48	137	35.0
Endometrial Cancer	15	86	17.4
Gastric Cancer	15	96	15.6

TABLE II

Elevated CEA Levels Among Malignant Patients

Malignant Disease	CEA+	Total	%Positive
Breast Cancer			
Stage 1	5	41	12.1
Stage 2	4	54	7.4
Stage 3	11	38	28.9
Stage 4	66	98	67.3
Ovarian Cancer	11	76	14.5
Hepatoma	21	78	26.9
Lung Cancer	27	72	37.5
Endometrial Cancer	17	79	21.5
Gastric Cancer	5	17	29.4

Tables I and II clearly indicate that both Hybri-BREScan antigen and CEA are often elevated in the sera of breast cancer patients and, as indicated by the staging data, that both increase in frequency of elevation in more advanced cases. Combining the Stage 3 and Stage 4 data, the incidence of elevation for Hybri-BREScan and CEA in metastatic breast cancer are 55% and 56%, respectively. In addition, both antigens are sometimes elevated in a number of other malignant conditions. Particularly striking are the ovarian cancer data where Hybri-BREScan antigen is elevated in 46% of the cases as compared to only 14.5% of the same 76 cases for CEA.

In addition to the malignant disease cases, Hybri-BREScan antigen was also measured in the sera of healthy individuals and patients with a variety of benign diseases of the organs also tested in the malignancy experiments. Table III presents data for the number and percentages of elevation of Hybri-BREScan antigen in these healthy and benign disease patients.

TABLE III

Elevated Hybri-BREScan Levels Among Benign Disease Patients

Benign Disease	Hybr-Br.+	Total	%Positive
Breast	15	230	6.5
Ovarian	6	49	12.2
Liver	23	74	31.1
Lung	21	64	32.8
Endometrial	7	85	8.2
Gastric	1	13	7.7
Healthy	30	484	6.2

Prospective Monitoring of Breast Cancer Patients

In order to evaluate the clinical utility of Hybri-BREScan in monitoring course of breast cancer, four groups of patients are being serially monitored prospectively for Hybri-BREScan and CEA: 1. Normal, presumed healthy females; 2. Females with biopsy proven breast disease; 3. Females with metastatic, Stage 4 breast cancer; 4. Females with resected, but poor prognosis breast cancer. In addition, CA 15-3 levels are also available for some patients. Figures 1 and 2 present selected patients from this ongoing study.

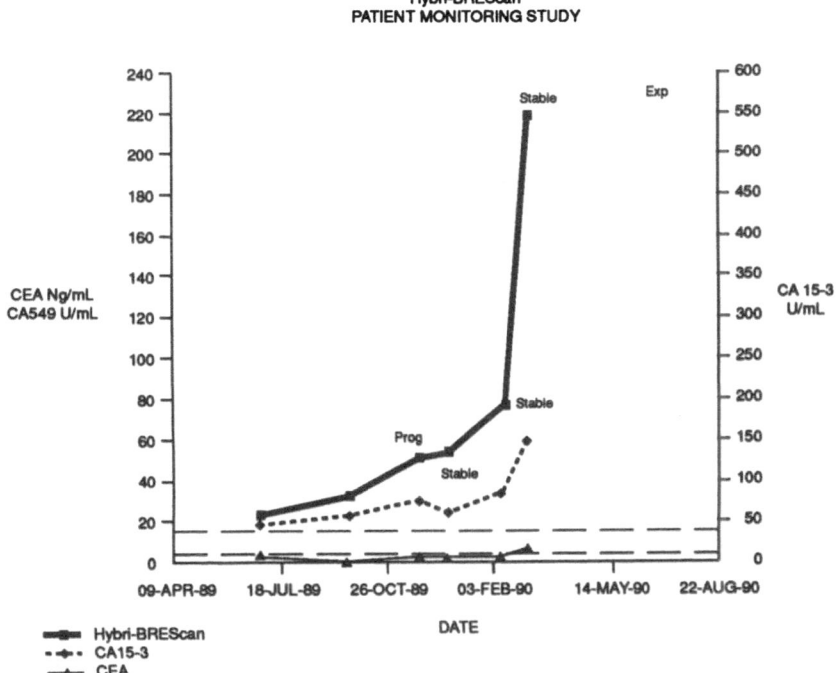

Figure 1. Rise in Hybri-BREScan levels in patient with apparently stable breast cancer. CA 15-3, but not CEA, also predicted progression. Patient expired June 1990.

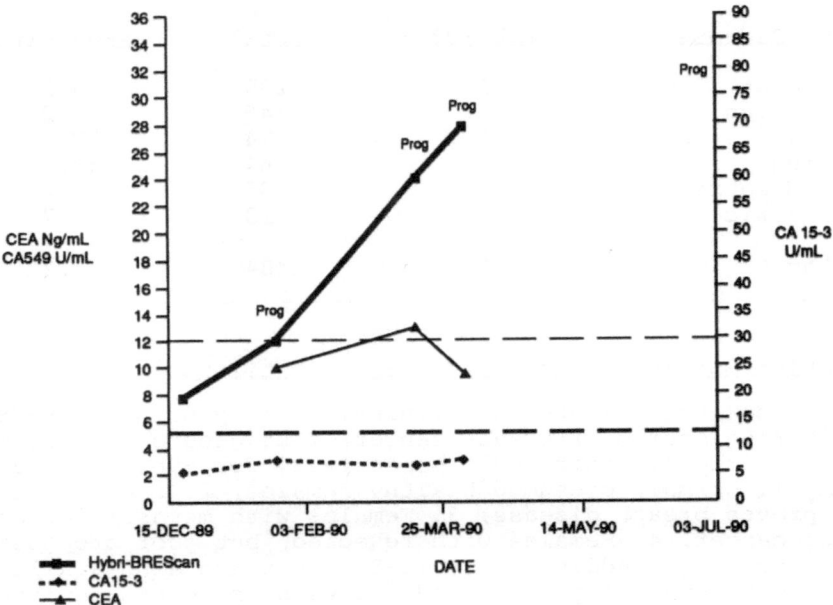

Figure 2. Rise in Hybri-BREScan levels in a patient with breast cancer progression. Note that neither CA 15-3 nor CEA levels reflect this apparent progression.

Figures 1 and 2 present two interesting cases that suggest that Hybri-BREScan assay may be valuable in monitoring the course of breast cancer. In figure 1, the patient was thought by the examining physician to have stable disease, but Hybri-BREScan antigen levels were increasing sharply. The accuracy of the Hybri-BREScan levels in predicting the course of disease was confirmed when the patient died of breast cancer about 6 months after the disease was declared stable but the Hybri-BREScan levels began their sharp rise.

In the case shown in figure 1, the CA 15-3 levels generally parallel those of Hybri-BREScan but do not rise as sharply. However, in other cases the converse is sometimes true where the levels are again parallel but CA 15-3 shows the sharper rise (data not shown). In the case in figure 2, The Hybri-BREScan antigen levels correspond with the clinical observation of breast cancer progression. Interestingly, both the CEA and CA 15-3 failed to detect this trend. In most patients, the Hybri-BREScan and CA 15-3 trends mirrored one another. Not so with CEA. Many patients have been observed in which the Hybri-BREScan assay detected a clinical trend that was missed by CEA (data not shown).

Enzyme Immunometric Assay

In order to assess the ability of a newly developed enzyme immunometric assay to also detect levels of Hybri-BREScan antigen in the sera of breast cancer patients, sera from 62 patients with active disease were assayed by both the immunoradiometric (IRMA) and enzyme immunometric assays. Figure 3 shows a linear regression of the paired antigen values provided by the two tests.

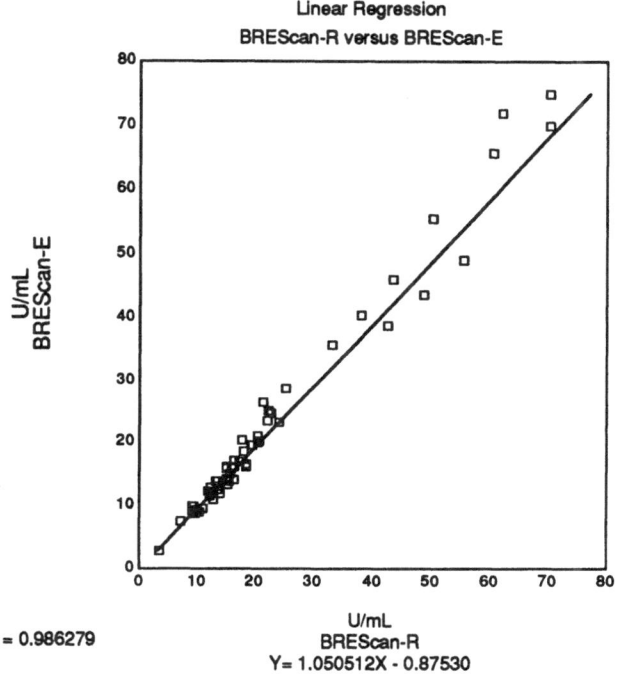

Linear Regression
BREScan-R versus BREScan-E

r = 0.986279

U/mL
BREScan-R
Y= 1.050512X - 0.87530

Figure 3. Linear regression of Hybri-BREScan antigen levels for 62 breast cancer patient sera as measured by the IRMA and enzyme immunometric assay. The two assays have excellent correlation.

DISCUSSION

Initial studies of the Hybri-BREScan assay indicated that it had approximately a 50% sensitivity for advanced breast cancer and a 98% specificity (6). The larger retrospective clinical trial reported herein has produced similar results with a 55% sensitivity for metastatic breast cancer (Table 1). When considering the incidence of antigen elevation in all benign conditions tested, the specificity is calculated to be 89.7% (Table 3). However, the benign disease populations tested were chosen with an eye to testing those diseases in which MFGM antigens were known to be elevated (2) and may thus provide a false positive rate somewhat higher than would otherwise be expected. When only healthy and benign breast disease patients are considered, as was done in the preliminary study (6), the clinical specificity rises to 93.7% (Table 3).

157

One unexpected finding of this study was the equivalent performance of CEA and Hybri-BREScan in the retrospective study of malignant diseases (Tables 1 and 2). As mentioned above, the clinical performance of Hybri-BREScan was consistent with earlier findings; the unusual observation of this study was the performance of CEA in the breast cancer patients (Table 2). Previous studies had shown CEA to be far less sensitive in detecting breast cancer than was shown in this study (1,9). The reason why CEA performed so well in this study as compared to previous studies is not clear.

The main advantage revealed in this study of Hybri-BREScan over CEA is in the prospective monitoring of breast cancer patients. In several patients, including those examples shown in Figures 1 and 2, the CEA was not elevated and/or displayed no concomitant rise and fall in level as the disease progressed or remissed. This study is currently ongoing and when complete should provide the clearest data for optimizing a strategy for the clinical use of Hybri-BREScan.

Previous studies have biochemically characterized the epitopes recognized by Hybri-BREScan and contrasted them with those recognized by CA 15-3 (6). While these two tests apparently both recognize MFGM, they probably react with overlapping but somewhat different subsets of this heterogeneous family of antigens. This difference probably explains why some patients (Figure 2) have been observed that react with one test but not the other. Although the degree of elevation can often vary (Figure 1), in the majority of patients tested, CA 15-3 and Hybri-BREScan levels paralleled each other, suggesting that both are useful for breast cancer monitoring.

In addition to the clinical trial results, we also report the development of an enzyme immunometric assay for Hybri-BREScan antigen. It uses a simple protocol similar to but faster than that of the IRMA, and has the additional advantages of longer shelf life and no radioactivity-related safety concerns. This Hybri-BREScan E assay showed excellent correlation to the IRMA in measuring antigen levels in 62 breast cancer sera (Figure 3). This new assay expands the availability of Hybri-BREScan testing to a wider group of clinical and hospital laboratories and opens the door for possible future automation of this test.

ACKNOWLEDGEMENTS

The authors gratefully acknowledge the fine technical assistance of Claudine Batton, Pascale Cheront, Carla D'Agostino, Laura Dollar, Tim Hill, Oliva McWilliams, Pat Nunnelly, and Karen Suchocki, as well as the discussion and helpful suggestions of the scientific staff at Hybritech. We are also grateful to Mark Rosney for help with the graphics.

REFERENCES

1. R. Lamerz, Serial carcinoembryonic antigen (CEA) determinations in the management of metastatic breast cancer. Oncodev Biol Med. 1:123 (1980).

2. D. F. Hayes, M.T. Sekine, M. Abe, K. Keefe, and D.W. Kufe, Use of murine monoclonal antibody for detection of circulating plasma DF3 antigen in breast cancer patients, J Clin Invest. 75:1671 (1985).

3. J. Hilkens, V. Krozen, J.M.G. Bonfrer, M. DeJong-Baker and P.F. Bruning, MAM-6 antigen, a new serum marker for breast cancer monitoring, Cancer Res. 46:2582 (1986).

4. M. R. Price, G. Crocker, S. Edwards, I.S. Nagra, R.A. Robins, M. Williams, R.W. Blamey, D.M. Swallow, and R.W. Baldwin, Identification of a monoclonal antibody-defined breast carcinoma antigen in body fluids, Eur J Cancer Clin Onc, 23:1169 (1987).

5. P. S. Linsley, V. Ochs, S. Laska, D. Horn, D.B. Ring, A.E. Frankel, and J.P. Brown, Elevated levels of a high molecular weight antigen detected by antibody W1 in sera from breast cancer patients, Cancer Res, 46:5444 (1986).

6. K. R. Bray, J.E. Koda, and P.K. Gaur, Serum levels and biochemical characteristics of cancer associated antigen CA-549,a circulating breast cancer marker, Cancer Res, 47:5853 (1987).

7. K. R. Bray, and P.K. Gaur, Correlation of serum levels of CA-549 to disease status in post-treatment serial samples from breast cancer patients, J Clin Lab Anal, 2:134 (1988).

8. R. A. Beveridge, D.W. Chan, D. Bruzek, D. Damron, K.R. Bray, P.K. Gaur, D.S. Ettinger, R.C. Rock, S. Shurbaji, and F.P. Kuhajda, A new biomarker in monitoring breast cancer: CA 549, J Clin Oncol, 6:1815 (1988).

9. D. K. Werneke, R.L. Elliot, and K.R. Bray, The comparative incidence of elevation of tumor markers CA-549, CA-15.3, and CEA in the sera of breast cancer patients, Clin Chem, 35:1078 (1989).

EVALUATION OF SEVERAL TUMOR MARKERS (MCA, CA 15.3, BCM AND CA 549) IN TISSUE AND SERUM OF PATIENTS WITH BREAST CANCER

Rafael Molina, Antonio M. Ballesta

Laboratory of Clinical Biochemistry, Medical School
Hospital Clinico, Barcelona, Spain

INTRODUCTION

The diagnosis and accompanying staging of human breast cancer has traditionally relied on clinical and radiologic criteria. The demostration of antigens shed by tumor cells into the circulation offers a possibility of assessing the clinical course of disease in cancer patients by monitoring changes in serum levels of such tumor associated-antigens (TAA's). Using monoclonal antibody technology a number of monoclonal antibodies directed against breast tumor-associated antigens have been characterized. Many well known monoclonal antibodies are reactive with high molecular weight cell surface glycoproteins identified as mucins or mucin-like structures (CA 15.3, CA 125, MCA, CA 549, BCM). These antigens are secreted by normal epithelia in substantial amount into the external environmet. Transformed epithelia may shed these substances into the blood circulation. This allows the measurement of levels ot these proteins in serum of cancer patients, using monoclonal antibodies.

CA 15.3 is an heterologous double determinant assay for determination of circulating levels of a large mucin-antigen, It is based on two different anti-mucin monoclonal antibodies: DF-3 as a tracer and 115D8 as the catcher antibody. (1, 2). Mam-6 antigen is a polypeptide with an estimated mw of over 400 kD, recently identified as heavily sialytated mucus glycoprotein with predominantly, but not exclusively, O-linked carbohydrates (3). Ställ et al (4) described a mouse monoclonal antibody, MAb b-12 that react with a 350 kD glycoprotein with a mucin-like characteristics expressed in cytoplasm and on the surface on human breast carcinoma cell lines. This antigen, calles MCA (Mucinous Carcinoma associated Antigen) can be evaluated in serum or tissue by a enzymoimmunoassay (5).

Bray et al (6) have identified a circulating breast cancer associated glycoprotein, called CA 549. This antigen is defined by two monoclonal antibodies, directed against distinct membrane epitopes from a breast cancer cell line (BC4E 549) and the milk fat globule membrane (BC4N). Demers et al (7), using a sandwich immunoradiometric assay reported abnormal CA 549 serum levels in 83% of patients with advanced breast cancer. Likewise, Anderson et al reported a new mucin-like antigen, calles BCM and defined by the monoclonal antibody M85, wich recognizes a family of N-acetyllactosamie (LacNAc) oligosaccharides (8).

All the new antigens described above had in common the organ-specificity, with highest levels in breast and ovarian carcinomas, that they are reactive with mucin-like structures and that several of them are defined by two monoclonal antibodies obtained against milk fat globule

Breast Epithelial Antigens, Edited by R.L. Ceriani
Plenum Press, New York, 1991

membrane antigens and/or enriched cytoplasmic membrane of breast cancer cell lines (1-8, 9). This study investigates the distribution of four tumor-associated antigens (CA 15.3, MCA, CA 549, BCM) in serum and tissue (CA 15.3, MCA) of patients with breast disease, correlates their levels with other tumor markers (CEA) or pathological parameters of known prognostic value and evaluates their possible application in patients with breast cancer.

TUMOR ASSOCIATED ANTIGENS IN PATIENTS WITH BENIGN BREAST DISEASES

Serum levels of TAA's in benign breast diseases did no differ significantly from those measured in apparently healthy subjects. Using 40 U/ml as cut-point for CA 15.3, 11 U/ml for MCA, 12 U/ml for CA 549, 40 U/ml for BCM and 5 ng/ml for CEA, abnormal levels of these markers were found in less than 3% of the evaluated patients: CA 15.3 in 2% (n=65), MCA in 2% (n=41), CA 549 in 0% (n=20), BCM in 3% (n=37) and CEA in 2% (n=164) of patients with non-tumoral diseases. There were not differences in the TAA's serum levels in relation to the etiology of the disease, with similar results in patients with fibroadenomas or with fibrocystic disease.

TUMOR ASSOCIATED ANTIGENS SERUM LEVELS IN LOCO-REGIONAL BREAST CANCER

Table 1 shows the TAA's preoperative serum levels found in patients with loco-regional breast cancer. TAA's sensitivity in patients without metastasis is rather low, and the serum levels in those stages do not significantly differ from those found in patients with benign breast disease. The low sensitivity of these TAA's in primary breast cancer and the slighty elevations found in the majority of positive patients, indicate that these tumor markers are not useful in diagnosis.

Table 1. Preoperative TAA's serum levels in patients
with loco-regional breast cancer subdivided
in relation to tumor-size and nodal status.

	CEA		CA 15.3		MCA		CA 549		BCM	
	nº	rate*	nº	rate	nº	rate	nº	rate	nº	rate
Total	668	22%	243	18%	106	21%	100	25%	105	18%
T1-2	415	16%	152	13%	62	16%	68	19%	66	15%
T3-4	201	29%	60	25%	31	26%	29	38%	39	23%
Infla.**52		42%	27	33%	8	38%	3	33%	–	–
Node –	267	12%	89	11%	41	15%	42	17%	40	3%
Node +	338	25%	128	18%	63	21%	58	31%	65	28%
1-3	199	27%	82	13%	34	15%	20	40%	34	18%
3	139	22%	46	26%	29	28%	38	26%	31	35%

* Proportion of patients with abnormal serum levels
** Inflammatory

TAA's serum evaluation show two different populations in patients with loco-regional disease, those with positive and those with negative results. Table 1 shows the relationship between preoperative serum levels of these antigens and well known prognostic factors as tumor size or nodal involvement. In general all the TAA's are related to the presence of more advanced disease i.e. large tumors or nodal involvement. These resuts seem to indicate that TAA's results in loco-regional breast cancer reflected indirectly the number of malignant cells (tumor size) and/or tumor extension (nodal involvement) as an indicator of an easier shedding into the blood site where they are evaluated. Moreover, tumor size and nodal involvement are two factors related themselves. Table 2 correlates the TAA's serum levels and the tumor size subdivided according to the nodal status. CA 15.3 is a tumor marker mainly associated with tumor size. In contrast, MCA and CA 549 had relationship to tumor size only in node positive patients, with significantly higher levels in those cases with node invasion and tumors larger than 5 cm of diameter. BCM and CEA seems mainly related to nodal involvement, with significantly higher values in patients with nodal involvement. No relationship between TAA's serum levels and steroid receptor status were found.

Table 2. Preoperative TAA's serum levels in patients
with loco-regional breast cancer according
to nodal involvement and subdvided in relation
to tumor size.

	CEA		CA 15.3		BCM		MCA		CA 549	
	nº	rate*	nº	rate	nº	rate	nº	rate	nº	rate
N-T1-2	215	12%	74	9%	31	3%	34	15%	35	17%
T3-4	44	9%	20	21%	9	0%	7	14%	7	14%
N+T1-2	196	20%	75	16%	35	26%	26	12%	33	21%
T3-4	101	31%	52	25%	30	30%	18	28%	25	44%

* Proportion of patients with serum levels higher than the cut-point

Tumor size and nodal involvement are the most used prognostic factors in patients with loco-regional breast cancer. TAA's relationship to these parameters suggest their possible application as prognostic factors. Patients with preoperative abnormal MCA serum levels show a lower DFI (p=0.017) than tose with normal levels (n=97) (10). CEA also had prognostic value but only in patients with T 3-4 (p 0.025) (n=144). No differences in the DFI according to the CA 15.3 (n=217) or BCM (n=92) serum levels were found (11).

To compare the sensitivity (proportion of patients with abnormal levels)of TAA's serum levels in patients with locoregional breast cancer the same samples from 91 patients were evaluated. Sensitivity of these antigens were 15% for CA 15.3, 15% for MCA, 16% for BCM and 22% for CEA. sensitivity of use two of these TAA's antigen was: 20% for CA 15.3-MCA, 21% for CA 15.3-BCM and 20% for MCA-BCM. Higher sensitivity was when CEA was used with combination to one of the mucin-associated antigens 31% for CEA-CA 15.3, 29% for CEA-MCA and 33% for CEA-BCM. Abnormal serum levels of one or another TAA's were found in 36% of these patients with loco-regional breast cancer.

TUMOR-ASSOCIATED ANTIGENS IN PATIENTS WITH METASTASIS

TAA's in patients with cancer are mainly used in disease monitoring of patients with advanced disease. In patients with abnormal levels of these TAA's increasing serum levels correlate with the progression of disease while partial or complete response to therapy was accompanied by decreasing serum levels. Table 3 shows the sensitivity ot the TAA's evaluated in patients with advanced disease. Patients with metastatic breast cancer showed elevations of TAA's ranging from 65% to 85%.

Overall survival in patients with metastatic breast cancer is mainly related to tumor response, steroid receptors and site of metastasis.

Table 3. TAA,s serum levels in patients with
metastatic breast cancer, according
to site of recurrence.

	CEA		CA 15.3		MCA		CA 549		BCM	
	nº	rate*	nº	rate	nº	rate	nº	rate	nº	rate
Bone	116	73%	104	79%	54	76%	10	-	59	73%
Lung	40	42%	23	78%	6	67%	-	-	10	80%
Liver	20	90%	10	90%	6	100%	-	-	3	100%
Loco-regional	28	25%	22	18%	12	42%	-	-	8	38%
Others	24	50%	8	63%	-	-	-	-	5	80%
Multiple	52	75%	31	84%	19	74%	-	-	30	83%
Total	300	65%	198	73%	97	76%	35	80%	115	75%

* Proportion of patients with serum levels higher than the cut-point

Table 3 correlates the sensitivity of the different TAA's studied with
the site of metastasis. Lowest serum levels were found in patients with
loco-regional recurrences and highest results were associated with liver
or bone metastasis. Similar results has been reported by other authors
(5,10,12,13). Likewise, some of these TAA's had relationship to the steroid receptor status, as shows table 4.

Table 4. TAA's serum levels in patients with
metastatic breast cancer subdivided
to ER status.

	CEA		CA 15.3		MCA		BCM	
	nº	rate*	nº	rate	nº	rate	nº	rate
ER+	100	75%	55	82%	18	72%	36	86%
ER-	60	33%	23	52%	10	60%	11	91%

* Proportion of patients with serum levels higher than the cut-off

CEA and CA 15.3 had significantly higher serum levels in patients with
ER+ tumors than in those with ER- tumors. This relationship between CEA
and CA 15.3 serum levels and ER status as well as site of recurrence,
suggest their possible application as prognostic factors in patients with
advanced disease. Patients with abnormal CA 15.3 at diagnosis of relapse
had a lower overall survival than those with normal levels $(p < 0.001)$ (14)
In contrast, no differences in the overall survival of these patients
were found in relation to the CEA results at diagnosis. CEA relationship
with two opposite prognostic factors in patients with metastasis as ER
status and site of recurrence, may explain the lack of application of
this antigen as prognostic factor.
    Mucin-associated antigens had a similar sensitivity in patients with
metastasis as well as a similar relationship to site of recurrence.
Likewise, all these TAA's are antigens related to mucin-like structures.
In order to know if these mucin-associated antigens, are related themselves, we evaluate all of them in 140 samples from 57 patients with metastatic disease. Sensitivity of these antigens were : 79% for CA 15.3, 77%

for MCA, 79% for BCM, and 67% for CEA. Eighty eigh percent of these 57 patients with metastasis had abnormal levels of one or another of these TAA's. Table 5 compares the adventages of using one or more of these mucin-associated antigens in relation to the sensitivity obtained in patients with metastasis. Considering the results as positive or negative (above or below the cut-points) a good overall concordance between the

Table 5. Comparison of TAA's sensitivity
in patients with metastatic breast
cancer.

	Rate*	overall concordance	overall concordance in positive samples
CEA-CA 15.3	50/57 (88%)	70%	66%
CEA-MCA	49/57 (86%)	72%	67%
CEA-BCM	50/57 (88%)	72%	67%
CA 15.3-MCA	46/57 (81%)	95%	93%
CA 15.3-BCM	46/57 (81%)	96%	96%
MCA-BCM	47/57 (82%)	91%	89%
MCA-BCM-CA 15.3	47/57 (82%)	-	-
All TAA's	50/57 (88%)	-	-

*Proportion of patients with abnormal serum levels

mucin-related antigens were found. These results indicate that sensitivity using the three mucin-associated antigens was not better than using only two of them. Likewise, the best sensitivity was obtained using CEA and one of the mucin-associated antigens.

TUMOR MARKERS IN TISSUE

Breast tissue for TAA's evaluation in pellet and in cytosol were homogenized and ultracentrifuged at 105.000 g. The cytosol was used for steroid receptor evaluation and CEA, CA 15.3 and MCA assays (cytosol). The pellet was resuspended with 4 ml of TEM buffer and 0,3 ml of 5% Triton X, sonicated and centrifuged at 2.000 g. for 10 minutes and the supernatant used for TAA's assays (pellet). Table 6. shows the TAA's concentrations found in primary breast tumors. A wide range of values were found in both cytosols and pellets. All TAA's evaluated had higher concentrations in pellet and cytosol from malignant tumors than in those from nonmalignant tissues. Higher CEA values in pellet than in cytosol were found. In contrast, CA 15.3 and MCA had significantly higher levels in cytosol than in pellet. These results are in concordance with the location of these TAA's reported by different authors. Kufe et al (15) using immunohistochemistry with one of the two monoclonal antibodies (DF-3) which define CA 15.3 antigen, mainly found a cytoplasmic reaction in the breast cancer cells. Likewise, Zenklusen et al (16) reported that MAb b-12 (MCA) had intracytoplasmic reaction.

A trend toward higher CA 15.3 tissue levels in more advamced cancers were found, but statistically significance differences were obtained when number of axillary nodes were evaluated ($p < 0.03$). In contrast no relationship between CEA or MCA concentrations and tumor size or nodal involvement were found. All tumor markers studied had relationship to steroid receptor status. Higher TAA's concentrations in ER+ tumors than in ER- tumors, were found (Table 7).

The relationship between TAA's concentrations in tissue and well

Table 6. CEA, CA 15.3 and MCA concentrations in
pellet and cytosol from breast tissue.

		CEA*		CA 15.3*		MCA*	
	nº	Median	nº	Median	nº	Median
Pellet Benign	37	3.5	29	10.0	8	7.4
Cancer	277	18.0	144	18.5	65	18.3
T 1-2	136	17.5	69	16.0	30	11.0
T 3-4	51	14.0	26	31.0	8	7.5
Node+	109	16.1	52	24.5	22	8.5
Node-	88	18.6	43	10.0	20	13.0
Cytosol Benign	11	0.9	-	-	-	-
Cancer	233	5.6	86	21.0	43	21.6
T 1-2	106	3.7	36	15.0	33	10.0
T 3-4	44	2.1	15	15.0	9	22.0
Node+	79	3.5	27	27.0	24	18.7
Node-	77	3.1	25	6.0	19	20.0

* Concentrations in ng/mg (CEA) and U/mg (CA 15.3, MCA)

known prognostic factors in breast cancer, suggest their possible prognostic interest. Patients with cytosolic CA 15.3 values above 20 U/mg of protein had a shorter DFI than those with levels lower than this cut-point. (p 0.05) (n=59). Similar results were found using 30, 40 or 50 U/mg as cut-points. In contrast, no relationship between CA 15.3 concentrations

Table 7. TAA´s concentrations in pellet and
cytosol from malignant breast tumors,
in relation to steroid receptor status.

		CEA*		CA 15.3*		MCA *	
	nº	Median	nº	Median	nº	Median
Pellet ER+	184	27.5	102	26.5	47	15.2
ER-	59	3.8	22	1.5	13	3.4
Cytosol ER+	154	9.1	37	32.0	39	23.3
ER-	54	1.7	11	1.0	17	7.7

* Concentrations in ng/mg (CEA) and U/mg (CA 15.3, MCA)

in pellet and prognosis was found(n=142) (11). CEA in tissue had prognostic value only when the results obtained in serum (preoperative) and in pellet are used combined. Using both results (CEA pellet/serum) (n=100) it is possible to stablish different prognostic groups, independently of axillary nodal status: if CEA was positive in the mastectomy specimen (pellet) those patients with presurgical normal CEA serum levels experienced 15% of relapse, while those with positive serum CEA experienced 72% of relapse in 4 years. Patients with low CEA concentrations in pellet and normal CEA serum levels experienced 47% relapses (17,18).

CONCLUSIONS

TAA´s in serum or in tissue are not useful in diagnosis, but can be employed as prognostic factors and in disease monitoring of patients with metastatic breast cancer. Mucin-associated antigens had a good overall concordance with a similar sensitivity and relationship to well known prognostic factors. Sensitivity using CA 15.3, MCA and BCM was not better than using only two of them.

ACKNOWLEDGEMENTS

We are grateful to Judith Jo, Celia Aparicio and Francisca Coca for their expertise in performing the immunoradiometric assays. We would like to thank to Eulalia Salom for her preparation of this manuscript.

REFERENCES

1) Tobias R, Rothwell C, Wagner J, Green A, Liu YS. Development and evaluation of a radioimmunoassay for the detection of a monoclonal antibody defined breast tumor associated antigen 115D8/DF-3. Clin Chem 31: 986 (1974).

2) Hilkens J, Buijs F, Hilgers J, et al. Monoclonal antibodies against human milk-fat globule membranes detecting differentiation antigens of the mammary gland and its tumors. Int J Cancer 34:197 (1984).

3) Hilkens J, Bonfrer J, Kroezen V, et al. Comparison of circulating MAM-6 and CEA levels and correlation with the estrogen receptor in patients with breast cancer. Int J Cancer, 39:431 (1987).

4) Stäli C, Takacks B, Miggiano V, Staechelin T, Carman H. Monoclonal antibodies against antigens on breast cancer cells. Experientia, 41:1377 (1985).

5) Bombardieri E, Gion M, Mione R, Dittadi R, Bruscagin G, Buraggi GL. A mucinous-like carcinoma associated antigen (MCA) in the tissue and blood of patients with primary breast cancer. Cancer, 63:490 (1989).

6) Bray KR, Koda JE, Gaur PK. Serum levels and biochemical characteristics of cancer-associated antigen CA-549, a circulating breast cancer marker. Cancer Res, 47:5853 (1987).

7) Demers LM, Harvey HA, Glenn JD, Gaur PK. CA 549: A new tumor marker for patients with advanced breast cancer. J Clin Lab Analysis, 2:168 (1988).

8) Anderson B, Slota J, Kundu S, et al. Characterization of monoclonal antibody to pargloboside and sialosyl-PG, and improved chromatogram binding assay for rapidly identifying antibodies to tumor antigens. J Cell Biochem (Suppl), 110 (1987).

9) Colomer R, Ruibal A, Genolla J, Salvador L. Circulating tumor marker levels in advanced breast carcinoma correlate with the extent of metastatic disease. Cancer, 64:106 (1989).

10) Molina R, Filella X, Mengual P, Prats M, Zanon G, Daniels M, Ballesta AM. MCA in patients with breast cancer: correlation with CEA and CA 15.3. Int J Biol Markers, 5: 14 (1990).

11) Molina R, Filella X, Rivera-Fillat F, Jo J, Bedini JL.Ballesta AM. Carbohydrate antigen 15.3 (CA 15.3) in tissue and serum of patients with breast diseases. Brest C Res Treat, (Submitted).

12) Hayes DF, Zurawski VR, Kufe DW. Comparison of circulating CA 15.3 and carcinoembryonic anitigen levels in patients with breast cancer J Clin Oncol 4:1542 (1986).

13) Beveridge RA, Chan DW, Bruzek D, Damron D, Bray KR, Gaur PK, Ettinger DS, Rock RC, Shurbaji MS, Kuhajda FP. A new biomarker in monitoring breast cancer: CA 549. J Clin Oncol 6:1815 (1988).

14) Molina R, Filella X, Rivera-Fillat F, Bedini JL, Jo J, Gallart A, Ballesta AM. Carbohydrate antigen 15.3 (CA 15.3) in tissue and serum of patients with breast diseases. Breast Cancer Res Treat. Submitted.

15) Kufe DG, Inghirami M, Abe D, Hayes H, Justi-Wheeler H, Schlom H. Differential reactivity of a novel monoclonal antibody (DF-3) with human malignant versus benign breast tumors. Hybridoma 3:223 (1984).

16) Zenzlusen HR, Stäli C, Gudat F, Overebeck J, Rolink J, Heitz PhU. The immunohistochemical reactivity of a new anti-epithelial monoclonal antibody (Mab b-12) against breast carcinoma and other normal and neoplastic human tissues. Virchows Archiv A Pathol Anat 413:3 (1988).

17) Molina R, Ballesta AM, Rivera-Fillat F, Balague A. Study of the carcinoembryonic antigen (CEA) and estrogen receptors (ER) in breast cancer tissue, in: "Marker tumorali in Ginecologia", Coli AC, Torre GC, Vecchione A, Zacutti Jr A, eds., CIC Edizioni Internazionali, Roma (1985).

18) Molina R, Rivera-Fillat F, Filella X, Prats M, Zanon G, Ballesta AM. Carcinoembryonic antigen in tissue and serum from breast cancer patients: relationship with steroid receptors and clinical applications in prognosis and in early diagnosis of relapse. Cancer Res. Submitted.

IMx BCM: A NOVEL MONOCLONAL BASED SYSTEM FOR THE DETECTION OF

BREAST CANCER ASSOCIATED MUCIN

John G. Konrath, Louise W. Przywara, Donald M. Lynch,
Kim K. Borden, Albert L. Sorrell, Ruthie L. Thillen,
Colleen C. McInerney, Anne C. Black, George L. Manderino

Abbott Laboratories
Diagnostic Division
North Chicago, Illinois 60064

INTRODUCTION

Breast Cancer Mucin (BCM) is a cancer associated mucin often found in the sera of breast cancer patients. Mucin and mucin like molecules have been shown to be expressed by malignant epithelial cells [1,2].

The role of mucins as markers of malignant conditions [3,4] and the association of blood group substances i, I, and paragloboside with cancer have been reported [5,6]. Recent findings suggest that some breast cancer tumors demonstrate diminished ABH expression but an increase in precursor substances such as the I oligosaccharide [7]. These findings have been the basis for the development of serum assays to monitor breast cancer patients and evaluate response to therapy [8-13].

The BCM M85 monoclonal antibody recognizes a family of mucins containing O-linked oligosaccharides with lactosamine structures. Further studies indicate the primary target for M85 to be terminal Gal(B1-4) GlcNAc [14]. This sequence is present in i, I, and paragloboside blood group structures.

An automated immunoassay for the detection of Breast Cancer associated Mucin (BCM) in serum has been developed for use on the Abbott IMx[R] Immunoanalyzer. The specificity of the assay is based on the reactivity of an IgM monoclonal antibody M85 with high molecular weight glycoproteins found in the sera of many breast cancer patients. IMx[R] BCM uses a two-step, enzyme mediated, competitive inhibition format. An on-line neuraminidase (*C. perfringens*) treatment of patient samples takes place during the assay. Desialylation of the BCM glycoprotein significantly increases the immunoreactivity of M85 for BCM resulting in a rapid (35 minute), sensitive immunoassay.

Clinical studies indicate that serum of breast cancer patients often contains elevated levels of BCM compared to serum of normal individuals. BCM serum levels also often reflect changes in disease status of breast cancer patients.

Breast Epithelial Antigens, Edited by R.L. Ceriani
Plenum Press, New York, 1991

MATERIALS AND METHODS

Sera specimens were supplied by a serum bank operated at Abbott Laboratories. These sera were obtained from various sources and were used for in-house clinical evaluations. Specimens were also supplied by clinical sites: Dr. Robert Bast, Duke University Medical Center, Durham, N.C., Dr. Herbert Fritsche, M.D. Anderson Medical Center, Houston, Texas, Dr. Arvind Bhargava, Roswell Park Memorial Institute, Buffalo, New York, Aaron Malkin, The Sunnybrook Medical Center, Toronto, Canada. These specimens included serum specimens from individuals classified as normal and patients with breast cancer, other cancers, and various other non-malignant conditions.

M85 is a murine monoclonal IgM antibody. It was generated by immunizing BALB/c mice with mucin enriched fractions of pooled ascites from patients with urogenital cancers. M85 reacts with mucins of sera, ascites fluids and xenografts. M85 recognizes glycoproteins in serum with molecular weights exceeding 750 Kd (HPLC and S-400 chromatography). Reactivity from ascites fluids or tumor cell lines remains near, or at the origin of 4%-20% SDS polyacrylamide gels. M85 reacts with O-linked but not N-linked oligosaccharides. Immunoreactivity is enhanced in most incidences by pre-treatment with neuraminidase. This suggest that the epitope is siaylated. Binding studies of specific oligosaccharides indicate the primary epitope is Gal(Beta 1-4)GlcNAc1-R. These data will be published elsewhere.

An IMx^R BCM retrospective study was conducted at clinical sites and at Abbott Laboratories. Analysis of 2425 serum specimens was completed. 1334 of these specimens were also tested with the CA ·15-3 RIA (Centocor, Malvern, Pennsylvania). 72 serial specimens from breast cancer patients were evaluated with IMx^R BCM, CA 15-3, and in some cases CEA. All IMx specimens were run in duplicate, but only the first replicate was used for the CA 15-3 comparison. All IMx values were read off calibration curves stored the first day of a study and adjusted for each subsequent assay run. Three controls were included with each run (30, 50, 70 U/ML). Manufacture's instructions were followed for testing with the commercially available CA 15-3 and CEA assays. IMx^R BCM kits (2202/3A43) and CEA kits were supplied by Abbott Laboratories, North Chicago, Illinois.

The IMx Immunochemistry Analyzer System

The IMx^R Immunochemistry Analyzer is a benchtop instrument that automates BCM testing in the clinical laboratory. This microprocessor based instrument uses a robotic arm to process specimens using Microparticle Enzyme ImmunoAssay (MEIA) technology.

IMx^R BCM MEIA technology relies upon a very small (0.2u-0.4u) high surface area microparticle solid phase. The suspension of microparticles used in the BCM assay has a significantly larger surface area than a quarter inch polystyrene bead or tube solid phase. The total area available for reaction with a bead is approximately 1.27 cm^2 compared to 12.22 cm^2 for the BCM assay microparticles. This 10 fold increase in surface area increases reaction kinetics and reduces incubation times. The use of MEIA technology allows processing of a 24 specimen BCM carousel in only 35 minutes. Use of a microparticle suspension solid phase also facilitates relatively simple automation of the BCM protocol. A microparticle solid phase is easily and precisely manipulated by the IMx robotic aspirate/dispense probe. The IMx performs all BCM immunoassay procedures required for the analysis of patient serum samples: sample counting, identifying assay type, warming the reagent pack, pipetting reagents and samples, timing critical assay steps, washing the aspirate/dispense probe, quantifying fluorescent signals, performing data reduction, and printing out BCM U/mL results. The IMx[R] also automatically evaluates assay validity based upon reagent availability, calibrator characteristics, 0 U/mL Calibrator Counts/Sec/Sec (Ct/S/S), linearity of the fluorescent reads (NRMSE) and other assay/instrument performance factors 15.

Assay quantitation is accomplished through the use of a front surface fluorometer which measures fluorescence of methylumbelliferone (MU). MU production results from M85 alkaline phosphatase mediated turnover of the assay substrate 4-methylumbelliferyl phosphate (MUP). The rate of the fluorescence production is calculated (Ct/S/S) and is inversely proportional to the BCM concentration in U/mL.

BCM calibration is accomplished by running a calibration curve (0, 15, 35, 55, 75, 125 U/mL calibrators) and then using this stored curve for determining specimen values on subsequent assays. Each specimen assay includes a Mode 1 Adjustor (35 U/mL Calibrator) which is used, if necessary, to adjust the original calibration curve Ct/S/S to reflect run variation.

The IMx BCM Immunoassay

The IMx[R] BCM assay employs BCM coated microparticles in a two-step competitive inhibition format. The IMx BCM reagents are added to a reaction cell in the following sequence. A serum sample is added to the specimen well of the reaction cell through a non-precision pipetting step.

Step 1. Upon initiation of the IMx[R] BCM assay, specimen and Specimen Diluent Buffer containing heuraminidase are transferred into the predilution well. BCM disialylation occurs at this time.

Step 2. The Anti-BCM Alkaline Phosphatase Conjugate is delivered to the predilution well. The Anti-BCM Alkaline Phosphatase Conjugate binds to the desialylated BCM forming an antibody-antigen complex.

Step 3. Microparticles are delivered to the incubation well along with an aliquot of the Anti-BCM Conjugate:BCM Complex and unbound Anti-BCM Conjugate. Any unbound Anti-BCM Conjugate binds to the BCM coated Microparticles.

Step 4a. An aliquot of this reaction mixture is transferred to the glass fiber matrix to which the Microparticles are irreversibly bound.

Step 4b. The matrix is washed to remove unbound materials.

Step 5. The substrate, 4-Methylumbelliferyl Phosphate (MUP) is added to the matrix. Fluorescent product is measured by the MEIA optical assembly. The rate of fluorescence production is inversely proportional to the concentration of BCM in the sample. BCM U/mL values are derived by reading specimen Ct/S/S off adjusted calibration curves. Point to point data reduction is used. A typical cal curve is shown in Fig. 1.

ASSAY PROTOCOL

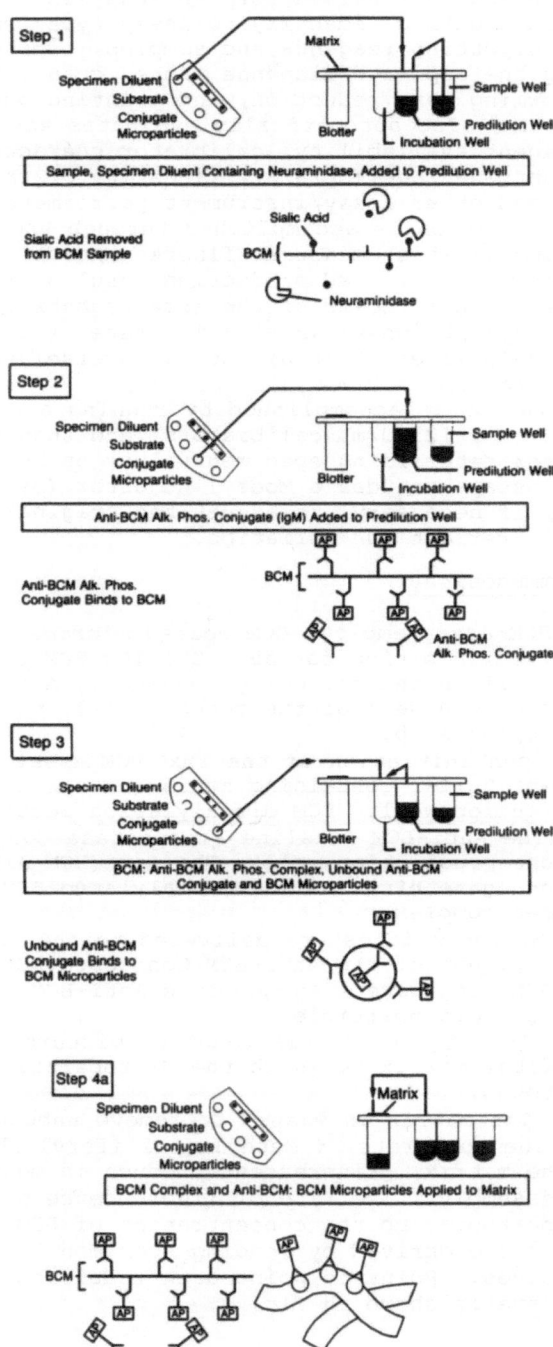

Step 1

Specimen Diluent
Substrate
Conjugate
Microparticles

Matrix

Sample Well
Predilution Well
Incubation Well

Blotter

Sample, Specimen Diluent Containing Neuraminidase, Added to Predilution Well

Sialic Acid Removed from BCM Sample

Sialic Acid

BCM

Neuraminidase

Step 2

Specimen Diluent
Substrate
Conjugate
Microparticles

Sample Well
Predilution Well
Incubation Well

Blotter

Anti-BCM Alk. Phos. Conjugate (IgM) Added to Predilution Well

Anti-BCM Alk. Phos. Conjugate Binds to BCM

BCM

Anti-BCM Alk. Phos. Conjugate

Step 3

Specimen Diluent
Substrate
Conjugate
Microparticles

Sample Well
Predilution Well
Incubation Well

Blotter

BCM: Anti-BCM Alk. Phos. Complex, Unbound Anti-BCM Conjugate and BCM Microparticles

Unbound Anti-BCM Conjugate Binds to BCM Microparticles

Step 4a

Specimen Diluent
Substrate
Conjugate
Microparticles

Matrix

BCM Complex and Anti-BCM: BCM Microparticles Applied to Matrix

BCM

172

RESULTS

IMx^R BCM Assay Performance

Reproducibility

IMx^R BCM assay reproducibility was determined by assaying five
samples in replicates of 3 on 10 independent runs at 3 clinical
sites and at Abbott Laboratories (n=120 for each sample). All
specimens at a site were read off of a stored calibration curve.
The coefficients of variation (CV) for between run, and between
lab, were determined from the components of variance of replicates
of one for the multiple assay runs. Within run CV's were
calculated for replicates of 3 (Fig. 2). The stability of a stored
curve is illustrated by control tracking over 160 days at one of
the clinical sites (Fig. 3). The between run CV for the high,
medium, and low control was 3.6%, 5.7%, and 6.5%, respectively.
The %CV between the duplicate values for 1444 specimens was also
compared. A high degree of correlation (r.=0.997) was observed
between replicate 1 and the mean of replicate 1 and replicate 2
(Fig. 4).

Dilution Linearity

The linearity of patient sample dilutions was evaluated using
the IMx^R BCM Assay. Acceptable linearity was observed in all
specimens. Two representative specimen dilution curves are found
in Fig. 5.

Recovery

Known amounts of BCM were added to normal human serum. The
concentration of BCM was determined using the IMx^R BCM Assay. The
resulting recovery was calculated (Fig. 6). Recovery ranged from
91% to 113%.

Sensitivity

The sensitivity of the IMx^R assay was determined. This was
defined as two standard deviations above the IMx^R BCM Calibrator A
(0 U/mL) mean. IMx^R BCM sensitivity was calculated as better than
5.1 U/mL. This represents the lowest measurable concentration of
BCM that can be distinguished from zero.

ASSAY PROTOCOL

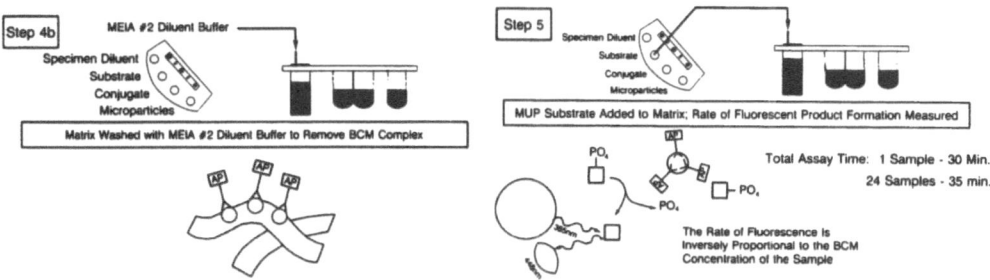

STANDARD CURVE
CT/S/S (Thousands)
U/mL
0
15
35
55
75
125

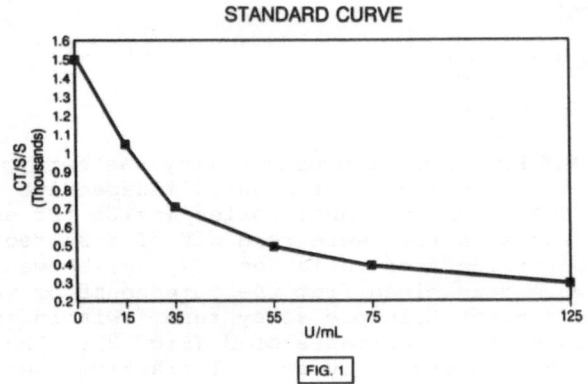

FIG. 1

REPRODUCIBILITY

Sample	Mean BCM (U/mL)	Within Run % CV	Between Run % CV	Between Lab % CV
1	25.39	6.8	11.1	12.4
2	46.71	4.0	6.3	6.4
3	68.48	3.0	4.2	4.5
4	89.91	3.4	6.5	6.7
5	509.16	2.8	6.0	7.0

5 Samples Were Assayed in Replicates of 3 in 10 Independent Runs at 4 Labs (N = 120 for Each Sample)

FIG. 2

IMx BCM CONTROL TRACKING
Read Off Of Stored Curve
BCM Concentration (U/mL)
% CV
3.56
5.70
6.53
Days
Low
Medium
High

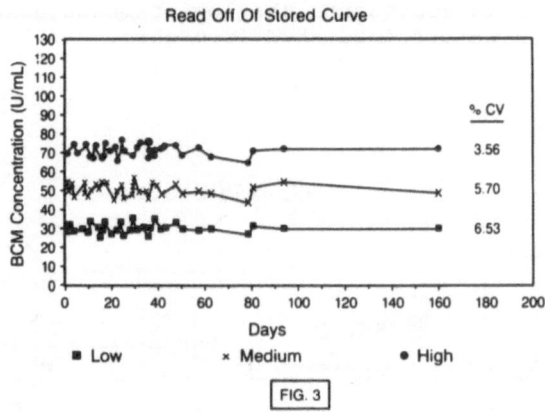

FIG. 3

Specificity

Lipemic specimens up to 3000 mg/dL triglycerides, icteric specimens up to 25.6 mg/dL bilirubin, and hemolyzed specimens with up to 1250 mg/dL hemoglobin were tested and did not interfere in the IMx^R BCM assay. Concentrations of protein in specimens from 5.0 to 11.0 g/dL showed no significant interference.

Clinical Results

Comparison of BCM and CA 15.3 values

A description of the 1334 specimens tested with both assays is provided in Fig. 7a and 7b. The correlation coefficient between the two assays for these specimens was r=0.97 and the slope m-1.51 (Fig. 8).

WITHIN RUN SPECIMEN PRECISION

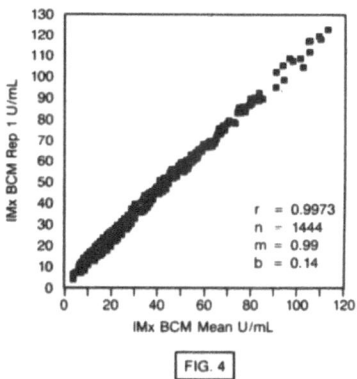

FIG. 4

Levels in Normal Individuals: 411 apparently normal individuals were tested for BCM and CA 15-3 levels. Seventeen (4.1%) normal controls had BCM values >35 U/mL compared to 21 (5.1)% of the values >30 U/mL for CA 15-3. The correlation coefficient between the two assays was r=0.72 and slope m=0.91. A 35 U/mL cutoff between normal and elevated BCM levels was chosen based upon the 95% normal specificity achieved with this value.

Levels in Individuals with Benign Conditions: 394 individuals with various benign conditions were tested. BCM values were >35 U/mL in 73 (18.5%) incidences compared to 79 (20.1%) of the CA 15-3 values. The coefficient of correlation was r=0.77 and slope m=0.88 (Fig. 9). Individuals with benign breast disease had elevated BCM values in 8 of 117 (6.8%) cases compared to 9 (7.7%) for CA 15-3. Benign liver conditions resulted in the highest percentage of non-malignant elevations. Both BCM and CA 15-3 values were elevated in 18 of 40 (31%) individuals. Some individuals with benign lung disease, autoimmune conditions, benign kidney disease, and infections also were found to have marker elevations (Fig 7b).

Levels in Breast Cancer Patients: 354 patients with breast cancer were tested with both IMx BCM and CA 15-3. BCM levels were elevated in 129 (36.1%) patients compared to 127 (35.6%) for CA 15-3. Analysis of 142 patients with active breast cancer found that 80 (56.3%) of these patients had elevated values compared to 75 (52.8%) for CA 15-3. The correlation coefficient for these two groups was r=0.99 and the slope m=1.44 (Fig. 10). Thirty-three pre-operative breast cancer patients also had elevated values. Eleven (33.3%) had elevated BCM values and 10 (30.3%) had elevated CA 15-3 values. The correlation coefficient was r=0.84 and slope m=0.53 (Fig 11).

Distribution of BCM values by stage is found in Fig. 12. This analysis uses 35 U/mL (95% normal specificity) as the cutoff between normal and elevated BCM values. Of 151 stage IV breast cancer patients 72.2% were found to have elevated values. This compares with 47.7%, 41.4% and 27.3% for stages III, II, and I respectively.

For breast cancer patients with serial samples, BCM level changes were generally consistent with changes in disease status. Also comparison of BCM and CA 15-3 values for these patients found consistent parallelism. An example of BCM, CA 15-3, and CEA tracking is found in Fig. 13. Serial specimen data will be published elsewhere.

DILUTION OF BREAST CANCER SAMPLE

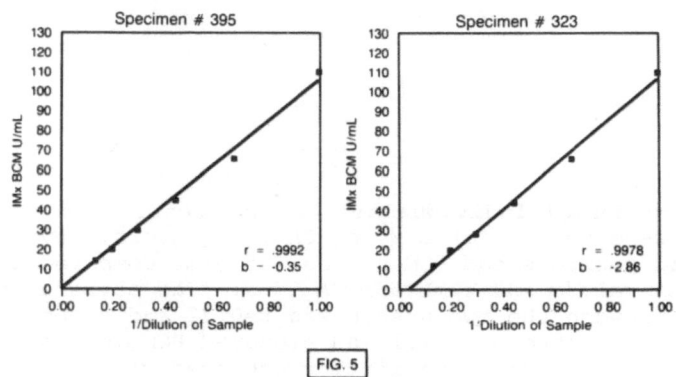

FIG. 5

Receiver Operating Characteristic Curve: A Receiver Operating Characteristic (ROC) Curve was constructed using normal specimen results versus active breast cancer specimen results. This was done in order to determine the BCM value which best discriminates between normal and elevated BCM values. The ROC curve plots sensitivity on the ordinate and 1 minus specificity on the abscissa (Fig. 14). Sensitivity is defined as the number of active breast cancer specimens which have a result above a selected cutoff as a percentage of the total active breast cancer specimens (TP/TP+FN). Specificity is defined as the number of normal plus benign specimens which have a result below a selected cutoff as a percentage of the total normal (TP/TN+TP).

From this study 30 U/mL was found to be the optimum cutoff value to discriminate between normal and elevated BCM values. This value results in the maximum clinical sensitivity (TP) and clinical specificity (TN).

RECOVERY

Sample Serum	Endogenous Level (U/mL)	BCM Added (U/mL)	Value Obtained (U/mL)	% Recovery*
A	19.2 19.2	38.5 81.4	56.1 104.7	96 105
B	22.8 22.8	36.7 77.7	56.1 110.5	91 113
C	19.5 19.5	34.5 81.5	52.2 100.0	95 99

$$*\% \text{ Recovery} = \frac{\text{BCM Value Obtained (U/mL) - Endogenous BCM Level (U/mL)}}{\text{BCM Added (U/mL)}} \times 100$$

FIG. 6

Levels in Individuals with Other Malignant Conditions: 529 patients with non-breast cancers were tested with both IMx BCM and CA 15-3. BCM levels were elevated in 201 (38.0%) patients compared to 184 (34.8%) for CA 15-3. The correlation coefficient for these two groups was r=0.97 and the slope m=1.5 (Fig. 15). Colorectal, pulmonary, ovarian, and uterine/cervical cancer patients were found to have elevated BCM and CA 15-3 levels in many incidences (Fig. 7a). BCM values and CA 15-3 values usually were elevated in the same patients. Specimens from patients with colorectal cancer demonstrated the largest difference between the assays. BCM values were elevated in 19 of 25 (39.6%) colorectal patients compared to 12 % elevated for CA 15-3.

Summary of Elevated Specimens
IMx BCM Vs CA 15-3

Category/Range	N	IMx BCM*			CA 15-3 –		
		# Below	# Above	% Elev.	# Below	# Above	% Elev.
Overall	1334	1043	291	21.8	1050	284	21.3
Malignant	529	328	201	38.0	345	184	34.8
Benign	394	321	73	18.5	315	79	20.1
Normal	411	394	17	4.1	390	21	5.1
Normal Female	311	299	12	3.9	304	7	2.3
Normal Male	100	95	5	5.0	86	14	14.0
Breast CA	357	228	129	36.1	230	127	35.6
No History	38	25	13	34.2	24	14	36.8
History	319	203	116	36.4	206	113	35.4
Inactive	171	138	33	19.3	135	36	21.1
Active	142	62	80	56.3	67	75	52.8
Pre-Op	33	22	11	33.3	23	10	30.3
Unknown	6	3	3	50.0	4	2	33.3

* Using a Cutoff of 35 u/mL
+ Using a Cutoff of 30 u/mL

FIG. 7a

Summary of Elevated Specimens
IMx BCM Vs CA 15-3

Category/Range	N	IMx BCM*			CA 15-3 +		
		# Below	# Above	% Elev.	# Below	# Above	% Elev.
Colorectal CA	48	29	19	39.6	36	12	25.0
Lung CA	50	23	27	54.0	26	24	48.0
Ovarian CA	30	12	18	60.0	13	17	56.7
Uter./Cerv. CA	24	19	5	20.8	22	2	8.3
Other CA	20	17	3	15.0	18	2	10.0
Benign Kidney	41	32	9	22.0	34	7	17.1
Benign Liver	58	40	18	31.0	40	18	31.0
Benign Lung	60	50	10	16.7	44	16	26.7
Infectious	39	27	12	30.8	29	10	25.6
Autoimmune	79	63	16	20.3	60	19	24.1
Benign Breast	117	109	8	6.8	108	9	7.7

* Using a Cutoff of 35 u/mL
+ Using a Cutoff of 30 u/mL

FIG. 7b

CORRELATION FOR BENIGN SPECIMENS

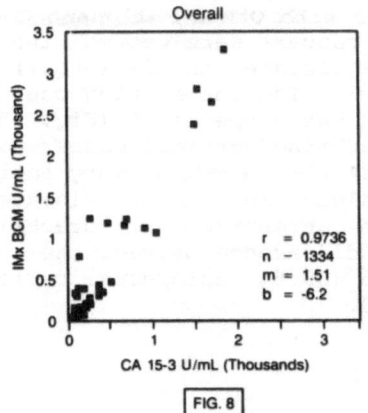

r = 0.9736
n – 1334
m = 1.51
b = -6.2

FIG. 8

CORRELATION FOR ALL SPECIMEN TYPES

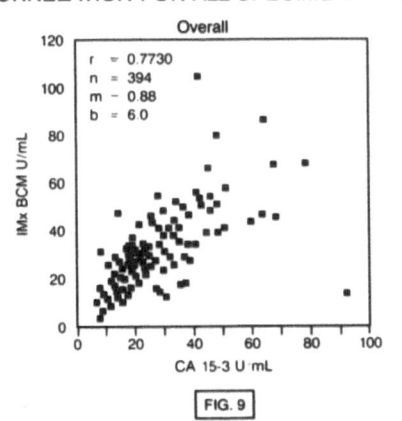

FIG. 9

CORRELATION FOR ACTIVE BREAST CANCERS

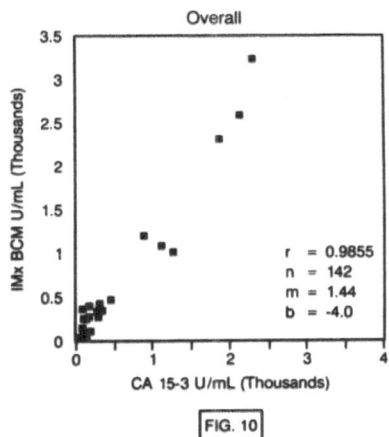

FIG. 10

CORRELATION FOR PRE-OP BREAST CANCERS

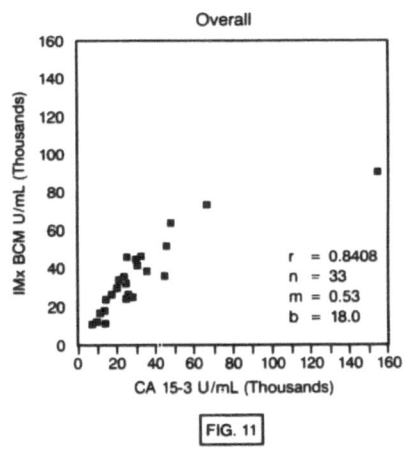

FIG. 11

Percent Distribution
BCM Values by Stage

Stage	Number of Subjects	U/mL 0.0-35.0	U/mL 35.1-50.0	U/mL 50.1-125	U/mL >125
I	132	72.7	14.4	7.6	5.3
II	254	58.7	14.6	11.0	15.8
III	86	52.3	20.9	12.8	14.0
IV	151	27.8	19.2	14.6	38.4

FIG. 12

179

SEQUENTIAL SPECIMEN

Pt. ID # 201523 (MAU)
Breast Cancer: Stage IV (not otherwise specified) Grade Unknown (10/24/85)
Demographics: 63 Year Old Postmenopausal Caucasian Female, Non-Smoker

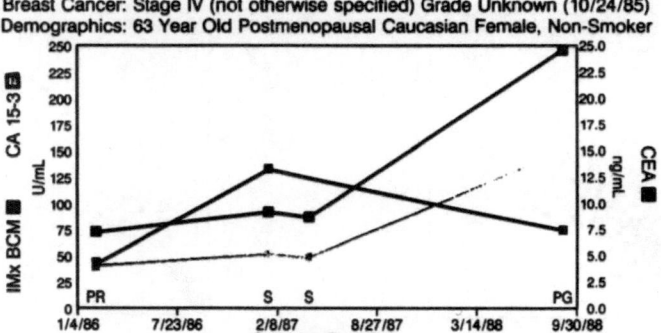

FIG. 13

COMBINED IMx BCM RESULTS
ROC CURVE ANALYSIS

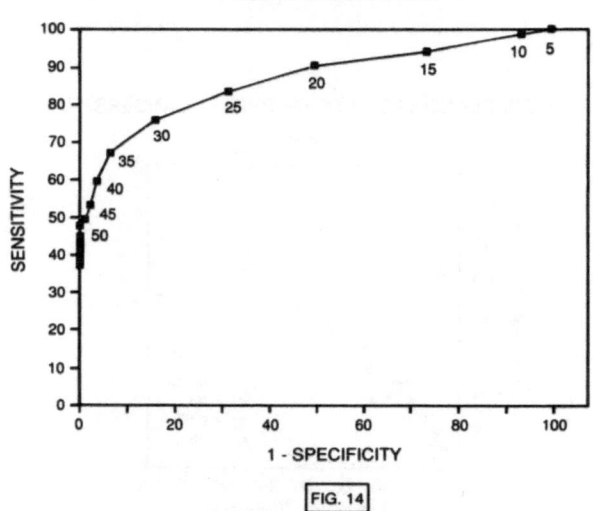

FIG. 14

CORRELATION FOR MALIGNANTS

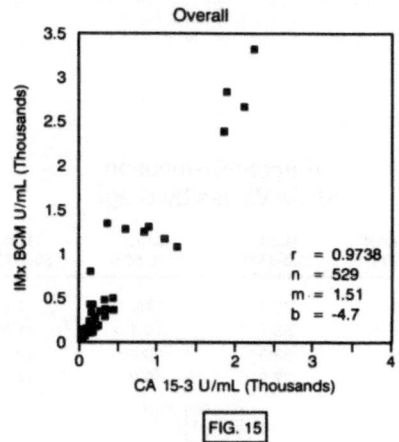

FIG. 15

180

SUMMARY

It has been demonstrated that IMx^R BCM serum levels often are
elevated in patients with active breast cancer compared to normal
individuals and those with benign breast conditions. The
reliability of singlet testing and feasibility of using adjusted
stored calibration curves has been shown. Acceptable
reproducibility, dilution linearity, recovery, and specificity
was demonstrated.

The IMx^R BCM assay has been compared to the commercially
available Centocor CA 15-3 RIA. Comparison using a retrospective
clinical panel (n=1334) indicates a high degree of correlation
(r=0.97) between these two assays. IMx values were elevated in 17
(4.1%) of 411 apparently normal individuals compared to 21 (5.1%)
for CA 15-3. Of 142 active breast cancer patients, 80 (56.3%%) had
elevated BCM levels compared to 75(52.8%) for CA 15-3. Evaluation
of serial serum specimens from breast cancer patients determined
that the BCM and CA 15-3 levels trend in parallel and often reflect
changes in disease status.

The IMx BCM assay may be useful in monitoring disease status
in breast cancer patients.

BIBLIOGRAPHY

1. Taylor-Papadimitriou J, Burchell J, Chang SE. "Use of
   Antibodies to Membrane Antigens in the Study of Differentiation
   and Malignancy in the Human Breast." IN: Monoclonal Antibodies
   and Cancer. Academic Press, Inc. 1983:227-238.

2. Griffiths AB, Burchell J, Gendler S, Lewis A, Blight K,
   Tilly R, Taylor-Papadimitriou J. Immunological Analysis of
   Mucin Molecules Expressed by Normal and Malignant Mammary
   Epithelial Cells. *Int J Cancer* 1987;40:319-327.

3. Feizi T. Demonstration by Monoclonal Antibodies that
   Carbohydrate Structures of Glycoproteins and Glycolipids are
   Onco-Developmental Antigens. *Nature* 1985;314(7):53-57.

4. Marcus DM. A review of the Immunogenic and Immunomodulatory
   Properties of Glycosphingolipids. *Molecular Immunology*
   1984;21(11):1083-1091.

5. Childs RA, Pennington KU, Scudder P, Goodfellow PN, Evans MJ,
   Feizi T. High Molecular Weight Glycoproteins are the Major
   Carriers of the Carbohydrate Differentiation Antigens I, i
   and SSEA-1 of Mouse Teratocarcinoma Cells. *Biochem J*
   1983;215:491-503.

6. Myoga A, Taki T, Arai K, Sekiguchi K, Ikeda I, Kurata K,
   Matsumota M. Detection of Patients with Cancer by Monoclonal
   Antibody Directed to Lactoneotetraosylceramide (Paragloboside).
   Cancer Research 1988;48:1512-1516.

7. Hrzenjak T, Matausic-Bratkovic M, Buguljic A, Muic V,
   Djelajlua-Kostic M. Gangliosides and Glycoproteinic Antigen
   from Human Breast Tumours. *Iugoslav Physiol Pharmacol Acta*
   1987;23:47-53.

8. Konrath JG, Manderino GL, Przywara LW. IMx Breast Cancer Marker. *Breast Cancer Research and Treatment* 1988;12(1):134.

9. Konrath JG, Manderino GL, Przywara LW. IMx BCM an Automated Microparticle Enzyme Immunoassay for Breast Cancer Mucin. *Clin Chem* 1989;35(6):1078.

10. Rittenhouse HG, Manderino GL, Hass GM. Mucin Type Glycoproteins as Tumor Markers. *Lab Med* 1985;16(9):556-560.

11. Kim YD, Robinson DY, JK Tomita JD. Monoclonal Antibody PR92 with Restricted Specificity for Tumor Associated Antigen in Prostrate and Breast Carcinoma Cancer Research 1988;48: 4543-4548.

12. Tondini C, Hayes DF, Gelman R, Henderson IC, Kufe DW. Comparison of CA 15-3 and Carcinoembryonic Antigen in Monitoring the Clinical Course of Patients with Metastatic Breast Cancer. *Cancer Research* 1988;48:4107-4112.

13. Thor A, Weeks MO, Schlom J. Monoclonal Antibodies and Breast Cancer. *Seminars in Oncology* 1986;13(4):393-401.

14. Anderson B, Slota J, Kundu S, Patrick J, Manderino G, Rittenhouse H, Tomita J. Characterization of Monoclonal Antibodies to Paragloboside (PG) and Sialosyl-PG (2,6-SPG), and an Improved Chromatogram Binding Assay for Rapidly Identifying Antibodies to Tumor Antigens. *J Cell Biochem* 1987;Supplement S11D:157.

15. Fiore M, Mitchell J, Doan T, Nelson R, Winter G, Grandone C, Zeng K, Haraden R, Smith J, Harris K, Leszczynski J, Berry D, Safford S, Barnes G, Scholnick A, Ludington K. The Abbott IMx Automated Benchtop Immunochemistry Analyzer System. *Clin Chem* 1988;34(9)1726-1732.

16. Turner DA. An Inituitive Approach to Receiver Operating Characteristic Curve Analysis. *The Journal of Nuclear Medicine* (1978) 19:213-220.

A NOVEL SERUM ASSAY USING RECOMBINANT BREAST EPITHELIAL MUCIN ANTIGEN

Roberto L. Ceriani[1], David Larocca, Jerry A. Peterson, Richard Amiya, Sean Enloe and Edward W. Blank

John Muir Cancer and Aging Research Institute, 2055 North Broadway, Walnut Creek, CA 94596

[1] To whom correspondence should be addressed.

ABSTRACT

Serum levels of breast epithelial antigens are presently used in the follow-up of breast cancer patients. Available assays do not have optimal sensitivity and rely on reagents that could vary in their source and purity. A novel competitive solid-phase radioimmunoassay was developed for BrE-Ags that consists of a fusion protein, (comprising a β-galactosidase fusion partner and a polypeptide sequence obtained from a breast carcinoma cell line cDNA library), and anti-human milk fat globule monoclonal antibody Mc5, both produced biosynthetically. The fusion protein carries a modified epitope sequence (mimotope) similar, but not identical, to the one found in the native antigen. This new competitive assay configuration has two features, a solid-phase affinity step that purifies the fusion protein carrying the mimotope for Mc5, and a competitive step that provides quantitation. Serum values for this assay show high specificity and sensitivity for breast cancer patients when compared to normal subjects and post-surgical breast cancer patients during their disease-free period.

The detection in serum of the breast epithelial mucin, although proposed as a means to follow the breast cancer patient, is mired with uncertainty. First, it is now known that anti-breast epithelial mucin monoclonal antibodies (MoAbs) used in the serum immunoassays bind a restricted area of the tandem repeated sequence of the mucin (J. A. Peterson, et.al, this volume), and that the tandem repeat is polymorphic

---

Abbreviations used: MoAb, monoclonal antibody; HMFG, human milk fat globule; BrE-Ags, breast epithelial antigens; NPGP, nonpenetrating glycoprotein; CEA, carcinoembryonic antigen; SDS-PAGE, sodium dodecylsulfate-polyacrylamide gel electrophoresis; Met-BSA, methylated bovine serum albumin; PBS, phosphate buffered saline; SEM, standard error of mean; NED, no evidence of disease; RIA buffer, radioimmunoassay buffer (PBS, 0.3% Triton X-100, 0.02% sodium azide); SD, standard deviation.

Breast Epithelial Antigens, Edited by R.L. Ceriani
Plenum Press, New York, 1991

(1) and can vary several fold from individual to individual. This makes the number of epitopes on the mucin a factor in altering the reading of the assays. Further, it was recently found that cellular mechanisms of release of the mucin can alter the levels found in circulation (2). To make the situation even more complex, 2 catabolic processes have been reported which, depending on their level of activity, could alter serum levels of the mucin. These catabolic processes comprise the hepatic uptake and destruction of this mucin (3), and fragmentation of the mucin by serum itself (4). Serum levels of the mucin result from all these intervening factors which undoubtedly could be different from patient to patient and even changing from time to time in the same patient.

Breast epithelial antigens (BrE-Ags) were first found in the circulation of breast cancer patients with a polyclonal antiserum raised against human milk fat globule (HMFG) (5). Approximately 75% of sera of patients with disseminated breast cancer disease were positive with this double determinant assay, while only 25% of those at Stage I of the disease had detectable serum values (5). Later, with radio assays, using MoAbs also created against the HMFG components were used (6,7), a similar percentage of positivity was found. An alternative assay, using MoAb Mc8, detecting breast antigens of lower molecular weight had an even higher percentage of positive serum values in disseminated breast cancer; however, at high tumor load in the patients, the antigens were detected in immune complexes (8).

The polyclonal assay (5) detected HMFG components with molecular weights 150, 70 and 45 kilodaltons, while CA15-3 and Mc5 detect the breast epithelial mucin ($\geq$ 400 kilodaltons antigen) (7) and the Mc8 assay, the 45 kilodaltons antigen (9). (For a review of breast epithelial markers see 4).

Assays measuring the breast epithelial mucin, also called non-penetrating glycoprotein (NPGP) when it was first detected immunologically (9), have as an advantage the ease of their performance, possibly due to the polyepitopic nature of this antigen (1). However, a large percentage of patients with disseminated breast cancer disease (approximately 30%) do not show detectable values over controls (4). This lack of detection could be explained by mechanisms of release and catabolism by liver and serum itself (3). Thus, assays with higher sensitivity are required because they could detect threshold levels of this antigen in circulation. This type of sensitivity, without compromising specificity, can be found in competitive assays (4).

The use of competitive immunoassays has a double advantage over that of immunometric assays. First, in competitive immunoassays the probing antibody can choose between the putative antigen in the serum and the true native antigens used to compete for the same epitope region, thus assuring specificity. Second, it is possible to manipulate the sensitiveness of the immunoassay by varying the ratio of the participating native antigen to Ab. With this approach, it was found that a competitive assay using Mc5 (7), a MoAb against NPGP, had a higher sensitivity than the CEA assay.

Recently, it has been possible to obtain fusion proteins produced by cDNA clones isolated from breast epithelial cell λgt11 cDNA libraries, that are bound by anti-HMFG antibodies. We have previously described a cDNA encoding an epitope for Mc5 that was selected from a MCF-7 λgt11 expression library (11). This epitope is not recognized by other MoAbs [(Mc1 (9), BrE-1, BrE-2, BrE-3 (11)] against the breast mucin, NPGP. This cDNA, NP5, was subcloned into the expression vector pEX2 (12) to achieve high levels of expression of a fusion protein consisting of

Cro/β-galactosidase fused to the NP5 cDNA. The entire nucleotide sequence of NP5 was determined (10), and when the derived amino acid sequence of the reading frame of the cDNA insert portion of fusion protein was compared with that of the tandem repeat of the breast mucin (13), only a short region of homology was found. Four out of 6 amino acids were identical (10). Epitope mapping of the entire NP5 amino acid sequence using the method of Geysen (14) revealed that this region of homology was the only linear peptide sequence which bound Mc5. The sequence for binding of Mc5 on NP5 was Asp-Leu-Arg-Pro-Gly-Pro, where the underlined residues correponded to those that are identical to the continuous epitopic sequence of the tandem repeat. The amino acid sequence of the Mc5 epitope in the native tandem repeat was Thr-Arg-Pro-Ala-Pro (15). Fusion protein NP5 contains an altered epitope for the MoAb Mc5 that could be referred to as a mimotope.

A novel breast cancer serum assay is described in this study. It employs a mimotope for MoAb Mc5 found in the fusion protein NP5. This mimotope participated in a competitive assay with a novel configuration that had greater sensitivity and specificity for circulating mucin antigen than others previously reported (6). This configuration purifies in situ the fusion protein carrying the mimotope and then presents it to MoAb Mc5 in competition with the circulating antigen. Here the preparation of the fusion protein carrying the mimotope is presented, as well as the configuration of the competitive immunoassay created with it and details of its procedure. Serum values of the BrE-Ags obtained with this immunoassay in breast cancer patients with and without active disease, and in normal subjects are reported.

MATERIALS AND METHODS

Antibodies and antigen: MoAb Mc5 was created as described (7). Purified MoAb Mc5 was obtained from Coulter Immunology (Hialeah, FL), polyclonal rabbit anti-β-galactosidase (3 mg/ml) (Organon Teknika, West Chester, PA), and goat-anti-mouse immunoglobulin (Sigma, St. Louis, MO).

The mimotope for MoAb Mc5 was contained in a fusion protein produced by a cDNA clone originally isolated from a λgt11 cDNA library made from human breast carcinoma cell line MCF-7 (16). The λgt11 NP5 clone (16) was separated by polyethylene glycol precipitation, cut with EcoR1 restriction enzyme to excise the NP5 insert, which was then subcloned into the expression plasmid pEX2 (12) at the EcoR1 restriction enzyme recognition site downstream from the lac z gene. The fusion protein produced by bacteria infected with pEX2 containing the NP5 insert bound only the Mc5 MoAb, and not Mc1 (9), DF3 (17), BrE-1 (11), BrE-2 (11), BrE-3 (11), all anti-NPGP MoAbs; not to Mc13 (9), McR2 (9), Mc3 (5) and Mc8 (11) MoAbs, which recognize smaller components of HMFG. The NP5 cDNA was also subcloned into pGEM3 according to standard methods, and sequenced (10) by the dideoxy method (18) using a modified T7 DNA polymerase (sequenase) directly on the plasmid DNA using T7 or SP6 promoter sequence primers (Promega, Madison, WI) according to the manufacturer's protocol (United States Biochemicals, Cleveland, OH).

Under appropriate conditions these bacteria carrying pEX2 recombinant NP5 produced fusion proteins comprising β-galactosidase plus the NP5 peptide sequences. The fusion protein, NP5, was concentrated in inclusion bodies in these pEX2 bacteria. Inclusion bodies were then separated and the fusion protein purified.

Preparation of fusion-protein from bacterial culture: Fusion protein was prepared according to the method of Marston (19). Fifty milliliters of inoculant culture were added to 1000 ml of LB medium (1:20

ratio) containing 0.1 mg/ml ampicillin. The culture was maintained at 28-30°C in a shaker incubator until an $A_{600}nm=0.5$ was obtained with LB medium [Bacto-Tryptone, 10 gm/l; Bacto-Yeast extract, 5 gm/l; NaCl, 10 gm/l; 0.01 M Tris (pH 7.5)]. The culture was heat induced in a 42°C shaker water bath for 1 h, then incubated for 1 h at 37°C, chilled on ice, and centrifuged at 3290xg for 15 min at 4°C. The supernatant was discarded, the pellets weighed and lysozyme (0.8 mg/gm cells) in lysis buffer (0.05 M Tris, pH 8.0, 1 mM EDTA, 0.1 M NaCl) was added.

The suspension was thoroughly mixed by drawing up and down with a pipet, deoxycholate (sodium salt, 4 mg/gm of cells) in lysis buffer was added and the viscous suspension mixed again using a pipet. Each tube was sonicated on ice three times for 30 sec (30 sec on and 30 sec off) and centrifuged for 15 min at 12,000g at 4°C. The supernatant was removed and the pellet weighed and a volume of lysis buffer with 10 mM EDTA and 0.5% Triton X-100 of 9 times the pellet weight was added and mixed using a pipet. The last two steps were repeated twice and supernatants voided. One gram of pellet was weighed and then dissolved in 68 mM Tris pH 7.5, 2% SDS, 2% β-mercaptoethanol and 1 mM of phenylmethylsulfonyl fluoride in 90 ml. When the solution was completely clear, it was dialyzed against PBS and 0.3% Triton X-100 at room temperature.

The dialyzed preparation was run on a 7.5% SDS-PAGE and, a Western blot was performed to confirm the presence and purity of the fusion protein and to detect any protein degradation occurring during dialysis. The total protein concentration was estimated by the BCA method (20). The dialyzed preparation was used directly in the serum assay.

Coated plate preparation: Microtiter plates (Dynatech, Alexandria, VA) were coated as previously described (21) using methylated BSA (Met-BSA) (22). A 0.01% solution of Met-BSA was made by dissolving it in 0.3% Triton-X-100 in PBS, and 0.1% sodium azide (RIA buffer). Fifty μl of the Met-BSA solution were added per microtiter plate well, and then dried overnight at 37°C. The next day 200 μl of PBS were added per well, kept for 30 minutes at room temperature and pipetted out. This was repeated once more, this time letting PBS stand for 5 min, the microtiter plate was let dry and stored at room temperature in a dry place for up to 6 months.

Solid Phase preparation: Two hundred μl of PBS were added to the Met-BSA coated plates, then left 5 min at room temperature, the PBS aspirated and 50 μl of buffered glutaraldehyde solution (0.25% glutaraldehyde in PBS pH 7.0) were added. Next, the microtiter plate was left at room temperature for 1 h, then washed twice with 200 μl/well of PBS. These plates can now be stored for up to 6 months at room temperature. To bind the anti-β-galactosidase to the plate 50 μl of a 2 μg/ml solution of the antibody was pipetted per well coated with Met-BSA and glutaraldehyde and let bind overnight. At the end of the incubation, the wells were washed 3 times with PBS and 200 μl of 0.5% solution of glycine in PBS were added for 1 h at room temperature to block remaining protein binding sites of the glutaraldehyde-Met-BSA, then washed once and the wells let dry. To bind the fusion protein to the microtiter plate coated now with anti-β-galactosidase, 50 μl/well of a 13 μg/ml solution of NP5 in and RIA buffer were added and incubated for 24 h at 37°C, then washed three times with PBS and let dry.

NPGP affinity chromatography: Affinity chromatography purification of NPGP from HMFG was performed using activated Sepharose-4B (Pharmacia, Uppsala, Sweden) coated with MoAb Mc5. For this purpose 1 mg of MoAb Mc5 per ml of swollen activated Sepharose-4B were reacted using procedures prescribed by the manufacturer. The Mc5-coated beads were suspended in RIA buffer. Delipidated HMFG (22) was dissolved in RIA buffer and sonicated. One mg protein of dissolved HMFG [dissolved using a double step sonication sequence with a microtip horn at 25 watts on a Sonifier Cell Disrupter 185 (Branson, Danbury, CT) for 4 min (10 sec sonication, 10 sec silent period at 4°C)] per ml of packed beads was added. It was then left overnight with mild agitation at 4°C. After this incubation the beads were washed 3 times alternately with 0.1 M acetate, 1 M NaCl pH 4.0 and 0.1 M Tris, 1 MNaCl pH 8.0 and then 2 times with PBS plus 0.3% Triton X-100, packed by slow centrifugation and the supernatant decanted. To each ml of the packed beads 3 ml of a solution of 3 M sodium isocyanate in PBS were added, and let stand for 30 min. After this step the beads were packed, the supernatant collected and washed again with 3 ml of 3 M sodium isocyanate. Then the beads were packed and the supernaturant again collected. Both 3 M sodium isocyanate supernaturants were then dialyzed at 4°C against 2000 ml of PBS and 0.05% sodium azide 3 times. The final dialysis took place against 0.1X PBS and 0.005% Na azide. Then the dialysate was concentrated by evaporation with an electrical fan. A protein determination was obtained using the BCA method (20).

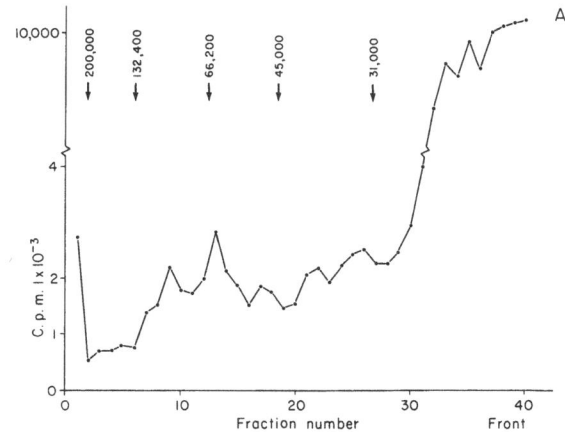

Figure 1. (A) Electrophoretic profile of ^{125}I-HMFG run on a 10% continuous SDS-PAGE.

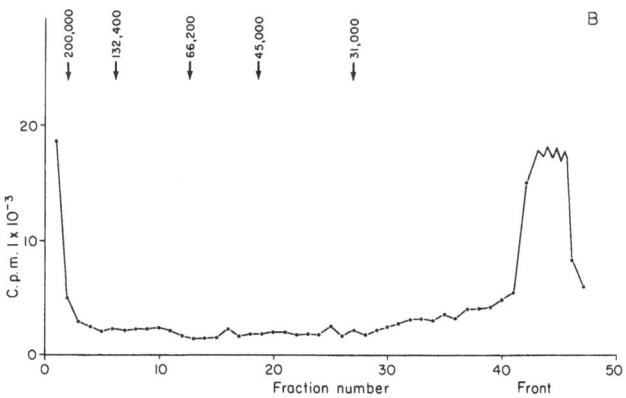

Figure 1. (B) Electrophoretic profile of Mc5 affinity purified NPGP labeled with ^{125}I on a 10% continuous SDS-PAGE.

Determination of binding constants: They were performed by displacement on microtiter plates which had NP5 and NPGP as a solid phase; K_a was calculated as reported by Sheldon (23).

NP5 fusion protein competititon serum assay: To prepare the standard curve a 500 μg protein/ml solution of NPGP was prepared in RIA buffer and sonicated for 4 min (10 sec sonication, 10 sec. silent period at 4°C). This standard solution, was then diluted in decreasing concentration in RIA buffer containing 14% normal female serum, (concentrations of NPGP prepared were 30, 10, 3, 1, 0.3, 0.1, and 0 μg NPGP protein/ml). Thirty μl of the decreasing NPGP content solutions were pipetted in quadruplicate into wells of a microtiter plate, and then to every well 20 μl of a 0.6 μg/ml solution of MoAb Mc5 diluted in RIA buffer was added. The wells were covered with a non-porous tape and left on a rotatory shaker overnight at room temperature. The next day the wells were washed 4 times with 200 μl of RIA buffer, and the last wash was left for 5 min. Then 200,000 cpm of ^{125}I-goat-anti mouse immunoglobulin in 50 μl of RIA buffer were added. The wells were covered with a non-porous tape and placed on a gyratory agitator for 3 h at room temperature. After this, the wells were washed 5 times with 200 μl of RIA buffer, leaving the last wash 5 min. Then wells were cut out and counted for radioactivity.

To determine serum levels of NPGP patient serum samples were diluted (1:6) in RIA buffer and 30 ul in quadruplicate were pipetted, (in place of the NPGP solution used to prepare the standard curve), into microtiter plate wells coated with Met-BSA, plus anti-β-galactosidase plus NP5. Then 20 μl of a 0.6 μg/ml MoAb Mc5 solution were added (as for the NPGP standard curve) and the procedure was continued as above.

The results were plotted using the mean value of cpm of each quadruplicate of the standard dilutions versus the NPGP concentration. Counts obtained for the subjects' serum samples were used to determine the μg/ml concentration of NPGP (1:6 dilution) from the standard curve. The cut off line for this assay was obtained by multiplying the standard deviation of the mean of the normal subject samples times 3, and adding it to this mean. Specificity and sensitiveness were calculated as reported (24). Standard error of the mean (SEM) was calculated as reported (25).

RESULTS

In order to assess purity of the Mc5 affinity purified NPGP it was radiolabeled with ^{125}I and run on a 10% SDS-PAGE. A radioactivity profile of HMFG species is shown in (Fig. 1A) with several smaller molecular weight bands while ^{125}I-NPGP is shown remaining at the top of the gel. In Fig. 1B, the purity of the heavy molecular weight NPGP after affinity chromatographic purification is shown. Small molecular weight denatured labeled products and free radioactive iodine are seen at the front of both gels.

To measure circulating NPGP in breast cancer patient serum, a radioimmunoassay was developed with the following configuration (See MATERIALS AND METHODS):

Met-BSA>>Rabbit-anti-β-galactosidase>>NP5>>Mc5>>[125]I-Goat-anti-mouse Ab
  I   II   III IV   V

MoAb Mc5 was added to the coated microtiter plates at a stoichiometric concentration with the solid bound NP5-fusion protein carrying Mc5's mimotope to increase sensitivity. Thus, when either NPGP (the native antigen) or patient sera containing circulating NPGP were added at this step (step IV), they competed for Mc5 binding. The quantity of MoAb Mc5 remaining attached to the solid phase after washings was identified by ^{125}I-labeled anti-mouse Ig. Using the affinity purified NPGP, a standard curve was constructed (Fig. 2). The sensitivity of the standard curve reached 0.1 μg/ml.

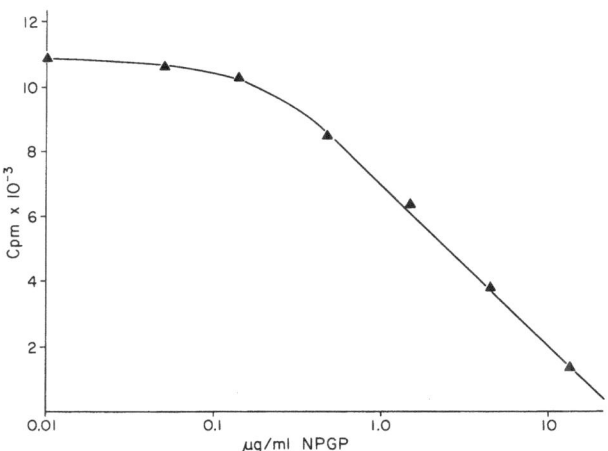

Figure 2. Standard curve of MoAb Mc5-NP5 fusion protein competitive assay using different concentrations of the breast mucin NPGP (See MATERIALS AND METHODS).

To explore the ability of Mc5 to bind either NPGP, (the native antigen carrying the tandem repeated sequence natural epitope), or NP5, carrying the mimotope, their respective binding constants (23) were obtained. MoAb Mc5 was shown to have a higher binding avidity for the epitope on NPGP ($K_a = 3.98 \times 10^{-9}$ 1/M) than for the mimotope on NP5 ($K_a = 7.4 \times 10^{-8}$ 1/M).

Serum values were obtained from 40 normal healthy women, 31 patients who had undergone surgery for breast cancer who at the time were clinically free of disease, that is with no evidence of disease (NED), and 24 patients with active breast cancer disease. Clinical diagnosis had been obtained by at least 3 consecutive visits for NED patients, and in the case of active disease by direct confirmation of the recurrent lesion.

Normal patients had levels of NPGP by the NP5 assay ranging from 0 to 0.960 μg/ml, with a mean of 0.136 μg/ml and a standard deviation (SD) of ± 0.278 μg/ml and SEM of ± 0.044.

A cut-off value of 0.97 μg/ml NPGP for this assay (Fig. 3) was determined as described in MATERIALS AND METHODS. Using this cut-off value, it was found that no NED patient had values above it and that both normal and NED patients clearly separate from patients with active breast cancer (Fig. 3). The mean for NED patients was 0.030 μg/ml, the S.D. was ± 0.078 μg/ml and the SEM was ± 0.014 μg/ml. The NED patients did not differ statistically from the normal subjects in terms of circulating NPGP levels. The patients with active breast cancer disease had values ranging from 4.26 to >83 µg/ml, all of them above the cut-off line. Some of the patients with active breast cancer disease reached very high values of NPGP in circulation which were not directly proportional to tumor mass present in these patients.

Both sensitivity and specificity for this serum assay using the NP5 fusion protein were established to be 100%.

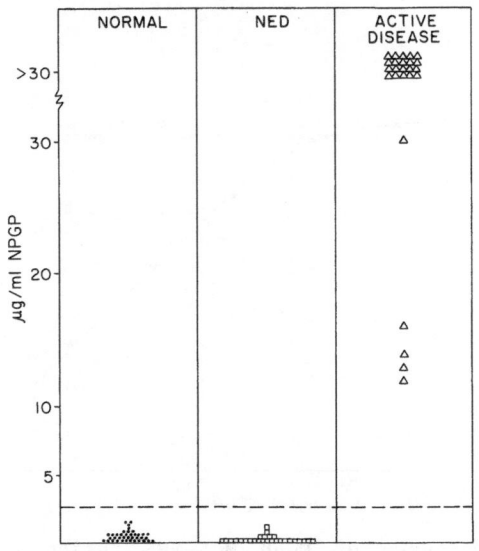

Figure 3. Levels of the breast mucin NPGP in sera from normal women, breast cancer patients with no evidence of disease (NED), and breast cancer patients with active detectable disease.

DISCUSSION

A novel serum assay for the breast epithelial mucin NPGP is described that takes advantage of the availability of NP5, a fusion protein obtained from a cDNA library of MCF-7 cells. It contains a mimotope for MoAb Mc5. The unique assay configuration described here achieves two goals: First it purifies the fusion protein while firmly binding it to a solid phase (step III); and second, it presents the mimotope to MoAb Mc5 (step IV) allowing for the performance of a competitive assay for circulating NPGP in breast cancer patient sera with high specificity and sensitivity.

The solid phase binding of the NP5 fusion protein was obtained via a polyclonal anti-β-galactosidase antibody that was in turn covalently bound to a firmly adhered layer of Met-BSA (21). Thus, the anti-β-galactosidase further purifies, by a step of affinity chromatography, the fusion protein NP5 that had already been partially fractionated in the form of inclusion bodies from E. coli. Another advantage of the configuration presented is that a very sensitive competitive assay is used to determine levels in circulation of NPGP. This is first demonstrated by the very steep standard curve obtained and by the clear cut separation in terms of NPGP levels, between active breast cancer disease vs. NED and normal subjects. The reason for such sensitivity could be found in the fact that the MoAb Mc5, has a higher binding affinity when binding to the native epitope on NPGP (that is a fully glycosylated mucin) than when binding the mimotope found on NP5. Impaired glycosylation of the mucin in breast cancer (26, 27) as well as possible catabolism of this molecule in circulation (4) could provide an altered epitope to the assay that could compete less effectively with NPGP for binding of anti-mucin MoAbs like Mc5, and also for those present in the CA15-3 assay. However, in our present assay configuration, the altered circulating antigen could be more successful when competing with NP5 for binding of MoAb Mc5, since Mc5 has a lower binding affinity for NP5 than for NPGP. It can also be added that the mimotope is present only once on each NP5 molecule, thus quantitative calculations can be arrived at, in contrast to the use of NPGP where it is not clear how many repeated sequences containing the Mc5 epitope are expressed or exposed on this polymorphic molecule (1).

NP5 is a permanent source of antigen for the assay as well as a standard reagent free from the risk of denaturation during NPGP purification and will not represent possible glycosylation differences depending on the source of this native antigen. The assay includes two pure reagents: the NP5 fusion protein and MoAb Mc5, allowing for rigid standardization of the assay.

The sensitivity of the assay allows for early detection of increased serum values of NPGP that are known to be useful in the follow-up of the breast cancer patient (4). It is the patient in disease-free period that could benefit the most from the early recognition of recurrence or relapse. As for the ability of assays like the present one to be useful for screening, the high sensitivity obtained here has to be weighed against negative factors related to the antigen itself. The breast epithelial mucin or NPGP is cleared very rapidly from the circulation (4 and R.C. unpublished results); therefore, catabolic factors play an important role in controlling NPGP serum levels, especially at low tumor load when the amount of NPGP released is small. Obviously, in the presence of fast serum clearance of NPGP it will be necessary to arrive to a certain secretion level of the antigen for it to be able to emerge above the assay's cut-off level. It will only be the true positives out of those results that will be of significance, since false negatives are possible with the specificity kept at 100%.

Comparison of other tests (6,7) vs. the present one using paired samples will permit choosing the most accurate. These assays (6,10) have as a common denominator the presence of some false positives among normal subjects and that only approximately 70% of active disease patients show elevated NPGP levels. Moreover, one of these previously employed tests (6) have as an inconvenience the lack of standardization of values, (usually quoted in units) and the varying and limited availability of antigen. The use of the fusion proteins as an antigen source provides a solution to those issues. In addition, substitution of the NP5 fusion-protein with others carrying the epitopes of alternative breast cancer markers could allow for the performance of a panel of assays keeping to a minimum the changes in assay reagents other than the recognizing MoAb.

In this study the development of a novel assay for a serum breast tumor marker is described. This competitive assay employs as antigen a fusion protein carrying an artificial epitope or mimotope of an anti-HMFG MoAb (Mc5). High sensitivity and specificity can be obtained with this serum test which lends itself to become the basis for a multi-breast epithelial antigen assay panel having as the only variable component the fusion protein carrying the epitope, with the anti-breast cancer marker MoAbs provided as a mixture.

ACKNOWLEDGEMENTS

Supported by PHS Grants RO1-CA39932 and PO1-CA42767 from the National Cancer Institute and BRSG Grant S07-RR05929.

REFERENCES

1. Gendler, S.J., Burchell, J.M., Duhig, T., Lamport, D., White R., Parker, M., and Taylor-Papadimitriou, J. (1987) Proc. Natl. Acad. Sci. USA **84**, 6060-6064.

2. Ceriani, R.L. and Chan, C. (1990) Breast Cancer Res. Treat. 17, 55-58.

3. Ceriani, R.L., and Blank, E.W. (1991) Breast epithelial mucin serum clearance, Cancer Letters, accepted for publication.

4. Ceriani, R.L., and Rosenbaum, E.H. (1991) In: Immunodiagnosis of Cancer. (Herberman,R.B., Mercer, D., Eds.) Vol. 2, pp. 223-241, Marcel Dekker Inc.

5. Ceriani, R.L., Sasaki, M., Sussman, H., Wara, W.M., and Blank, E.W. (1982) Proc. Natl. Acad. Sci. USA 79, 5420-5424.

6. Hayes, D.F., Sekine, H., Ohno, T., Abe, M., Keefe, K., and Kufe, E.W. (1985) J. Clin. Invest. 75, 1671-1678.

7. Ceriani, R.L., Blank, E.W., Rosenbaum, E.H., Ben Zeev, D., Lowitz, R.S., Johansen, L. and Trujillo, T. (1990) Breast Cancer Res. Treat., 15, 161-174.

8. Salinas, F.A., Wee, K.H., and Ceriani, R.L. (1987) Cancer Res. 47, 907-913.

9. Ceriani, R.L., Peterson, J.A., Lee, J.Y., Moncada, R., and Blank, E.W. (1983) Somat. Cell Genet. 9, 415-427.

10. Larocca, D., Peterson, J.A., Walkup, G. and Ceriani, R.L. (1990) High level expression in E. Coli of an alternate reading frame of pS2 mRNA that encodes a mimotope of human breast epithelial mucin tandem repeat. Submitted for publication.

11. Peterson, J.A., Zava, D.T., Duwe, A.K., Blank, E.W., Battifora, H. and Ceriani, R.L. Hybridoma, (1990) 9, 221-235.

12. Stanley, K., and Luzio, J. (1984) EMBO. J. 3, 1429-1434.

13. Lichtenberg, M.J.L., Voss, H.L., Gennissen, A.M.C. and Hilkens, J. J. Biol. Chem. 265:5573-5578, 1990.

14. Geysen, H.M., Meloen, R.J., and Barteling, S.J. (1984) Proc. Natl. Acad. Sci. USA 81, 3998-4002.

15. Peterson, J.A., Larocca, D., Walkup, G., and Ceriani, R.L. (1990) Monoclonal antibodies to overlapping polypeptide epitopes of the breast mucin have different specifications to the native antigen. Submitted for publication.

16. Walter, P., Green, S., Greene, G., Krust, A., Pornert, J.M., Jeltsch, J.M., Staub, A., Jensen, E., Scrace, G., Waterfield, M., and Chambon, P. (1985) Proc. Natl. Acad. Sci. USA 82, 7889-7893.

17. Kufe, D., Inghirami, G., Abe, M., Heyes, D., Justi-Wheeler, H., and Schlom, J. (1984) Hybridoma, 3, 223.

18. Sambrook, et al, (1989) Molecular Cloning: A laboratory manual, Second Edition, Cold Spring Harbor Press, New York.

19. Marston, F. (1987) In: DNA Cloning. A practical approach. (Glover, P., Ed.) IRL Press, Oxford, U.K. 3, 59-88.

20. Smith, P.K., Krohn, R.I., Hernonson, G.T., Mallia, A.K., Gartner, F.J., Provenzano, M.D., Fujimoto, E.K., Goeke, N.M., Olson, B.J., and Klenk, D.C. (1985) Anal. Biochem. 150, 76-85.

21. Ceriani, R.L.: (1984) In: Monoclonal Antibodies and Functional Cell Lines. Progress and Applications. (Bechtol, K.B., McKern, T.J. and Kennett, R., Eds.) pp. 398-402, Plenum Press, New York, .

22. Ceriani, R.L., Thompson, K., Peterson, J.A., and Abraham, S. (1977) Proc. Natl. Acad. Sci. USA 74, 582-586.

23. Sheldon, K., Marks, A., and Baumal, R. (1987) Biochem. Cell Biol. 65, 423-428.

24. Galen, R., and Gambino, R. (1975) Beyond Normality: The Predictive Value and Efficiency of Medical Diagnosis. John Wiley & Sons, New York.

25. Snedecor, G.W., and Cochrane, W.G. (1967) Statistical Methods, 6th Edition, Iowa State University Press, Ames, Iowa.

26. Parodi, A.J., Blank, E.W., Peterson, J.A., and Ceriani, R.L. (1982) Breast Cancer Res. Treat., 2, 227-237.

27. Hull, S.R., Bright, A., Carraway, K.L., Abe, M., Hayes, D.F., and Kufe, D.W. (1989) Cancer Communications 4, 261-267.

RADIOIMMUNOGUIDED™ SURGERY IN BREAST CANCER

Carol A. Nieroda* and Edward Martin, Jr.**

*National Cancer Institute
Laboratory of Tumor Immunology and Biology
Bethesda, Maryland

**Ohio State University
College of Medicine
Department of Surgery
Columbus, Ohio

ABSTRACT

Twenty-two patients with biopsy proved carcinoma of the breast received radiolabeled MAb B72.3 (5-1 mCi of ^{125}I per 2.0-0.25 mg MAb, Iodo-Gen™ method) intravenously 6-30 days prior to definitive surgery (mastectomy or lumpectomy/axillary dissection). Using the Neoprobe 1000™ gamma detecting probe, gamma counts of breast and axillary tissues were obtained preoperatively, intraoperatively, and on ex vivo specimens. In breast tissue, the RIGS™ system identified tumor that was histopathologically confirmed in 11 of 14 patients. There were two false positives each having a histopathologic diagnosis of apocrine metaplasia and hyperplasia. Unsuspected tumor was histopathologically documented in 3 of 6 breast biopsies performed based on the preoperative presence of high external gamma counts in the countralateral breast or in a quadrant other than that of the original primary. All six patients had negative mammograms and physical exams. The 3 false positives had diagnoses of aprocrine metaplasia and hyperplasia. In axillary tissue, probe counts identified metastatic disease in 3 of 8 patients and verified absence of disease in 10 of 14 patients. False positive counts were obtained in 4 having histopathologic diagnoses of sinus histocytosis or reactive nodes. RIGS appears to be able to identify residual, subclinical, and multicentric carcinoma of the breast and delineate the pattern of antigenic drainage of tumor into lymph nodes.

INTRODUCTION

Carcinoma of the breast is a leading cause of death in American women second only to lung cancer.[1] Over the past two decades with increased public education, improved mammagraphic techniques, and an ever increasing appreciation of the psychological issues surrounding the treatment of carcinoma of the breast, surgical therapy has progressed from radical mastectomy to modified radical mastectomy to the present trend toward conservation of the breast using lumpectomy or quadrantectomy and axillary dissection. These conservative surgical approaches emphasize resection of the tumor with a thin rim of grossly normal mammary tissue, which in many instances, leaves behind a subclinical burden of cancer cells,[2] which must be dealt with using another treatment modality, for example, radiation therapy. There are now, however, several studies which support the findings of equal disease free survival and long term survival in women receiving this conservative treatment compared with those more radical treatments of the past.[2,3]

As well as utilizing new surgical approaches, the recent advances in hybridoma-monoclonal antibody technology and gene cloning procedures are revolutionizing our understanding and investigative approach to carcinoma. MAbs which are reactive with breast carcinoma have been developed using several different categories of immunogens: membrane enriched extracts of breast carcinoma metastases, milk fat globule membrane, mammary carcinoma cell lines, lymph nodes from mastectomy patients, intermediate filaments, and surface receptors.[4-8] Many such MAb's are being used successfully in immunohistopathologic and immunohistochemical as well as radioscintigraphic diagnosis of breast cancer. In an attempt to incorporate these new scientific advances into the surgeon's armamentarium of tools to identify subclinical disease, a new surgical detection method that uses a hand-held gamma detecting probe has been developed. The probe is used preoperatively and intraoperatively after injection of a radiolabeled monoclonal antibody (MAb) to assist the surgeon in more accurately localizing, intraoperatively staging, and optimally resecting the tumor. In addition, the pathologist scans ex vivo specimens using the probe to assist in determining the best location from which to obtain specimens for histopathologic analysis.

MATERIALS AND METHODS

Antibody

B72.3, a first generation murine whole IgG1 antibody, is reactive with a 220-1,000,000 K glycoprotein complex (TAG-72). This antibody has been described in detail elsewhere.[9-11] TAG-72 is found in abundance in several epithelially derived cancers including colonic adenocarcinomas, invasive ductal carcinomas, non-small cell lung carcinomas, and ovarian carcinomas. The antibody was provided through the courtesy of Jeffrey Schlom, Ph.D. (National Cancer Institute, Bethesda, Maryland).

The purified whole antibody was radiolabeled with ^{125}I using the 1,3,4,6-tetrachloro 3 alpha, 6 alpha diphenylglycouril (Iodo-Gen™) method.[12] Unbound iodine was removed by gel-filtration chromatography on a cross linked dextran (Sephadex™) column. Resulting specific activities ranged from 1- 5 millicuries of ^{125}I per milligram of antibody depending on the particular dosage each patient received. Radiolabeled MAb B72.3 was sterilized by filtration through a 0.22 micrometer filter. The percentage of ^{125}I bound to protein was determined regularly after purification by trichloracetic acid precipitation and instant thin-layer chromatography. A level of greater than or equal to 90 per cent protein-bound ^{125}I was required before injection into humans.

Probe

The Neoprobe 1000 instrument is a hand-held gamma detecting probe (GDP) consisting of an 18 cm long x 2 cm in diameter stainless steel tube with an angled tip (120^0) for better maneuverability. The tip of the probe contains a 12 millimeter radiosensitive cadmium telluride crystal that relays through a preamplifier and signal processor to emit a digital readout and an auditory signal. The intensity of the auditory signal is directly proportional to the level of radioactivity detected. The detector element is housed in a chromium plated copper housing about the size of a pocket flashlight. The audible signal threshold level may be adjusted to eliminate the sound (squelch) when the probe is over normal tissues.

Patients

Twenty-two patients with a biopsy proved histopathological diagnosis of carcinoma of the breast were studied. The study was conducted in accordance with the Ohio State University human subjects review committee guidelines, and a consent form was signed by each patient.

A history and physical examination were done and a diagram of the breasts and axillae made. The cytologic or histologic diagnosis was verified, and a recent mammogram was obtained.

Antibody Injection

Prior to the injection of the radiolabeled B72.3, each patient was begun on a saturated solution of potassium iodide, ten drops two times per day until the time of surgery or for 21 days, whichever was greater, to block thyroid uptake of ^{125}I. The research protocol nurse reviewed the purpose, risks, and benefits of the study with the patients and obtained informed consent.

Subcutaneous forearm hypersensitivity testing was performed by the physician using unlabeled B72.3 antibody (0.1 milligram in 0.1 milliliter of sodium chloride). After 15 minutes, a reaction two times that of the sodium chloride control or an erythma 20 millimeters in diameter was considered positive. No positive reactions were observed. Each patient then received the radiolabeled antibody (5-1 mCi ^{125}I per 2.0-0.25 mg MAb) intravenously over two to three minutes and was observed by a physician and nurse for at least 15 minutes. No adverse reactions were observed.

Tumor Activity Determination

Patients returned as outpatients at 6-10 day intervals for an external probe survey of the breasts and axillae and were scheduled for definitive operation (modified radical mastectomy or lumpectomy/quadrantectomy and axillary dissection) when external precordial gamma counts reached 20 counts per two seconds. Patients underwent final external probe survey on the day of surgery and diagrams were made noting any reactive areas. In addition, intraoperative (in vivo) probe survey was carried out to assess the site of the biopsy and nodal drainage (axillary and internal mammary). Postoperatively, ex vivo probing of specimens was done by the pathologist to aid in localizing areas for histologic sections. Each reading was taken for two seconds and repeated three times. The mean and standard deviation of counts for each site was calculated. The degree of localization of B72.3 in tumor at each site was derived from the ratio of radioactive count activity in tumor compared with background (nontumor) counts taken on clinically normal mammary tissue of a quadrant other than that containing the known carcinoma. Ratios of 2:1 were defined as positive, 1.5-2.0:1.0 as suspicious and ratios of less than 1.5:1.0 as negative. All individual numerical counts were converted to this positive, suspicious, or negative grading system for convenience.

Squelching Technique

The probe was placed over mammary tissue in a quadrant other than that containing the carcinoma proved by biopsy, and the threshold level was adjusted to eliminate the sound produced by the "background" normal tissue. After squelching in this manner, the system becomes extremely sensitive to small increments in count rate. The audible signal goes from no sound to an almost continuous sound with an increase in count rate by a factor of two. As the background activity may vary from one type of tissue to another, it is important to reset the squelch level when scanning a new tissue type.

RESULTS

The numerical GDP counts were converted to a positive, suspicious, or negative grading system to simplify presentation. In Table I, the correlation between tissue gamma counts (in in vivo as well as ex vivo specimens) and final histopathological diagnosis for twenty-two patients (27 specimens) is shown for breast tissue. In Table 2, the correlation between tissue gamma counts (in in vivo as well as ex vivo specimens) and final histopathological diagnosis for twenty-two patients (19 specimens) is shown for axillary tissues.

Table 1. CORRELATION OF TISSUE GAMMA COUNTS WITH HISTOPATHOLOGY

BREAST BIOPSY SITE TISSUE

Pt. No.		IN VIVO	EX VIVO	PATH
1		+	+	+
2		+	+	+
3		+	+	+
4		+	+	+
5		+	+	+
6		+	+	+
7	(R)	+	+	+
*	(L)	+	+	A
*8	(R)	+	+	+
	(L)	+	+	+
9		+	+	A
10		+	+	A
*11	(R)	+	-	A
	(L)	+	+	+
*12	(R)	+	N	+
*	(L)	+	N	+
13	(R)	+	+	+
*	(L)	+	-	A
14		-	-	+
15		-	-	-
16		-	-	-
17		-	-	-
18		-	-	-
19		-	-	-
20		-	N	-
21		-	-	+
22		-	+	+

* Probe-directed biopsy   A= atypical hyperplasia/fibrosis
+ Positive
- Negative
N= not examined

Table 2. CORRELATION OF TISSUE GAMMA COUNTS WITH HISTOPATHOLOGY

AXILLARY TISSUE

Pt. No.		IN VIVO	EX VIVO	PATH
1		-	-	r
2		-	-	r
3		-	-	-
4		-	-	-
5		-	-	-
6		+	+	+
7	(R)	-	+	r
*	(L)	N	N	N
* 8	(R)	N	N	N
	(L)	+	N	+
9		+	+	r
10		N	N	r
* 11	(R)	N	N	N
	(L)	-	N	+
* 12	(R)	N	N	N
*	(L)	N	N	N
13	(R)	-	-	-
*	(L)	N	N	N
14		-	-	+
15		-	-	-
16		-	-	r
17		-	-	-
18		-	+	SH
19		+	+	+
20		-	-	+
21		-	+	+
22		N	N	+

* Probe-directed biopsy   A= atypical hyperplasia/fibrosis
+ Positive   SH= Sinus histiocytosis
- Negative   r= reactive
N= not examined

Actual examples of gamma probe counts for mammary tissue for each patient are given for breast biopsy site tissue in Table 3 and normal breast tissue in Table 4. Actual examples of gamma probe counts for axillary tissues for each patient are given for clinically suspicious nodal tissues in Table 5 and for normal appearing nodal tissue in Table 6. Any patient having negative in vivo counts and positive ex vivo counts was classified based on in vivo counts.

Eleven of 14 patients with histopathologically documented residual tumor of the breast at the site of the previous excisional biopsy of the primary tumor were identified by high probe counts (true-positives). Of these, 10 (Nos.1-6,7R,8L,11L,13R) (Table 1) had high probe counts both in vivo and ex vivo , while 1 (No.12R) had no ex vivo counts done. Three patients (Nos.14, 21, 22) had residual tumors not identified by the probe counts (false-negatives). One patient (No. 22), not counted amoung the positive localizations but instead included with the false negatives, had negative in vivo counts but positive ex vivo counts. Six patients (Nos. 15-20) had histopathologically negative biopsy site specimens and low probe counts (true negatives), while 2 patients (No. 9, 10) had specimens negative for tumor but diagnoses of fibrosis, atypical intraductal hyperplasia, and cystic changes with high probe counts (false-positives).

Six probe-directed biopsies were performed based on the presence of high preoperative probe counts. One was the original biopsy site of patient No.12; however, the five others (7L,8R,llR,12L,13L) were performed in the opposite breast from the site of the original primary tumor. Of these 6 specimens, there were 3 noted to be carcinoma on histopathological examination and 3 noted to contain atypia, aprocrine metaplasia,or hyperplasia.

The probe identified metastases in axillary nodes by in vivo counts in 3 patients (Nos. 6, 8, 19) (true positives). One patient (No. 12) did not have axillary tissues removed for examination. Two patients (Nos. 10, 22) had tissues removed, but these were not examined either in vivo or ex vivo using the gamma probe. Eleven patients (Nos. 1-5,7,13,15-18) had low probe counts in histopathologically negative specimens (true negatives). Two of these eleven patients (Nos.7, 18) had negative in vivo counts, but positive ex vivo counts. There were 4 patients classified as false negatives (Nos. 11, 14, 20, 21). One of these patients (No.11) had low in vivo counts but no ex vivo counts were performed. Another patient (No. 21) had low in vivo counts, but positive ex vivo counts. There was one patient with high probe counts and negative pathology (false positive) (No. 9).

In summary, in mammary tissue, the RIGS system identified tumor that was histopathologically confirmed in 11 of 14 patients and confirmed absence of tumor in 6 of 8 patients giving a sensitivity of 79% and a specificity of 76%. Probe counts were suspicious for tumor but not documented histopathologically in two patients having atypia, aprocrine metaplasia, and hyperplasia. In axillary tissues, probe counts identified tumor in 3 of 8 patients with histopathologically documented tumor metastases. However, one patient with positive nodes did not have tissue examined using the gamma detecting probe. Another patient had low in vivo counts, but high ex vivo counts. Probe counts were suspicious for tumor but not documented histopathologically in one patient.

DISCUSSION

The use of monoclonal antibodies in such technologies as radioimmunoscintography is becoming increasing popular as a diagnostic tool in oncology.[13-25] Polyclonal and monoclonal antibodies and a variety of isotopes including ^{131}I, ^{123}I, ^{111}In, and ^{99}Tc, have been used extensively with varied results. There are several disadvantages of scintigraphy for detection of tumor including the dose of radioactive material to which the patient is exposed and the fact that lesions less than 1 centimeter are difficult to detect.[26] For instance, using the MAb B72.3, Lastoria [27] reported that in 20 patients with primary breast tumors (size 0.8 to 9 cm) only 9 were demonstrated although 2 undetected lesions were negative for B72.3 on immunoperoxidase staining. In 12 patients with recurrent or metastatic disease with a total of 21 lesions, 15 were imaged (3/4 liver, 6/9 nodes). Since the ability to detect radioactivity is based on the inverse square law which, briefly stated, defines the ease of detection as proportional to the inverse square of the distance of the detector from the radioactive source, it stands to reason that an instrument which can be placed within millimeters of the source will be able to detect smaller masses of tumor. In addition, in patients who have lesions detected by scintigraphy, the same lesions can be detected with equal efficiency using smaller doses of radiation, consequently decreasing the patient's radiation exposure. The Neoprobe 1000 hand-held gamma detecting instrument (probe) has been developed with this in mind. The probe has been demonstrated to be able to identify tumor cell masses as small as 3.9×10^4 cells (<1mm).[28] The effectiveness of the probe in localizing primary and recurrent metastatic carcinoma of the colon and breast has been documented in the literature elsewhere.[29-34]

Although numerous variables ranging from tumor cell population heterogeneity to the specific antigenic properties [35-40] are involved in the ability of an antibody to recognize a specific antigen on a tumor cell, many antibodies that identify carcinomas of the breast have been described. At the outset of the present study, it was believed that B72.3 had the most selective degree of reactivity for tumor tissues of any of the MAb's available at the time, in that no reactivity with normal adult human tissues had been observed.[10] It was noted, however, that different mammary cell tumors varied in the amount of antigen detected by B72.3.[41] In a given tumor cell population, blood supply, necrosis, and antigen density will also help determine positive or negative scores depending on the dose of antibody used. The specific location and pattern of staining within the cell has been studied in detail with two-thirds of specimens demonstrating cell-associated membrane or diffuse cytoplasmic staining, or both, associated with the production of mucin as opposed to focal staining or apical and marginal staining.[4, 7,41] Infiltrating ductal, infiltrating lobular, comedo, and in situ breast carcinoma lesions have also been demonstrated to pick up B72.3 more efficiently than other histological types such as medullary.[41-43] Using fine needle aspiration of mammary masses and immunoperoxidase staining, it was found that 26 of 27 malignant specimens stained positive with B72.3, while 22 of 23 benign specimens stained negative with B72.3.[44]

The effectiveness of the Neoprobe 1000 gamma detecting probe is directly related to the availability of antibodies that identify mammary tumors and the expertise of the operating surgeon in using the probe. If used with precision and in combination with a specific and sensitive antibody, it is able to identify occult tumor. In the present study, 11 of 14 patients with histopathologically documented residual tumor of the breast at the site of a previous biopsy were identified by high probe counts in vivo (confirmed by immunohistochemical staining), demonstrating that these tumors had receptors (TAG-72) for B72.3. Three of eight patients with metastatic disease to axillary nodes were also identified (3 others did not have tissues examined) while two false negatives were found. Immunohistochemical staining of false negative tissues showed poor staining with B72.3. Even more importantly, in six instances, biopsies were performed based on the finding of high probe counts on external scanning in a quadrant other than that of the primary tumor and, in 3 of these cases, carcinoma was confirmed on histopathology. These tumors would have gone undetected until some later date since both physical examination and mammograms were negative in these patients. The finding of aprocrine metaplasia and hyperplasia in the other 3 biopsy specimens raises the question of the presence of shared antigens in these "benign" diagnoses which may be associated with an increase breast cancer risk.[45] Further studies are needed to delineate the exact nature and significance of these findings.

An absence or low concentration of receptors on the tumors of several patients would explain the lack of detection both in mammary tissue and axillary nodes. However, the inability of the probe to detect high rates in vivo, that is, before the specimen was surgically removed, in some instances can be attributed to a high blood pool background of MAb not attached to tumor, making the ratio of abnormal to normal tissue less obvious in vivo but easily demonstrated ex vivo. Several variables will affect the time required for blood pool background clearance of an antibody including the particular antibody used, the amount of isotope used, and whether a whole antibody of fragment is used.

Table 3. MEAN GAMMA COUNTS OF BREAST TISSUE

BREAST BIOPSY SITE TISSUE

Pt. No.		IN VIVO	EX VIVO	BPB
1		84	94	53
2		37	42	16
3		49	53	22
4		46	68	23
5		37	22	17
6		51	188	23
7	(R)	37	25	
*	(L)	88	43	15
* 8	(R)	120	72	
	(L)	85	58	36
9		72	125	48
10		61	40	16
* 11	(R)	75	33	
	(L)	60	38	15
* 12	(R)	100	N	
*	(L)	69	N	40
13	(R)	81	30	
*	(L)	67	26	14
14		4	6	6
15		26	6	24
16		29	25	25
17		25	25	16
18		6	6	4
19		30	24	20
20		14	N	19
21		16	23	4
22		13	37	13

BPB = Blood pool background
* = Probe-directed biopsy
N = Not examined

Table 4. MEAN GAMMA COUNTS OF BREAST TISSUES

NORMAL BREAST TISSUE

PT. NO		IN VIVO	EX VIVO	BPB
1		53	7	53
2		6	1	16
3		6	7	16
4		23	20	23
5		12	9	17
6		N	25	23
7	(R)	N	N	
*	(L)	22	17	15
* 8	(R)	25	N	
	(L)	N	N	36
9		7	4	48
10		18	10	16
* 11	(R)	N	4	
	(L)	21	N	15
* 12	(R)	N	N	
*	(L)	36	N	40
13	(R)	N	N	
*	(L)	10	15	14
14		2	3	6
15		4	4	24
16		24	15	25
17		15	14	16
18		4	4	4
19		25	13	20
20		15	N	19
21		11	23	4
22		10	N	13

BPB = Blood pool background
N = Not examined
* = Probe-directed biopsy

203

Table 5. MEAN GAMMA COUNTS OF AXILLARY TISSUES

CLINICALLY ENLARGED NODES

PT. NO		IN VIVO	EX VIVO	BPB
1		39	59	53
2		5	1	16
3		5	14	22
4		11	7	23
5		19	15	17
6		38	105	23
7	(R)	9	99	
*	(L)	N	N	15
* 8	(R)	N	N	
	(L)	29	N	36
9		20	47	48
10		68	39	63
* 11	(R)	N	N	
	(L)	17	N	15
* 12	(R)	N	N	
*	(L)	N	N	40
13	(R)	16	16	3
*	(L)	N	N	14
14		6	1	6
15		8	3	24
16		N	N	
17		21	27	14
18		6	21	4
19		37	48	20
20		10	16	15
21		10	24	4
22		N	N	13

BPB = Blood pool background
N = Not examined
* = Probe-directed biopsy

Table 6. MEAN GAMMA COUNTS OF AXILLARY TISSUES

CLINICALLY NORMAL NODES

PT. NO		IN VIVO	EX VIVO	BPB
1		N	2	53
2		6	7	16
3		14	10	22
4		22	8	23
5		10	5	17
6		18	N	23
7	(R)	N	N	
*	(L)	9	15	15
* 8	(R)	N	N	
	(L)	15	N	36
9		3	14	48
10		5	3	63
* 11	(R)	N	N	
	(L)	10	N	15
* 12	(R)	N	N	
*	(L)	32	N	40
13	(R)	N	N	
*	(L)	13	N	14
14		1	1	6
15		14	10	24
16		N	N	
17		22	3	14
18		1	3	4
19		N	12	20
20		N	N	15
21		N	N	4
22		N	N	13

BPB = Blood pool background
N = Not examined
* = probe directed biopsy

205

With experience, we have found that in the case of B72.3, this problem can be essentially eliminated by planning the operation at a time when external probe counts over the heart have reached 20 or less per two seconds. In the future, the availability of antibodies with higher affinities which also are cleared quickly from the blood (for example a fragment of a second generation antibody such as CC49 or CC83) may greatly reduce the present average 21 day waiting period between the time of antibody injection and the surgical procedure while maintaining high detection rates.

The confirmation of the absence of tumor in 6 breast specimens and 11 axillary node specimens is encouraging. However, the phenomenon of false localization (not confirmed with light microscopy) that occurred in 1 original biopsy site, 3 probe-directed biopsies, and 4 node specimens is of concern. All mammary specimens were noted to have fibrosis, atypical intraductal hyperplasia, and cystic changes, while the axillary specimens were diagnosed as reactive or sinus histocytosis. Although the presence of shed antigen or shared antigens may explain the high probe counts, the possibility of the presence of tumor cells undetectable by current light microscopic methods must be considered. If the problem lies with shed antigen in regional nodal drainage, it is felt that the use of second generation antibodies of B72.3[46], for example CC83, may eliminate many false positives in this regard. To evaluate these premises, a controlled study will need to be designed. However, the implications of the possibility of more accurate identification of occult disease, in particular for chemotherapy in patients with histopathologically negative nodes,is evident. Flow cytometry in combination with electron microscopy may be helpful in more clearly delineating this area in the future.

In summary, the technique of radioimmunoguided surgery is in its infancy. However, we believe that the ability to intraoperatively define tumor margins clearly and detect multicentricity may allow the surgeon to make more appropriate therapeutic decisions in the surgical management of carcinoma of the breast. With further investigation using more sensitive and specific antibodies, it may be possible to detect externally bilateral synchronous tumor or distinguish malignant from benign mammographic change. Furthermore, it may be possible to use the RIGS system to examine internal mammary and supraclavicular lymph nodes that are routinely not examined presently in the staging of breast cancer patients. The possibility of external identification of boney or skin metastases also needs to be studied, since any tool that could lead to an earlier and more precise staging of disease would allow earlier and more judicious use of additional or adjuvant therapies.

BIBLIOGRAPHY

1. J. R. Harris, I. Henderson, D. W. Kinne, and S. Hellman. Cancer of the breast. in: Cancer:Principles and Practice of Oncology, V.T. DeVita, Jr., ed., S. Hellman, and S.A. Rosenblum. 3rd edition, pp. 1197-1268. Philadelphia: J. P.Lippincott Co., (1989).

2. A. Recht and J. R. Harris. Conservative surgery and radiation therapy for early breast cancer. In: Cancer:Principles and Practice of Oncology, Updates. V.T. DeVita, Jr., ed., S. Hellman and S.A. Rosenberg, 2nd ed., Vol. 2,pp.1-12. Philadelphia: J.B. Lippincott Co. (1987).

3. Treatment of Early Breast Cancer: NIH Consensus Development Conference. June 18-21, 1990, NIH, Bethesda, MD.

4. A. Thor, M. Weeks, and J. Schlom. Monoclonal antibodies and breast cancer. Seminars in Oncology 13:393 (1986).

5. J. Schlom, J. Greiner, P. Horan Hand, D. Colcher, G. Inghirami, M. Weeks, D. Pestka, P. Fisher, P. Nobuchi, and D. Kufe. Monoclonal antibodies to breast human cancer-associated antigens as potential reagents in the management of breast cancer. Cancer 54:2777 (1984).

6. D. Colcher, P. Horan Hand, M. Nuti, and J. Schlom. A spectrum of monoclonal antibodies reactive with human mammary tumor cells.Proc. Natl. Acad. Sci. 78:3199 (1981).

7. E. E. Sterns. Monoclonal antibodies in the diagnosis and treatment of carcinoma of the breast. Surg. Gynec. and Obs. 169:81 (1989).

8. R. L. Ceriani, M. Sasaki, H. Sussman, W. M. Wara, and E. W. Blank. Circulating mammary epithelial antigens in breast cancer. Proc. Natl. Acad. Sci. 79:5420 (1982).

9. D. Colcher, M. Zalutsky, W. Kaplan, D. Kufe, F. Austin, and J. Schlom. Radiolocalization of human mammary tumors in athymic mice by a monoclonal antibody. Cancer Research 43:736 (1983).

10. M. Nuti, Y. A. Teramoto, R. Mariani-Costantini, P. Horan Hand, D. Colcher, and J. Schlom. A monoclonal antibody (B72.3) defines patterns of distribution of a novel tumor-associated antigen in human mammary carcinoma cell populations. Int. J. Cancer 29:539 (1982).

11. V. G. Johnson, J. Schlom, A. J. Paterson, J. Bennett, J. L. Magnani, and D. Colcher. Analysis of a human tumor-associated glycoprotein (TAG-72) identified by monoclonal antibody B72.3. Cancer Research 46:850 (1986).

12. B. Sickle-Santanello, P. J. O'Dwyer, C. Mojzisik, S. E. Tuttle, G. H. Hinkle, M. Rousseau, J. Schlom, D. Colcher, M. O. Thurston, C. Nieroda, A. Sardi, J. P. Minton, and E.W. Martin. Radioimmunoguided surgery using the monoclonal antibody B72.3 in colorectal tumors. Dis. Colon Rectum 30:761 (1987).

13. A. G. Siccardi, G. L. Buraggi, L. Callegaro, A. C. Colella, P. G. DeFilippi, G. Galli, G. Mariani, R. Masi, R. Palumbo, P. Riva, M. Salvatore, G. A. Scassellati, K. Scheidhauer, G. L. Turco, P. Zaniol, S. Benini, G. Deleide, M. Gasparini, S. Lastoria, L. Mansi, G. Paganelli, E. Salvischiani, E. Seregni, G. Viale, and P. G. Natali. Immunoscintigraphy of adenocarcinomas by means of radiolabeled F(ab')$_2$ fragments of an anti-carcinoembryonic antigen monoclonal antibody: a multicenter study. Cancer Research 49:3095 (1989).

14. A. A. Epenetos, S. Mather, M. Granowska, C. C. Nimmon, L. R. Hawkins, K. E. Britton, J. Shepherd, J. Taylor-Papadimitriou, H. Durbin, J. S. Malpas, and W. F. Bodmer. Targeting of iodine-123-labeled tumour-associated monoclonal antibodies to ovarian, breast, and gastrointestinal tumours. Lancet 2:999 (1982).

15. A. M. Keenan, J. C. Harbert, and J. M. Larson. Monoclonal antibodies in nuclear medicine. J.Nucl. Med. 26:531 (1985).

16. M. S. Larson. Radiolabeled monoclonal anti-tumor antibodies in diagnosis and therapy. J. Nucl. Med. 26:538 (1985).

17. D. M. Goldenberg, F. KeLand, E. Kin, S. Bennett, F. J. Primus, J. R. Van Nagell, N. Estes, P. DeSimone, and P. Rayburn. Use of radiolabelled antibodies to carcinoembryonic antigen for the detection and localization of diverse cancer by external photoscanning, New Engl. J. Med. 298:1384 (1978).

18. S. E. Order, J. L. Klein, D. Ettinger, P. Alderson, S. Siegelman, and P. Leichner. Use of isotopic immunoglobulin in therapy, Cancer Res. 40:3001 (1980).

19. J. P. Mach, F. Buchegger, M. Forni, J. Ritshard, D. Berche, J. D. Lumbroso, M. Screyer, C. Giradet, R. S. Accolla, and S. Carrel. Use of radiolabelled monoclonal anti-CEA antibodies for the detection of human carcinoma by external photoscanning and tomoscintigraphy, Immunol. Today 2:239 (1986).

20. B. Delaloye, A. Bischof-Delaloye, F. Buchegger, V. Von Fliedner, J. P. Grob, J. C. Volant, J. Pettavel, and J. P. Mach. Detection of colorectal carcinoma by emission-computerized tomography and after injection of ^{123}I-labelled Fab or F(ab')$_2$fragments from monoclonal anticarcinoembryonic antigen antibodies, J.Clin.Invest. 77:301 (1986).

21. R. M. Rainsbury. The localization of human breast carcinomas by radiolabelled monoclonal antibodies, Br. J .Surg. 71:805 (1984).

22. D. M. Goldenberg, E.E. Kin, and F. H. KeLand. Human chorionic gonadotropic radioantibodies in the radioimmunodetection of cancer and for disclosure of occult metastases, Proc. Natl. Acad. Sci. U.S.A. 78:7754 (1981).

23. D. M. Goldenberg, E. E. Kin, F. J. DeLand, S. Bennett, and F. J. Primus. Radioimmunodetection of cancer with radioactive antibodies to carcinoembryonic antigen. Cancer Res. 40:2984 (1980).

24. S. M. Larson, J. A. Carrasquillo, K. A. Krohn, B. W. Mc Griffin, D. Williams, I. Hellstrom, K. E. Hellstrom, and D. Lyster. Diagnostic imaging of malignant melonoma with radiolabeled antitumor antibodies. JAMA 249:811 (1983).

25. F. J. Primus and D. M. Goldenberg. Immunological considerations in the use of goat antibodies to carcinoembryonic antigen for the radioimmunodetection of cancer. Cancer Res.40:2979 (1980).

26. J. L. Murray, M. G. Rosenblum, and L. Laniki. Clinical parameters related to optimal tumor localization of Indium-111-labeled mouse anti melanoma monoclonal antibody ZME-D18. J. Nucl. Med. 28: 25-33 (1987).

27. S. Lastoria and M. Salvatore. Diagnostic value of B72.3 monoclonal antibody in epithelial cancers. in: "Nuklearmedizin: New Trends and Possibilities in Nuclear Medicine", H.A.E. Schmidt and L Cservay eds. Schattauer Verlag, p. 601 (1988).

28. O. A. Oredipe, R. F. Barth, S. E. Tuttle, D. M. Adams, I. Sautins, D. M. Bucci, C. M. Mojzisid, G. H. Hinkle, S. Jewell, Z. Steplewski, M. O. Thurston, and E. W. Martin Jr. Limits of sensitivity for the radioimmunodetection of colon cancer by means of a hand held gamma probe, Nucl. Med. Biol. 15: 595 (1988).

29. C. A. Nieroda, C. Mojzisik, G. Hinkle, M. O. Thurston, and E. W. Martin, Jr. Radioimmunoguided surgery (RIGS) in recurrent colorectal cancer, Cancer Detection and Prevention, In press. (1990).

30. C. A. Nieroda, C. Mojzisik, A. Sardi, P. Ferrara, G. Hinkle, M. O.Thurston, and E. W. Martin, Jr., Radioimmunoguided surgery in primary colon cancer. Cancer Detection and Prevention 14: 651 (1990).

31. C. Nieroda, C. Mojzisik, A. Sardi, W. Farrar, G. Hinkle, M. Siddiqi, P. Ferrara, A. James, J. Schlom, M. Thurston, and E. W. Martin, Jr. Staging of carcinoma of the breast using a hand-held gamma detecting probe and monoclonal antibody B72.3, Surg. Gynec. and Obst. 169:35-40 (1989).

32. C. A. Nieroda, C. Mojzisik, A. Sardi, P. Ferrara, G. Hinkle, M. O. Thurston, and E.W. Martin, Jr. The impact of radioimmunoguided surgery (RIGSTM) on surgical decision-making in colorectal cancer, Dis. Colon Rectum 32:927 (1989).

33. A. Sardi, C. Agnone, C. Nieroda, C. Mojzisik, G. Hinkle, P. Ferrara, W. B. Farrar, J. Bolton, M. O. Thurston, and E. W. Martin Jr. Radioimmunoguided surgery in recurrent colorectal cancer: the role of carcinoembryonic antigen, computerized tomography, and physical examination, Southern Medical Journal, 82:1235 (1989).

34. A. Sardi, M. Workman, C. Mojzisik, G. Hinkle, C. Nieroda, and E. W. Martin Jr. Intra-abdominal recurrence of colorectal cancer detected by radioimmunoguided surgery (RIGS system), Arch Surg. 124:55 (1989).

35. R. S. Kerbel. Implications of immunological heterogeneity of tumors, Nature. 280:358 (1979).

36. L. E. Hart, and I. J. Fidler. The implications of tumor heterogeneity for studies on the biology and therapy of cancer metastases, Biochem. Biophys. Acta 651:37 (1981).

37. R. T. Prehn. Analysis of antigenic heterogeneity within individual 3-methylcholanthrene-induced mouse sarcomas, J. Natl. Cancer Inst. 45:1039 (1970).

38. M. V. Pimm, and R. W. Baldwin. Antigenic differences between primary methyl cholanthrene-induced rat sarcomas and post-surgical recurrences, Int. J. Cancer. 20:37 (1977).

39. F. R. Miller, and G. H. Heppner. Immunologic heterogeneity of tumor cell subpopulations from a single mouse mammary tumor, J. Natl. Cancer Inst. 63:1457 (1970).

40. G. Poste, J. Dolla, and I. J. Fidler. Interactions amoung clonal subpopulations affect stability of the metastatic phenotype in polyclonal populations of B16 melanoma cells, Proc. Natl. Acad.Sci. U.S.A.78:6226 (1981).

41. J. Schlom, D. Colcher, P. Horan Hand, J. Greiner, D. Wunderlich, M. Weeks, P. B. Fisher, P. Noguchi, S. Pestka, and D. Kufe. Monoclonal antibodies reactive with breast tumor-associated antigens, Adv. Cancer Res.43:143 (1985).

42. S. C. Lottich, W. W. Johnston, C. A. Szpak, E. R. Delong, A. Thor, and J. Schlom. Tumor-associated antigen TAG-72: correlation of expression in primary and metastatic breast carcinoma lesions, Breast Cancer Res. Treatment. 6:49 (1985).

43. A. Thor, N. Ohuchi, C. A. Szpak, W. W. Johnson, and J. Schlom. Distribution of oncofetal antigen tumor-associated glycoprotein-72 defined by monoclonal antibody B72.3, Cancer Res. 46:3118 (1986).

44. J. Lundy, M. Lozowski, and Y. Mishriki. Monoclonal antibody B72.3 as a diagnostic adjunct in fine needle aspirates of breast masses, Ann. Surg. 203:399 (1986).

45. M. Castagna, M. Nuti, and F. Squartini. Mammary cancer antigen recognized by monoclonal antibody B72.3 in apocrine metaplasia of the human breast, Cancer Res. 47:902 (1987).

46. R. Muraro, M. Kuroki, D. Wunderlich, D. J. Poole, D. Colcher, A. Thor, J. W. Greiner, J. F. Simpson, A. Molinolo, P. Noguchi and J. Schlom. Generation and characterization of B72.3 second generation monoclonal antibodies reactive with the tumor-associated glycoprotein 72 antigen, Cancer Res. 48: 4588 (1988).

IMMUNO-PHARMACOLOGIC PURGING OF BREAST CANCER FROM BONE MARROW

E.J. Shpall, R.C. Bast, Jr., C.S. Johnson, W.P. Peters,
and R.B. Jones

From the Bone Marrow Transplant Program (EJS,CJ,RBJ),
Division of Oncology (EJS,RBJ), University of Colorado
Health Sciences Center, Denver, Colorado 80262 and the
Division of Hematology/Oncology (RCB,WPP), Duke University
Medical Center, Durham, North Carolina 27710

INTRODUCTION

High dose chemotherapy and autologous bone marrow support (ABMS) has
produced high complete response rates and durable remissions for patients
with advanced breast cancer[1,2,3]. Breast cancer commonly metastasizes to
bone marrow. Twenty eight percent of women with newly diagnosed breast
cancer and no evidence of metastases, had tumor cells detected in their
bone marrow using immuno-histochemical methods[4]. The possibility of
infusing clonogenic tumor into patients receiving autologous marrow-
supported chemotherapy regimens stimulated our efforts to develop a bone
marrow purging regimen for breast cancer.

We chose to develop a combined immunomagnetic plus pharmacologic purging
regimen because of our data[5] as well as other studies[6,7,8] demonstrating a
superior anti-tumor effect with combined modality purging regimens,
compared to the tumor cell depletion achieved with single modality regimens.

MATERIALS AND METHODS

Pre-clinical studies were performed initially to optimize the ex vivo
purging regimen which was used in the clinical study.

Pre-Clinical Studies

Immunomagnetic Purging (IMP)

One ml mixtures of breast cancer cells plus a 10-fold excess of human
marrow mononuclear cells (MNCs) were incubated with the a panel of anti-
breast cancer monoclonal antibodies[9] for 60 minutes at 4°C with frequent
rotation. The cells were washed and an aliquot of immunoglobulin-coated
microspheres (Dynal A.S. Corporation, Oslo, Norway and 45 North Station
Plaza, Great Neck, New York) was added for 60 minutes 4°C. The entire
suspension was then subjected to the magnetic field generated by a small
samarium-cobalt magnet. The magnetospheres were rapidly attracted to the
magnet, pulling the tumor cells with it[10]. Non-adherent cells were poured
off and assayed as described previously. Residual clonogenic breast cancer
cells were assayed in a limiting dilution assay[6]. Bone marrow progenitor
cell recovery was evaluated in tissue culture assays[6].

4-Hydroperoxycyclophosphamide (4-HC)

One ml mixtures of breast cancer cells plus a 10-fold excess of human
marrow MNCs were incubated for 30 minutes in a 37°C water bath with the

appropriate concentration of 4-HC (donated by M. Colvin, JHOC, Baltimore, MD.) and TC-199 tissue culture media (Gibco Laboratories Inc, Grand Island, New York). The final incubation concentration was 2×10^7 cells/ml. Following the 4-HC incubation, the cell suspension was rapidly cooled to 4°C, washed three times (centrifugation at 2900 rpm for 10 minutes) and re-suspended in TC-199 for further evaluation.

IMP plus 4-HC

The IMP and 4-HC purging procedures were performed sequentially, as described above. Both sequences of incubation (IMP first and 4-HC first) were evaluated in clonogenic tumor cell and bone marrow progenitor cell assays.

Once the small scale studies were completed, and the purging regimen optimized, a larger magnetic separation device to be used in clinical studies was constructed. The experiments were then repeated to optimize the purging regimen with large volumes of marrow that would be required for an autograft[11].

Clinical Study

The clinical marrow purging trial was designed for previously untreated stage IV breast cancer patients with significant bone or bone marrow metastases. The patients received three cycles of the Duke AFM (adriamy-cin, 5-fluorouracil and methotrexate) induction regimen[12]. Following hematologic recovery from the third cycle, if the tumor involvement was less than 10% of the total cells on bilateral iliac crest biopsies, the marrow was harvested. A MNC fraction of marrow was obtained using a ficoll-diatrozoate density gradient (Lymphocye Separation Medium, Organon Teknika, Durham, North Carolina) on the Cobe 2991 Marrow Processor, and purged as previously described[13]. Patients then received high-dose cyclo-phosphamide, cisplatin, and carmustine[1] with infusion of the purged marrow.

The end-point of this phase I trial was a significant prolongation in marrow reconstitution, defined by the number of days until a peripheral leukocyte count of 1000 cells per microliter was achieved. The first group of patients had their marrows purged with 4-HC alone, beginning with 20 ug/ml, followed by 40 ug/ml, 60 ug/ml and 80 ug/ml respectively. Once the maximally tolerated dose of 4-HC was reached, the second group had their marrows purged with the immunomagnetic technique alone. Currently accrual onto the third phase of the trial continues with combined immunomagnetic plus 4-HC marrow purging. The un-purged historical control population was a group of metastatic breast cancer
patients with no bone or bone marrow involvement. This group received the same high-dose chemotherapy regimen folllowed by autologous marrow infusion with an unpurged buffy-coat fraction of marrow[1].

RESULTS

Pre-Clinical Studies

In small scale studies two incubations with antibody and magnetospheres were required for optimal removal of 3-4 logs of clonogenic breast cancer cells. An additional two logs of breast cancer could be removed by sequentially purging the marrow/tumor cell suspensions by IMP followed by 4-HC (or the reverse sequence)[6]. The concentrations of monoclonal antibod-ies and magnetospheres determined in the small scale studies were found to be optimal in the upscale setting[12]. One incubation with the antibodies and magnetospheres respectively removed 3.2-3.6 logs of clonogenic tumor. Increasing the number of antibody and/or magnetosphere incubations did not result in greater clonogenic CAMA cell elimination. As in the small scale studies, the addition of 4-HC to the IMP technique increased the tumor cell depletion to five logs.

Clinical Study

There was no difference in time to engraftment at the first three 4-HC dose levels of 20, (19 days), 40 (21 days), and 60 (23 days) ug/ml respectively, compared to the unpurged historical control group (17 days). At 80 ug/ml the engraftment period of 28 days was significantly delayed (p=0.027).

212

Further escalation of 4-HC was not attempted. There was no difference in engraftment for the immunomagnetically purged group compared to the controls. The study continues with marrow purging using the combination of 4-HC plus IMP.

Fifty-three percent of evaluable patients have achieved a complete remission of their breast cancer.

DISCUSSION

Pre-clinical studies were performed to eliminate breast cancer cells from bone marrow. In small scale IMP purging experiments, two incubations with a panel of monoclonal antibodies and magnetic microspheres respectively, produced three logs of breast cancer cell elimination. The addition of the 4-HC increased the tumor cell elimination to five logs.

An IMP purging apparatus was then constructed which could handle the large volumes of bone marrow required for patients receiving high dose chemotherapy with ABMS. In the upscale studies, 3.0-4.0 logs of CAMA breast cancer cells were eliminated using one incubation of monoclonal antibodies and magnetospheres. The addition of the 4-HC again augmented tumor cell elimination to five logs. The major difference between the small scale and upscale experiments was that only one incubation with the antibodies and magnetospheres was required for optimal tumor cell elimination in the upscale setting.

With the pre-clinical data described above, a clinical phase I purging study was designed for patients with breast cancer metastatic to bone marrow. The maximally tolerated dose of 4-HC alone for a MNC fraction of marrow in patients with breast cancer is 80 ug/ml. IMP alone produces no delay in engraftment compared to unpurged historical controls. Evaluation of combined IMP plus 4-HC purging is progress.

The follow-up for patients on this trial is short, but to date the time to disease progression in this group is not different from our identically treated patients with advanced breast cancer that does not involve the bone marrow[14]. Longer follow up is needed to assess the ultimate therapeutic outcome in these patients.

REFERENCES

1. Peters WP, Shpall EJ, Jones RB, Olsen G, Bast RC Jr, Gockerman J, Moore J. High-dose combination alkylating agents with bone marrow support as initial treatment for metastatic breast cancer. J Clin Oncol 6:1368-1375, 1988.
2. Williams SF, Mick R, Dresser R, Golick J, Beschorner J, Bitran J. High-dose consolidation therapy with autologous stem cell rescue in stage IV breast cancer. J Clin Oncol 7:1824-1830, 1989.
3. Dunphy F, Spitzer G, Buzdar A, Hortobagyi G, Horowitz L, Yau J, Spinolo J, Jagannath S, Dicke K. High-dose therapy (HDT) with ABMT in metastatic breast cancer (BC): Clinical features of prolonged progression-free survival(PFS). Proc Amer Soc Clin Oncol 8:25, 1989.
4. Redding H, Monaghan P, Ormerod M, Gazet J, Coombes R, Clink H, Dearnaley D, Sloane J, Powles T, Neville, A. Detection of micrometastases in patients with primary breast cancer. The Lancet December 3:1271-1273, 1983.
5. Anderson I, Shpall EJ, Leslie D, Daly , Nustad K, Ugelstad J, Peters W, Bast RC. Elimination of malignant clonogenic breast cancer cells from human bone marrow. Cancer Research 49:4659-1989.
6. Uckun F, Kazimiera G, Meyers D Ramsay N, Kersey J, Colvin M, Vallera D. Marrow purging in autologous bone marrow transplantation for T-lineage ALL: Efficacy of ex vivo treatment with immunotoxins and 4-hydroperoxycyclophosphamide against fresh leukemic marrow progenitor cells. Blood, 69:361-366, 1987.

7. DeFabritiis P, Bregni M, Lipton J, Greenberger J, Nadler L, Rothstein L, Korbling M, Ritz J and Bast R. Elimination of clonogenic burkitt's lymphoma cells from human bone marrow using 4-HC in combination with monoclonal antibodies and complement. Blood 65:1064-1079, 1985.

8. Marchetti-Rossi M, Centis F, Talevi N, Manna A, Sparaventi G, Porcellini A. Decontaminating bone marrow with Merocyanine 540, Mafosfamide or both. In: ABMT, Proc. In: The Third Intntl Symposium on ABMT. eds Dicke K, Spitzer G, Jagannath S. The University of Texas M. D. Anderson Hospital and Tumor Institute at Houston:151-157, 1987.

9. Frankel A, Ring D, Tringale F, Hsieh-Ma S. Tissue distribution of breast cancer-associated antigens defined by monoclonal antibodies. Journal of Biological Response Modifiers 4:273-286, 1987.

10. Ugelstad J, Mfutakamba HR, Mork PC. Preparation and application of monodisperse polymer particles. J Polymer Science 72:225-240, 1985.

11. Shpall EJ, Bast RC, Joines WT, Jones RB, Anderson I, Johnston C, Eggleston S, Tepperberg M, Edwards S, Peters WP. Immunomagnetic Purging of breast cancer from bone marrow for autologous transplantation. Bone Marrow Transplantation, 1990 (in press).

12. Jones RB, Shpall, EJ, Shogan J, etal: The Duke AFM program: intensive induction chemotherapy for metastatic breast cancer. Cancer 66:431-436, 1990.

13. Shpall EJ, Jones RB, Bast RC, Rosner G, Vandermark M, Ross M, Affronti ML, Johnston C, Eggleston S, Tepperberg M, Coniglio D, Peters WP. 4-hydroperoxycyclophosphamide (4-HC) purging of breast cancer from the mononuclear cell fraction of bone marrow in patients receiving high-dose chemotherapy and autologous marrow support: A phase I trial. J Clin Oncol, 1990 (in press).

14. Shpall EJ, Bast RC, Joines W, Jones RB, Ross M, Johnston C, Eggleston S, Tepperberg M, Peters WP. Immunopharmacologic bone marrow purging in metastatic breast cancer patients receiving high-dose chemotherapy with autologous bone marrow support. Proc Amer Soc Clin Oncol 9:9, 1990.

IMAGING, PHARMACOKINETICS, DOSIMETRY AND ANTITUMOR EFFECTS OF

RADIOLABELED ANTI-BREAST CANCER ANTIBODIES IN MOUSE AND MAN

Paul A. Bunn, Jr.[1], David G. Dienhart[1], Rene Gonzalez[1],
Ravindra Kasliwal[2], Duane Bloedow[3], Claudia Hartmann[3], James
Lear[2], Timothy Johnson[2], Philip Furmanski[8], Gary J. Miller[4],
Stephan Glenn[5], Clifford Longley[7], Roberto Ceriani[6],
and Gregory Butchko[5]

From the Division of Medical Oncology, Dept. of Medicine[1],
Department of Radiology[2], School of Pharmacy[3], Department of
Pathology[4], University of Colorado Cancer Center, Denver, CO
Coulter Immunology, Hialeah, FL[5], John Muir Cancer and Aging
Research Institute, Walnut Creek, CA[6]
AMC Cancer Research Center and Hospital, Lakewood, CO[7]

INTRODUCTION

Breast cancer is the most common malignancy in females in the
United States and the third most common cause of cancer death in the
United States[1]. Despite the advances in molecular biology,
mammographic screening and therapy, the mortality rate from breast
cancer has not changed appreciably over the past 30 years[1]. There is
no known cure for metastatic breast cancer. In fact, survival time from
the diagnosis of metastatic disease in some recent trials is shorter
than survival time reported in series from a decade ago. This has been
attributed to the potential of increased drug resistance developing
during adjuvant therapy. New treatment approaches are sorely needed.
One new approach has been to use intensive doses of chemotherapy in
conjunction with autologous bone marrow support as described in this
monograph by Shpall[2,3]. Despite this treatment, the majority of
patients experience disease progression and death.

The hybridoma technology developed by Kohler and Milstein[4], has
led to the production of multiple murine monoclonal antibodies
recognizing antigens on the surface of cancer cells. Ceriani and others
showed that human breast cancer cells and human milk share a number of
antigens[5-9]. Many of these antigens are high molecular weight mucins
with complex carbohydrate side chains on a protein backbone. Several
laboratories have developed monoclonal antibodies recognizing various
epitopes of these complex glycoproteins[5-9]. The antigens are often
referred to as human milk fat globule antigens (HMFG) or human mammary
epithelial antigens (HME). The gene encoding the protein backbone has
been cloned by several laboratories. These antigens are often shed into
the circulation and there is interest in using monoclonal antibodies for
in vitro detection and monitoring[10]. We have chosen to study three of
these antibodies with differing characteristics, KC-4G3, Mc5 and BrE-3.

Breast Epithelial Antigens, Edited by R.L. Ceriani
Plenum Press, New York, 1991

Radiolabeled monoclonal antibodies offer the possibility of adding two new dimensions to breast cancer management: detection of sites of metastatic disease and antitumor therapy. For therapy, high dose radiolabeled antibodies could be used alone or in conjunction with other modalities including chemotherapy and autologous bone marrow transplantation. In this report we briefly summarize our preliminary studies with radiolabeled anti-breast cancer monoclonal antibodies in mouse and man.

METHODS AND MATERIALS

Antibodies and Radiolabeling

KC-4G3 is a murine monoclonal antibody of the IgG3 class developed and produced by Coulter Immunology. It reacts with a high molecular weight (M_r >400,000) mucin antigen present on >95% of human breast adenocarcinomas[7,11,12]. Mc5 is a murine monoclonal antibody of the IgG1 class developed by Ceriani and colleagues and reacts with a high molecular weight antigen found in human milk and on breast cancer cells and the binding is partially inhibited by KC-4G3[5,6]. BrE-3 is also a murine monoclonal antibody of the IgG1 class developed by Ceriani and colleagues[13] and produced for our mouse studies by Dr. Ceriani's laboratory and for human studies by Coulter Immunology. For studies involving KC-4G3 and BrE-3 with [111]Indium, antibody-DTPA conjugates were prepared by Coulter Immunology. Approximately 1 mg of the antibody was radiolabeled with 5mCi of [111]Indium, by mixing buffered [111]InCl$_3$ with the antibody chelate and incubating the mixture for 20 minutes at room temperature[12]. Mc5 and BrE-3 were iodinated with Na[131]I using the chloramine-T method[14] with 40 mg - 400 mg of the antibody being labeled with 5mCi of [131]I. The ^{131}I radiolabeled antibody was separated from free iodide by passage through a Sephadex G-25 column. The antigen reactive fraction of each radiolabeled antibody was determined using a linear extrapolation method at infinite antigen excess[15]. The antigen reactive fraction of iodinated Mc5 and BrE-3 approximated 80% and [111]In- KC-4G3 exceeded 60%.

Human Tumor Cell Line

ZR-75 a human breast adenocarcinoma cell line, was obtained from the ATCC (Rockville, MD). MX-1, a human breast adenocarcinoma cell line, was maintained in the laboratory of one of the authors (RLC). BALL-1 is a B-cell lymphoma derived cell line obtained from Dr. Ikuro Kimura, Cancer Institute Division of Pathology, Okayama University, Okayama, Japan that does not express the HME antigens to a measurable level.

Athymic nude mice were purchased from Life Sciences and housed in the Animal Resource Center at the University of Colorado Health Sciences Center or were raised and housed in the John Muir Colony. Mice were approximately 6 weeks of age when 10^7 cells were injected subcutaneously into the flank under sterile conditions.

Patients

Patients had histologically confirmed breast or lung cancers which bound the antibody investigated. All breast cancer patients had adenocarcinoma. The lung cancer patients studied comprised 13 adenocarcinomas, 6 squamous carcinomas, 3 adenosquamous carcinomas, 1 large cell undifferentiated carcinoma, and 1 bronchoalveolar carcinoma[12]. Prior to monoclonal antibody imaging clinical sites of disease were documented by physical examination, routine x-rays and scans. All patients had adequate renal, pulmonary, hepatic, cardiac and marrow function and gave informed consent on a human subject approved protocol. Most patients had received prior therapy but none within 4 weeks of study.

Antibody Administration

For human subjects, antibody infusions were administered over 1 to 5 hours depending on the dose (40-100 mg/hr). KC-4G3 antibody doses were 1 mg, 10 mg, 50 mg, 100 mg, 250 mg, and 500 mg. The antibody administered included 1 mg of KC-4G3 labeled with approximately 5mCi [111]Indium. Cold KC-4G3 (9-499 mg) was mixed with the labeled antibody prior to administration. For Mc5, 40 or 400 mg of antibody was labeled with 5mCi of [131]I. For mouse studies radiolabeled antibodies were injected into the tail veins.

Gamma Camera Imaging

Whole body and analogue images of regions of interest were obtained on a GE400 at gamma camera (General Electric, Milwaukee, WI) or Technicare Gemini 700 gamma camera (Johnson and Johnson, NJ). Data were stored in a DEC Scintigraphic data analyzer (Digital Electronics, Maynard, MN) or a Macintosh II computer[16]. There was no blood pool or organ subtraction.

Pharmacokinetics

Serial blood and plasma collections were obtained before infusion and at 1, 2, 4, 8, 12, 24, 48, 72 and 96 hours after infusion for determination of plasma ^{111}In radioactivity, ^{131}I radioactivity, KC-4G3 and Mc5 antibody levels. Radioactivity was measured by gamma counting of 1Aml aliquots which were corrected for isotope decay. Antibody levels were measured by radioimmunoassay. Plasma radioactivity and antibody data were fitted to a one or two compartmental model using the computer programs Rstrip and Minsq (MicroMath, Scientific Software, Salt Lake City, Utah). Rstrip provided initial estimates of slopes and intercepts for each exponential component. Statistical determination of the number of exponentials to describe the data[17], was based on the t-test ($p < 0.05$). Using these initial estimates the data sets were fitted with Minsq. The resulting fitted slopes and intercepts were used to calculate pharmacokinetic parameters describing radioactivity and antibody distribution and clearance according to standard methods[18].

Serum Antibody and Antigen Levels

Antibody levels were determined by competition assay using iodinated KC-4G3 (or Mc5 or BrE-3) tracer and cold KC-4G3 (or Mc5 or BrE-3) antibody[12]. Antigen coated beads were incubated with patient serum and ^{125}I-KC-4G3 (or Mc5 or BrE-3). The beads were washed, centrifuged and counted to determine the amount of bound antibody. Serum antigen levels were determined by a sandwich radioimmunoassay using antigen coated beads and serial distributions of patient sera. After overnight incubation ^{125}I antibody was added and the beads were washed, centrifuged and counted to quantitate serum free antigen.

Immunohistochemistry

Prior to study, immunoperoxidase staining was performed on formalin-fixed paraffin-embedded tissue using a modified avidin-biotin-complex procedure[12]. Tissues were scored by the percentage of positive cells where 0=negative, 1+=1-25%, 2+=26-50%, 3+=51-75%, and 4+=76-100%.

RESULTS

Radiolabeled Antibodies in athymic nude mice with human tumor xenografts

Preclinical studies were performed in athymic nude mice with each of the antibodies Mc5, BrE-3 and KC-4G3. Gamma camera images were optimal when obtained 48-72 hours after antibody administration due to clearance of blood pool activity, as shown in Figure 1[11]. A broad range of doses were administered from 10 µg to 1000 µg of antibody labeled with 30 to 500 µCi of radioisotope (^{111}In or ^{131}I). The gamma camera images and the biodistribution (tumor to normal organ ratios) were not substantially affected by the administered dose of antibody or radioisotope. The ^{111}In-KC-4G3 produced a greater percent injected dose per gram in the tumor than either ^{131}I-Mc5 or ^{131}I-BrE-3 (Table 1). The reasons for the decreased tumor localization with the iodinated antibodies are unknown but may reflect processing and dehalogenation by the tumor. The slow decline in the tumor radiolocalization for ^{111}In-KC-4G3 treated mice resulted in consistent and sometimes improved tumor/organ ratios over time. The persistence of radioactivity in tumor would lead to a much greater dose to tumor. Control animals bearing antigen negative tumor (BALL-1) had no specific tumor uptake but considerably more uptake in blood and normal organs than in tumor. There was also no tumor specific uptake of control antibodies (mouse IgG). Thus, subtraction of counts following co-administration of iodinated mouse IgG improved the tumor to background ratios. These data suggested the possibility that radiolabeled anti-mucin antibodies would have a useful role in tumor detection or therapy in humans.

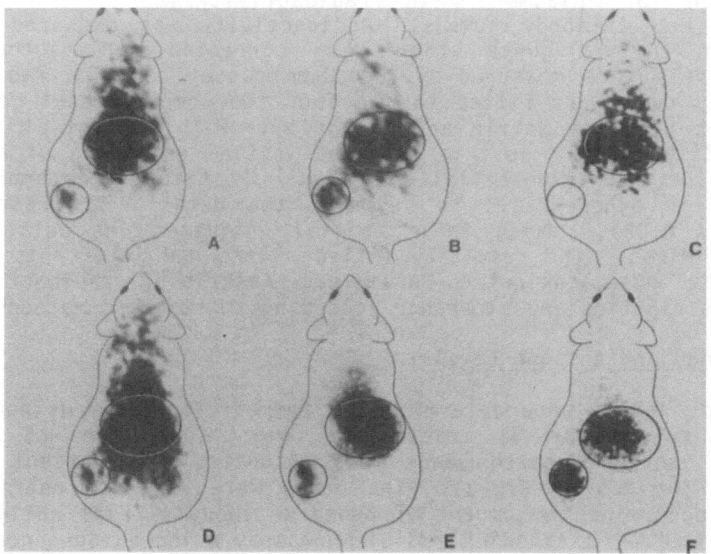

Figure 1. Gamma camera images of ^{111}In-KC-4G3-inoculated tumor-bearing nude mice. A,ZR-75 (breast adenocarcinoma), 10µg(37mCi), imaged at 24h post-inoculation; B,ZR-75, 10µg(37µCi), at 72h post-inoculation, C,BALL-1 (B-cell lymphoma), 100µg(90µCi), at 72h post-inoculation; D,ZR-75, 100µg(60µCi), at 24h post-inoculation; E,ZR-75, 100µg(60µCi), at 72h post-inoculation; F,ZR-75, 100µg(60µCi), at 8 days post-inoculation. Reprinted with permission, Ca Res. 50:5954,1990.

Table 1
Biodistribution of Radiolabeled MoAB in athymic nude mice with human breast adenocarcinoma ZR-75 xenografts

Isotope-Antibody	Time	% inj. dose/gm				
		Tumor	Blood	Liver	Thyroid	Lung
^{111}In-KC-4G3	72 Hrs	11.46	6.66	4.9	2.8	4.3
^{131}I-Mc5	72 Hrs	8	25	3.0	4.0	6.0
^{131}I-BrE-3	72 Hrs	4.8	5.1	1.5	1.75	3.0
^{111}In-KC-4G3	336 Hrs	12.3	1.1	3.8	2.0	1.1
^{131}I-Mc5	336 Hrs	1.0	1.2	0.1	0.1	0.3
^{131}I-BrE-3	336 Hrs	0.8	2.1	0.3	1.2	0.4

^{111}In-KC-4G3 in human breast and lung cancer patients

^{111}In-KC-4G3 was administered to 24 patients with advanced non-small cell lung cancer and 12 patients with advanced breast cancer. All patients had antigen positive tumors. Tumor uptake was seen in 22 of 24 (92%) lung cancer patients while lesions were imaged in 10 of 12 (83%) of breast cancer patients (Table 2)[12]. These numbers are less impressive when viewed as the percent of known clinical lesions successfully imaged. In lung cancer only 57% of the 91 clinical lesions were imaged. This was primarily due to poor visualization of small tumors. However, 81% of lesions greater than 3 cm in diameter were imaged. Infusions using the same antibody and isotope in similar doses given to breast cancer patients yielded a lower incidence of tumor visualization than that observed in lung cancer patients. Approximately 35% of known clinical lesions were detected in breast cancer patients. The reasons for these differences are not entirely clear. Serum antigen levels (Table 3) were similar. There was also no evidence that tumor antigen immunohistochemical expression (1+ to 4+) or serum antigen level affected imaging results. The most likely explanation is the site of lesions. In lung cancer patients pulmonary lesions were detected in 69% of instances(Figure 2)[12] and bone lesions were detected in only 32% of instances. Bone lesions were the most common site of metastases in breast cancer, and may partially account for the lower detection rate. The non-specific uptake in blood pool and marrow may obscure small lesions in these sites.

Table 2
Gamma Camera Imaging Detection of Tumor Sites

Isotype-Antibody	Cancer	% Pts imaged	%Lesions imaged				
			Total	Lung	Bone	# > 3cm	Primary
^{111}In-KC-4G3	Lung	92	57	69	32	81	69%
^{111}In-KC-4G3	Breast	83	35	40	25	68	100%
^{131}I-Mc5	Breast	12	5	0	10	NR	NR

Table 3
Pre-study Serum Antigen Levels

Antigen	Cancer	Median µg/ml (Range)
KC-4	Lung	0.11 (0 to 1.7)
KC-4	Breast	0.06 (0 to 1.6)
Mc5	Breast	42 (1.9 to >350)

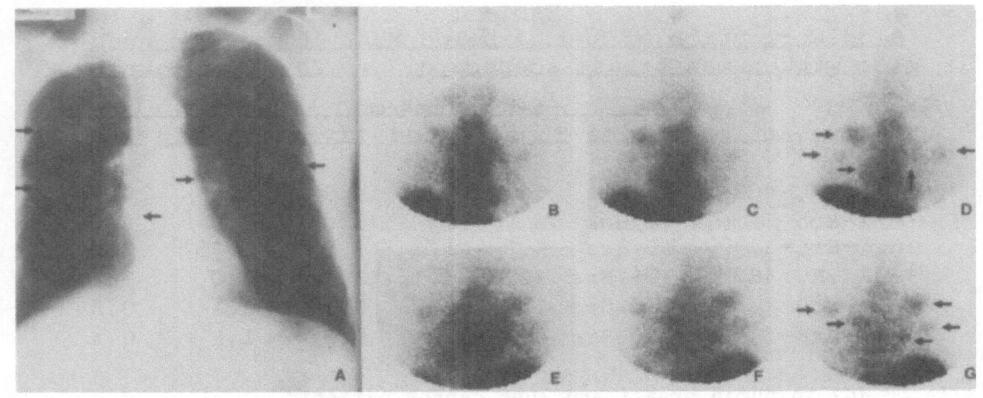

Figure 2. Gamma-camera chest images following 500 mg
^{111}In-KC-4G3. A, anterior chest X-ray; B, 24-h anterior
^{111}In-KC-4G3; C, 48-h anterior ^{111}In-KC-4G3; D, 72-h anterior
^{111}In-KC-4G3; E, 24-h posterior ^{111}In-KC-4G3; F, 48-h posterior
[111]-KC-4G3; G, 72-h posterior ^{111}In-KC-4G3. The arrows indicate the
sites of clinical lesions. Reprinted with permission, Cancer Research,
50:7068-7076, 1990.

The imaging results with ^{131}I-Mc5 were poor (Table 2). It is
unlikely that these results could be attributed to the use of [131]I
conjugates. The immunoreactivity of the ^{131}I-Mc5 was excellent
(average >80%) and the normal organs imaged well. The most probable
factor which may have adversely influenced lesion detection was the high
concentration of circulating Mc5 antigen. The median pretreatment level
was 42 µg/ml with a range of 1.9 to greater than 350 µg/ml (Table
3). In comparison, only 3 breast or lung cancer patients had a KC-4
antigen level exceeding 1 µg/ml and none had more than 2 µg/ml.
Thus, it may be likely that the high Mc5 antigen levels interfered with
the imaging results.

The pharmacokinetic results from these studies in mice and humans is
summarized in Table 4. In humans, the average volume of distribution for
the central compartment ranged from 2.8 to 6.5 L. The volume of
distribution and the plasma clearance of KC-4G3 appear to be smaller for
breast cancer patients compared to lung cancer patients. However, there

Table 4
Pharmacokinetic Data

Isotope Antibody	Dose	Cancer	Host	Volume Distrib	$t_{1/2\beta}$	Clearance ml/min
^{111}In-KC-4G3	1-500mg	Lung	Human	6.5(l)	29.0h	3.0
^{111}In-KC-4G3	1-500mg	Breast	Human	2.8(l)	26.5h	1.3
[131]I-Mc5	40-400mg	Breast	Human	5.7(l)	38.4h	2.2
^{111}In-KC-4G3	10-1000µg	Breast	Mouse	5.1(ml)	100h	4.6

Table 5
Relation of Dose to Imaging and Pharmacokinetics

Dose (mg)	% Imaged	Volume Distribution	t_{1/2β}hrs	Clearance
1	64	4.2	25.3	2.2
10	61	9.1	20.6	5.5
50	43	6.0	39.0	2.3
100	43	5.2	21.6	2.9
500	73	7.0	37.0	2.2

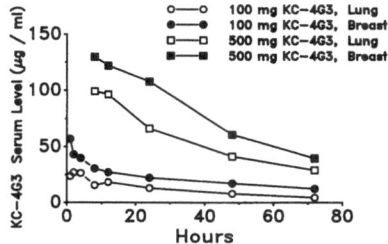

Figure 3. Serum disappearance of antibody KC-4G3 as determined by competition radioimmunoassay. Each point represents the mean value per dose.

are no statistically significant differences between the studies, probably related to high variability among patients. Examples of KC-4G3 antibody disappearance and achievable serum concentrations are illustrated in Figure 3.

For lung cancer patients, there was no obvious relationship between dose of administered antibody and the imaging results or the pharmacokinetics (Table 5)[12]. The fraction of lesions which successfully imaged, the volume of distribution, the t 1/2, and the clearance rates were similar at each dose.

In more recent studies we have collected serial region of interest data from gamma camera images of tumor and normal organs. An example of the results is shown in Table 6. These data were derived from the ^{131}I-Mc5 study and one of the tumor lesions which was successfully imaged. The calculated dose to the tumor following administration of a 100 mCi antibody dose was 739 cGy. This was at least 1.5 times higher than the dose to any normal organ. The liver, spleen and kidney received the highest doses to normal organs. The whole body and bone marrow were estimated to receive 66 and 57 cGy per 100 mCi, respectively. It is likely that the dose limiting toxicity would occur in the marrow (myelosuppression). These data would suggest that several thousand cGy could be delivered to the tumor at a dose to marrow which might not produce myelosuppression.

Table 6
^{131}I-Mc5 Dosimetry Calculations

Organ	Dose (cGy) 5 mCi	Dose (cGy) 100 mCi
Tumor	40	739
Marrow	2.9	57
Liver	22	426
Kidney	26	502
Lungs	2	40
GI	2.25	45
Thyroid	1.5	30
Spleen	62	1240
Total Body	3.3	66

BrE-3 is a potentially useful antibody because it reacts strongly with breast cancer tissues and has less normal tissue reactivity. In athymic nude mice it provides biodistribution similar to that of Mc5 when labeled with ^{131}I. In nude mouse therapy studies, it produced results equivalent to Mc5 when given alone. Combination studies show additive effects with Mc5. In humans, the serum BrE-3 antigen levels are considerably lower than Mc5 levels. Thus, it is logical to study BrE-3 in human breast cancer patients. These trials were recently instituted at the University of Colorado, and plans are to evaluate both an ^{111}In-BrE-3 and a ^{131}I-BrE-3 conjugate. In this way we can directly evaluate the effects of the radioisotope and chelate.

DISCUSSION

Monoclonal antibodies were heralded as the panacea to produce a "magic bullet" to deliver toxic substances to human tumors. Human breast cancers exhibit aberrant expression of mucin antigens[5-10]. Many of these antigens are shed into the circulation and may be useful for in vitro serum monitoring[10]. A number of monoclonal antibodies were developed which reacted with the antigens described as human milk fat globule (HMFG) or human mammary epithelial (HME) antigens. These antibodies were successfully radiolabeled and in vivo studies in athymic nude suggested that radiolabeled antibodies would be clinically useful in humans[11].

Our own studies in nude mice bearing ZR-75 or MX-1 human breast cancer xenografts demonstrated excellent tumor localization and cytotoxic effects when high doses of ^{131}I were delivered on Mc5 or BrE-3. Our preclinical studies in mice suggest that a antibody chelate with ^{111}In for imaging or Yttrium-90 (^{90}Y) for therapy might be preferable to ^{131}I labelled antibody.

Clinical trials in humans provided some unexpected results. While tumor uptake in large lesions was observed routinely, small lesions were poorly visualized. This was due, in part, to nonspecific uptake in blood pool, marrow, liver and spleen obscuring small lesions in liver and bone. The ^{131}I-Mc5 was particularly unsuccessful, most likely due to the binding of intravenously administered antibody to circulating free antigen and potential reticuloendothelial clearance of immune complexes. Despite the fact that small lesions were obscured, dosimetry studies suggested that uptake in normal organs would not lead to severe toxicity to any normal organ. These antibodies bind to the luminal aspects of cells in some normal glands. These antigens are at a distance from the capillaries and no excess accumulation was observed in normal organs. There are a number of potential ways to circumvent the problems found with _in vivo_ administration of murine monoclonal antibodies in humans. For example, circulating antigen can be removed by a plasma immunoadsorption anti-antibody column. For the administration of one or two antibody doses, the development of human anti-mouse antibody is not a significant factor. When multiple antibody doses are contemplated, chimeric antibodies could be used[19]. To overcome the problem of antigen heterogeneity, a combination of antibodies can be used and/or the use of radioisotopes which penetrate through several cell layers. Interferons may be used to increase the expression of HLA and tumor associated antigens and to increase killing by the normal immune system[20]. To increase the delivery of antibody to tumor, fragments can be used or pharmacologic methods may be utilized to increase vascular permeability. To increase the cytotoxic effects of the antibodies, investigators have studied antibodies conjugated with therapeutic radioisotopes (e.g.^{90}Y), with cytotoxic drugs, or in combination with immunologic agents such as IL-2.

The studies performed to date suggest that myelosuppression will be the dose limiting toxicity from high dose radiolabeled antibody therapy. Of course, one way to circumvent this toxicity is the use of autologous bone marrow transplantation. High dose chemotherapy with autologous bone marrow support produces a high rate of complete remission in patients with metastatic breast cancer[2,3]. Nonetheless, most of the patients relapse. Radiolabeled antibody therapy in combination with chemotherapy might increase the cell kill.

Another approach is to remove the long lived circulating antibody. As shown in Table 4, the biological half life of each antibody exceeds 24 hours. This circulating radiolabeled antibody adds considerable radiation dose to the marrow and whole body. Removal of the antibody may potentially decrease the dose to marrow and whole body without significantly affecting the dose to tumor. This approach could also improve the image results by lowering the blood pool background activity. An immunoadsorption system was developed by Cobe Laboratories[21,22]. A goat anti-mouse antibody column is connected in circuit to a plasma separator and the processed plasma is returned to the patient, thus creating an extracorporeal immunoadsorption device. We have initiated studies using this device in patients given infusions of 0.5 to 5.0 mg ^{111}In-KC-4G3. The data to date show that about 70% of the circulating radiolabeled antibody can be removed in 3-4 hours by immunoadsorption of 2-3 plasma volumes. The reduction in radiation dose to the marrow and whole body is dependent on the time at which the immunoadsorption procedure is initiated. When started 10 hours after antibody infusion the dose to marrow and whole body can be reduced by one-third. Further studies on the use of immunoadsorption and radiolocalization and therapy are in progress.

In summary, our studies suggest that radiolabeled anti-HMFG antibodies can be delivered safely to patients with metastatic breast cancer. To date, their effectiveness in tumor imaging is limited by non-specific distribution in blood pool and the reticuloendothelial system. The dose estimates to tumors when compared to normal organs suggests these antibodies could produce anti-tumor effects, especially if efforts to reduce or overcome myelosuppression are employed. We plan to continue our studies with alternative antibodies (BrE-3) and with ^{90}Y conjugated to KC-4G3. We will also consider the use of autologous bone marrow support or immunoadsorption removal of circulating antibody if preliminary studies show anti-tumor effects are limited by myelosuppression.

REFERENCES

1. Silverberg, E., Boring, C.C., and Squires, T.S., Cancer statistics, CA: A Cancer Journal for Clinicians, 40:9-26 (1990).
2. Shpall, E.J., Immunomagnetic purging of breast cancer from bone marrow of patients receiving high dose chemotherapy and autologous marrow support, Proceedings of this Workshop.
3. Jones, R.B., Shpall, E.J., Shogan, J., Affronti, M.L., Coniglio, D., Hart, I., Halperin, E., Iglehart, J.D., Moore, J., Gockerman, J., Bast, R.C., and Peters, W.P., The Duke AFM Program: Intensive induction chemotherapy for metastatic breast cancer, Cancer 66:431-436, (1990).
4. Kohler, G., and Milstein, C., Continuous cultures of fused cells secreting antibody of predefined specificity, Nature, 256:495-497 (1975).
5. Ceriani, R.L., Thompson, K.E., Peterson, J.A., Abraham, S., Surface differentiation antigens of human mammary epithelial cells carried on the human milk fat globule, Proc. Natl. Acad. Sci. (USA), 74:582-586 (1977).
6. Arklie, J., Taylor-Papadimitriou, J., Bodmer, W.F., Eagan, M., and Mills, R., Differentiation antigens expressed by epithelial cells in the lactating breast are also detectable in breast cancer, Int. J. Cancer, 28:23-29 (1981).
7. Hofheinz, D., Dienhart, D., Miller, G., Healy, S., Furmanski, P., Sedlacek, S., Longley, C., Bunn, P.A., Jr., and Kortright, K.,Monoclonal antibody, KC4G3, recognizes a novel, widely expressed antigen on human epithelial cancers, Proc. Am. Assoc. Cancer Res., 28:1552 (1987).
8. Ceriani, R.L., Peterson, J.A., Lee, J.Y., Moncada, F.R., and Blank, E.W., Characterization of cell surface antigens of human mammary epithelial cells with monoclonal antibodies prepared against human milk fat globule, Som. Cell Genet., 9:415-527 (1983).
9. Taylor-Papadimitriou, J., Peterson, J.A., Arklie, J., Burchell, J., Ceriani, R.L., and Bodmer, W.F., Monoclonal antibodies to epithelium-specific components of human milk fat globule membrane production and reaction with cells in culture, Int. J. Cancer, 28:17-24 (1981).
10. Ceriani, R.L., Rosenbaum, E.H., Chandler, M., Trijillo, T.T., Myers, B., and Sakada, M., Role of circulating human mammary epithelial antigens (HME-Ags) as serum markers for breast cancer. In "Tumor Markers and their Significance in the Management of Breast Cancer." IP, C., ed., A.R. Liss, ed., p 3-19 (1986).
11. Longley, C., Furmanski, P., Dienhart, D.G., Lear, J., Bloedow, D., Kasliwal, R., Bunn, P.A., Jr., Pharmacokinetics, biodistribution, and gamma camera imaging of ^{111}In-KC-4G3 murine monoclonal natibody in athymic nude mice with or without human tumor xenografts. Cancer Res., 50:5954-5961, (1990).
12. Dienhart, D.G., Schmelter, R.F., Lear, J.L., Miller, G.J., Glenn, S.D., Bloedow, D.C., Kasliwal, R., Moran, P., Seligman, P., Murphy, J.R., Kortright, K., and Bunn, P.A., Jr., Imaging of non-small cell lung cancers with a monoclonal antibody, KC-4G3, which recognizes a human milk fat globule antigen, Cancer Res., 50:7068-7076, 1990.
3. Peterson, J.A., Zava, D.T., Duwe, A.K., Blank, E.W., Battifora, H., and Ceriani, R.L., Biochemical and histological characterization of antigens preferentially expressed on the surface and cytoplasm of breast carcinoma cells identified by monoclonal antibodies against the human milk fat globule, Hybridoma, 9(3):221-235, (1990).

14. Greenwood, F.C., Hunter, W.M., and Glover, J.S., The preparation of 131-I-labeled human growth hormone of high specific activity, Biochem. J. 89:114-119 (1963).
15. Lindmo, T., Boven, E., Cuttitta, F., Fedorko, J. and Bunn, P.A., Jr., Determination of the immunoreactive fraction of radiolabeled monoclonal antibodies by linear ext Mary Ann Lieber, Inc., publisherrapolation to binding at infinite antigen excess, J. Immunol. Methods., 72:77-86 (1984).
16. Lear, J.L., Pratt, J.P., Roberts, D.R., Johnson, T., and Feyerabend, A., Gamma camera image acquisition, display, and processing with the personal microcomputer, Radiology, 175:241-245, (1990).
17. Boxenbaum, H., Riegelman, S., and Elashoff, R.M., Statistical estimation in pharmacokinetics, J. Pharmocokinet. Biopharm., 2:123-148 (1974).
18. Gibaldi, M., and Perrier, D., Pharmacokinetics, 2nd Ed. Marcel Dekker, Inc., New York, (1982).
19. LoBuglio, A.F., Wheeler, R.H., Trang, J., Haynes, A., Rogers, K., Harvey, E.B., Sun, L., Ghrayeb, J., and Khazaeli, M.B., Mouse/human chimeric monoclonal antibody in man: Kinetics and immune response, Proc. Natl. Acad. Sci. USA, 86:4220-4224, (1989).
20. Schlom, J., Hand, P.H., Greiner, J.W., Colcher, D., Shrivastav, S., Carrasquillo, J.A., Reynolds, J.C., Larson, S.M., and Raubitschek, A., Innovations that influence the pharmacology of monoclonal antibody guided tumor targeting, Cancer Res., 50:820-827, (1990).
21. Maddock, S., Bunn, P.A., Dienhart, D.G., Feyerabend, A., Glenn, S., Johnson, T., Kasliwal, R., Lear, J., Maddock, E., and McAteer, M., Strategy and use of immunoadsorption to improve RAIT. Presented at Third Conference on Radioimmunodetection and Radioimmunotherapy of Cancer, November 15-17, 1990, Antibody Immunoconjugates, and Radiopharmaceuticals, 4(1):34, 1991.
22. Dienhart, D.G., Kasliwal, R., Lear, J.L., Johnson, T.K., Bloedow, D.C., Hartmann, C., Seligman, P.A., Miller, G.J., Glenn, S.D., McAteer, M.J., Maddock, E.N., Maddock, S., and Bunn, P.A., Jr., Extracorporeal immunoadsorption of radiolabeled monoclonal antibody: A method for reduction of background radioactivity and influence on imaging, pharmacokinetic and dosimetric parameters. Manuscript in preparation.

Current address: Philip Furmanski, Ph.D., Department of Biology, New York University, New York, New York 10003.

Clifford Longley, M.Sc., Cytogen, 6000 College Road East, Princeton, NJ 08540.

RADIOIMMUNOTHERAPY WITH I-131 Chimeric L-6 IN ADVANCED BREAST CANCER

S.J. DeNardo[1], K.A. Warhoe[1], L.F. O'Grady[1], G. L. DeNardo[1], I. Hellstrom[2], K.E. Hellstrom[2] and S.L. Mills[1]

[1]Department of Internal Medicine, University of California, Davis Medical Center, Sacramento, CA, [2]ONCOGEN, Seattle, WA. Supported By National Cancer Institute CA47829 and Department of Energy DE FG03-84ER60233. Mouse and Chimeric L-6 were provided by ONCOGEN

SUMMARY ABSTRACT

Quantitative imaging studies with I-131 L-6 and I-131 Ch L-6 demonstrated that a target is present in the pulmonary vasculature readily available to the intravascular delivery of radiolabeled MoAb. An infusion of 200 mg of unlabeled L-6 covered this target so that subsequent injections of I-131 L-6 or Ch L-6 reached tumor.

Four patients have been treated with I-131 Ch L-6 at either 20, 50, 60 or 70 mCi/m^2, given after the infusion of 200 mg L-6. The 3 patients receiving 2 or more doses had greater than 70% reduction in tumor volume lasting 0.7 to 5.0 months. Transient marrow toxicity occurred with multiple doses. HAMA inhibited further therapy in 3 patients. These results suggest the potential for radioimmunotherapy to become clinically useful in breast cancer.

INTRODUCTION

Over the last 50 years, despite improved diagnostic technology, earlier detection of the disease, new chemotherapy agents and more aggressive regimens with autologous bone marrow rescue[1,2,3], there has been no change in survival rate of patients who develop metastatic breast cancer[4]. Although attempts at treating cancer patients with monoclonal antibodies (MoAb) have met with variable success, in breast cancer no benefit has been demonstrated[5-10]. We approached the treatment of breast cancer using I-131 Ch L-6 monoclonal antibody with the recognition that there was a need to administer unconjugated L-6 in amounts sufficient to cover nontumor targets and to activate biologic systems at the tumor so that subsequently administered I-131 Ch L-6 would achieve maximum targeting of the breast cancer. The treatment dose of I-131 Ch L-6 was preceded by an imaging study with the same radiopharmaceutical in order to assure tumor targeting and absence of normal tissue targeting. Each patient received an imaging and treatment dose of I-131 Ch L-6 following preload L-6. Using treatment doses of I-131 Ch L-6 given at 4 week intervals, responses were observed in 3 of the 4 patients with advanced breast cancer and were associated with acceptable toxicity. We believe that these promising results were made possible because of the capability of the pretreatment infusion of unconjugated L-6 and Ch L-6 to induce an inflammatory response thereby enhancing delivery and efficacy of the subsequently administered I-131 Ch L-6.

MATERIALS AND METHODS

The MoAb L-6, an IgG2a mouse antibody, targets a membrane bound antigen found on human adenocarcinoma cells of the lung, colon, ovary and breast[11]. It possesses tumoricidal activity manifested by antibody dependent cellular cytotoxicity (ADCC) in the presence of human peripheral blood mononuclear cells and complement dependent cytotoxicity (CDC) in the presence of human complement[12,13]. A chimeric human-mouse antibody was produced in which mouse constant domains C-G2a and C-kappa were replaced by the human C-G1 and C-kappa[14]. Chimeric and murine L-6 antibodies bind adenocarcinoma cells with the same avidity, but Ch L-6 is 50 to 100 times more effective at mediating ADCC[14].

The radiopharmaceutical was prepared using chloramine-T radioiodination with I-131 (ICN)[15,16] and the final products were sterile and pyrogen-free. HPLC TSK 3000 chromatography and cellulose acetate electrophoresis demonstrated that greater than 95% of the radioactivity was associated with the antibody. Immunoreactivity of each preparation had greater than 70% direct binding to a live human breast tumor cell line in vitro (HBT 3477)[17]. The radiopharmaceutical contained 10 mCi of I-131 per mg of antibody, 1 mCi of I-131 per ml, and human serum albumin 4% weight to volume.

Four patients with advanced metastatic breast cancer, L-6 positive by immunopathology, who had failed aggressive standard therapy, and demonstrated rapidly progressing disease were treated with I-131 Ch L-6. These patients were selected to assess the toxicity of the treatment approach and its potential for enhanced tumor uptake of the therapeutic radiopharmaceutical with little expectation of therapeutic responses in these patients with advanced disease.

Previous pharmacokinetic studies of murine L-6 in breast cancer patients revealed the need to give a 200 mg infusion of unconjugated antibody before the I-131 labeled antibody to achieve maximum tumor and minimal lung uptake[18]. All imaging and therapy doses were therefore given after slow infusion of 200 mg L-6 and 20 mg Ch L-6. In 3 of the 4 patients, an imaging study was performed the day prior to the therapeutic dose as final assessment for therapy. In the fourth patient the imaging study was performed one week prior to therapy. Therapy doses were given as close to monthly intervals as tolerated by the patient. Dose levels of 20, 50, 60 and 70 mCi/m^2 were used as the initial therapy dose and adjusted downward if necessitated by grade 3 or 4 hematologic toxicity. After each therapeutic dose, quantitative gamma camera imaging was performed for at least one week to obtain pharmacokinetics for dosimetric analysis (Fig. 1). Serum complement (C3) levels were obtained prior to and at the end of infusion of the unlabeled L-6/Ch L-6 and 24 hrs later.

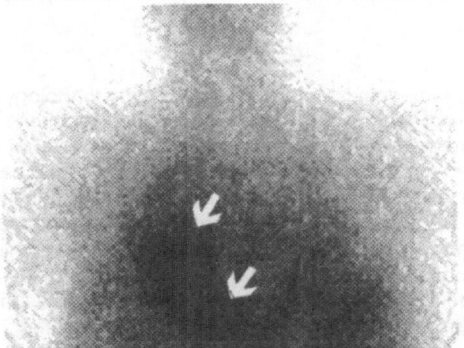

Figure 1. Radioimmunoimage of the anterior chest wall of a patient with metastatic breast cancer obtained 48 hrs. after therapy with I-131 Ch L-6. Concentration of the radioactivity in areas of metastatic tumor is demonstrated.

Chest wall lesions and all palpable lymph nodes were measured by calipers prior to the first treatment, at frequent intervals thereafter, and immediately prior to each subsequent treatment. Serial photographs were also taken of superficial disease on the chest wall. Computed tomography (CT) or Magnetic Resonance (MRI) images obtained at appropriate intervals were used to follow tumor size in nonpalpable areas (Fig. 2). Total tumor volume was calculated for data obtained pretherapy and posttherapy.

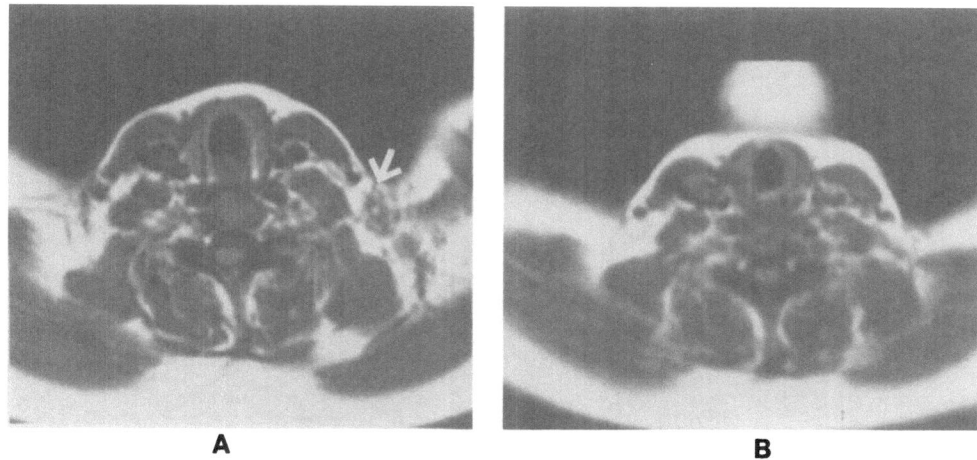

A   B

Figure 2. Regression of tumor in cervical nodes in response to I-131 Ch L-6 therapy demonstrated by Magnetic Resonance Imaging (MRI) before therapy (A), and one month after second therapy dose (B).

RESULTS

Three of the 4 patients were able to receive at least 2 and as many as 4 treatment doses. All 3 of these patients had measurable tumor regression calculated to be greater than 70% of tumor volume (Table 1). Tumor dose was calculated to range from 20 to 70 rads per administered mCi of I-131. Three of the 4 patients developed HAMA limiting further therapy. All 4 patients developed transient grade 3 or 4 hematologic toxicity during the course of therapy.

TABLE 1

BREAST CANCER THERAPY RESPONSE/TOXICITY				
PATIENT	DOSE #	HEMATOTOXICITY	RESPONSE	DURATION
1	1	0	PR	0.7 mos
	2	4		
2	1	0		
	2	3	PR	
	3	4		1.5 mos
3	1	0	PR	
	2	0		
	3	0		5 mos
	4	3		
4	1	3	NR (ascites)	0.7 mos

The patients noted pain and swelling at the sites of superficial disease, usually starting during the infusion of unlabeled antibody. Complement levels were also observed to decrease with antibody administration; C3 levels decreased from a mean of 145 mg/100 ml prior to infusion of unconjugated L-6 dose to 110 mg/100 ml at the end of the infusion, and 95 mg/100 ml at 24 hrs. postinfusion (Fig. 3).

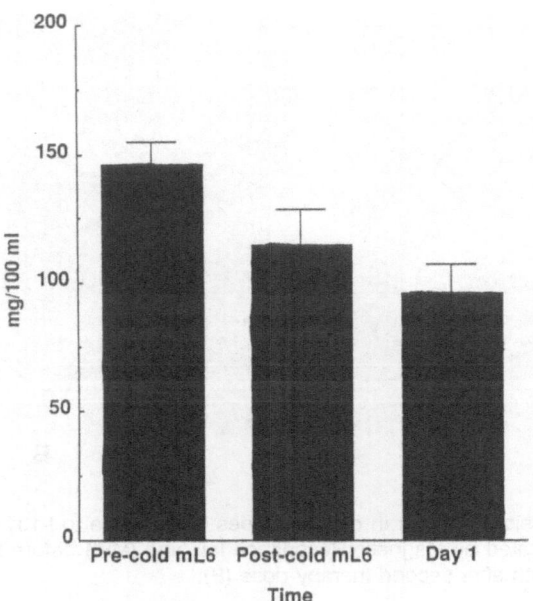

Figure 3. Mean complement (C3) levels immediately before, after and 24 hrs. after infusion of 200 mg L-6 and 20 mg Ch L-6 in 4 patients on 6 treatment occasions. The rapid complement consumption during the 4 hour infusion as well as evidence of continued activation during the ensuing 24 hours suggests significant in vivo biologic activity of this antibody. This activity seemed to enhance delivery of I-131 Ch L-6 to tumor.

DISCUSSION

These are the first instances of clinical responses to systemic radiolabeled MoAb seen in patients with breast cancer. Measurable tumor regression was observed in 3 of the 4 patients. The longest duration of response was 5 months and the patient could not be treated further because she developed a high HAMA titer. Her response included not only a reduction in tumor volume, but a significant reduction in bone pain allowing her to resume a normal life after being restricted to a wheelchair because of pain.

The mechanisms by which the tumoricidal effect was achieved may relate both to the biologic activity of the chimeric antibody and to the radiation received by the tumor. The preload of unconjugated L-6 followed by Ch L-6 was given to cover endothelial targets demonstrated by prior studies[18], as well as to utilize potential vasodilatory and capillary leak effects of this biologically active antibody to enhance delivery of the subsequently administered I-131 Ch L-6. The drop of serum complement (C3) levels during and immediately after the unlabeled antibody infusion indicates that complement was consumed by biologic activities initiated by this antibody. Inflammation at superficial tumor sites after each dose suggest antibody activation of immune effector cells and release of other cytokines. Although the evidence suggests that immunologic activity is occurring, it is unlikely that any major part of the therapeutic response seen in these patients was due to this activity, since studies using large amounts of unlabeled L-6 or Ch L-6 in patients with breast cancer have demonstrated little or no responses[19]. The radiation doses to tumor were calculated to be 20 to 70 rads per administered mCi I-131 and are considered to be the primary mechanism for these responses.

The hematologic toxicity observed at these doses of radiopharmaceutical were transient (Table 1). The radiation effect on the red blood cell line was minimal but the effect on platelet production was more obvious, as was expected, due to the short lifespan of the platelet and platelet effects previously observed in patients receiving radioimmunotherapy for lymphoma[20]. The patient receiving 70 mCi/m^2 had a grade 4 hematologic toxicity 12 days after the second therapy dose but 1 week later had recovered to grade 3 and at 5 weeks had recovered to well above normal. The fall and recovery of total white blood cells paralleled that of the granulocyte fraction. From this data, it appears that the nadir may vary in severity but is fairly constant in time of occurrence after administration of this radiopharmaceutical. Recovery time and level appeared to be more variable, probably relating to the patients' marrow reserve after previous radiation and chemotherapy.

We conclude that responses were achieved in 3 of 4 patients with advanced metastatic breast cancer after treatment with radiolabeled Ch L-6 MoAb. These patients were primarily selected to determine toxicity associated with this therapeutic approach therefore the clinical responses are more remarkable. Manageable levels of toxicity occurred in these patients. Unfortunately, due to initiation of treatment late in the disease course and onset of HAMA titers in most patients that limited further therapy, the responses were transient. Further therapeutic efforts will target metastatic breast cancer patients with less extensive disease using solely chimeric antibodies as well as support from marrow growth factors. In that setting we may better determine whether clinically useful radioimmunotherapy can be delivered to breast cancer.

REFERENCES

1. I. C. Henderson, J. R. Harris, D. W. Kinne, and S. Hellman. Cancer of the breast. In: V. T. Devita, S. Hellman, S. A. Rosenberg, eds. Cancer Principles and Practice of Oncology. Philadelphia: JB Lippincott, 1197-1261 (1989).
2. E. Silverberg, C. C. Boring, and T. S. Squires. Cancer statistics. CA 40:9-26 (1990).
3. I. C. Henderson, D. F. Hayes, and R. Gelman. Dose-response in the treatment of breast cancer: A critical review. J Clin Onc 6:1501-1515 (1988).
4. M. S. Fox. On the diagnosis and treatment of breast cancer. JAMA 241:489-494 (1979).
5. J. J. Tjandra and I. F. C. McKenzie. Murine monoclonal antibodies in breast cancer: an overview. Br J Surg 75:1067-1077 (1988).
6. A. Thor, M. O. Weeks, and J. Schlom. Monoclonal antibodies and breast cancer. Seminars in Oncology 13:393-401 (1986).
7. L. M. Weiner, J. O'Dwyer, J. Kitson, R. L. Comis, A. E. Frankel, R. J. Bauer, M. S. Konrad, and E. S. Groves. Phase I evaluation of an anti-breast carcinoma monoclonal antibody 260F9-recombinant ricin A chain immunoconjugate. Cancer Res 49:4062-4067 (1989).
8. J. Malamitsi, D. Skarlos, S. Fotiou, P. Papakostas, G. Aravantinos, D. Vassilarou, J. Taylor-Papadimitriou, K. Koutoulidis, G. Hooker, and D. Snook. Intracavitary use of two radiolabeled tumor-associated monoclonal antibodies. JNM 29:1910-1915 (1988).
9. B. J. Gould, M. J. Borowitz, E. S. Groves, P. W. Carter, D. Anthony, L. M. Weiner, and A. E. Frankel. Phase I study of antibreast immunotoxin by continuous infusion. J of the NCI 81:775-781 (1989).
10. B. P. Avner, S. K. Liao, B. Avner, K. DeCall, and R. K. Oldham. Therapeutic murine monoclonal antibodies developed for individual cancer patients. J of Biological Resp Mod 8:25-36 (1989).
11. I. Hellstrom, D. Horn, P. Linsley, J. P. Brown, V. Brankovan, and K. E. Hellstrom. Monoclonal antibodies raised against human lung carcinomas. Cancer Res 46:3917-23 (1986).
12. I. Hellstrom, P. L. Beaumier, and K. E. Hellstrom. Antitumor effects of L-6, an IgG2a antibody that reacts with most human carcinomas. Proc Natl Acad Science USA 83:7059-7063 (1986).
13. G. P. Adams, S. J. DeNardo, A. Amin, L. A. Kroger, G. L. DeNardo, I. Hellstrom, K. E. Hellstrom. Comparison of the pharmacokinetics in mice and the biological activity of murine L-6 and human-mouse chimeric Ch L-6 antibody. Antibody, Immunoconjugates, and Radiopharmaceuticals (In Press) (1991).

14. A. Y. Liu, R. R. Robinson, K. E. Hellstrom, E. D. Murray, Jr., C. P. Chang, and I. Hellstrom. Chimeric mouse-human IgG1 antibody that can mediate lysis of cancer cells. Proc Natl Acad Science USA 84:3439-3443 (1987).

15. W. M. Hunter and F. C. Greenwood. Preparation of iodine-131 labelled growth hormone of high specific activity. Nature 194:495 (1962).

16. S. L. Mills, S. J. DeNardo, G. L. DeNardo, A. L. Epstein, J.-S.Peng, and D. Colcher. I-123 radiolabelling of monoclonal antibodies for in vivo procedures. Hybridoma 5:265-275 (1986).

17. P. L. Beaumier, D. Neuzil, H.-M. Yang, E. A. Noll, R. Kishore, J. F. Eary, K. A. Krohn, W. B. Nelp, K. E. Hellstrom, and I. Hellstrom. Immunoreactivity assay for labeled anti-melanoma monoclonal antibodies. JNM 27:824-828 (1986).

18. S. J. DeNardo, L. F. O'Grady, D. J. Macey, L. A. Kroger, G. L. DeNardo, K. R. Lamborn, N. B. Levy, S. L. Mills, I. Hellstrom, and K. E. Hellstrom. Quantitative Imaging of Mouse L-6 Monoclonal Antibody in Breast Cancer Patients to Develop a Therapeutic Strategy. Nucl. Med. Biol. (In Press) (1991).

19. G. E. Goodman, I. Hellstrom, C. Nicaise, L. Brodzinsky, D. Hummel, and K. E. Hellstrom. Phase I trial of murine monoclonal antibody L-6 in breast, colon, ovarian,and lung cancer. J Clin Oncol 50:2449-2454 (1990).

20. S. J. DeNardo, G. L. DeNardo, L. F. O'Grady, N. B. Levy, S. L. Mills, D. J. Macey, J. P. McGahan, C. H. Miller, and A. L. Epstein. Pilot studies of radioimmunotherapy of B cell lymphoma and leukemia using I-131 Lym-1 monoclonal antibody. Antibodies, Immunoconjugates and Radiopharmaceuticals 1:17-33 (1988).

CONTRIBUTORS

D.C. Allred, Department of Pathology, University of Texas Health Science Center, San Antonio, TX 78284

R. Amiya, John Muir Cancer and Aging Research Institute, 2055 North Broadway, Walnut Creek, CA 94596

A.M. Ballesta, Laboratory of Clinical Biochemistry, Medical School, Hospital Clinico, Barcelona, Spain

R.C. Bast Jr., Division of Hematology/Oncology, Duke University Medical Center, Durham, North Carolina 27710

C.A. Beck, University of Colorado Health Sciences Center, Department of Pathology (B216), 4200 E. 9th Avenue, Denver, CO 80262

R. Beer, Genetic Systems, Seattle, Washington

A. Bistrain, John Muir Cancer and Aging Research Institute, 2055 North Broadway, Walnut Creek, CA 94596

A.C. Black, Abbott Laboratories, Diagnostic Division, North Chicago, Illinois 60064

E.W. Blank, John Muir Cancer and Aging Research Institute, 2055 North Broadway, Walnut Creek, CA 94596

D. Bloedow, Division of Medical Oncology, Department of Pharmacy, University of Colorado Cancer Center, Denver, CO

K.K. Borden, Abbott Laboratories, Diagnostic Division, North Chicago, Illinois 60064

K.R. Bray, Hybritech Incorporated, San Diego, California and Liege, Belgium

J.P. Brown, Genetic Systems, Seattle, Washington

A. Bui, Cell and Molecular Biology Division, Lawrence Berkeley Laboratory, University of California, Berkeley, CA 94720

P.A. Bunn Jr., Division of Medical Oncology, Department of Medicine, University of Colorado Cancer Center, Denver, CO

J. Burchell, Imperial Cancer Research Fund, P O Box 123, Lincoln's Inn Fields, London WC2A 3PX, U.K.

M. Burstein, Department Microbiology/Cell Biology, Tel Aviv University, Ramat Aviv, Israel 69978

G. Butchko, Coulter Immunology, Hialeah, FL

F. Buys, Department of Tumor Biology, The Netherlands Cancer Institute (Antoni van Leeuwenhoek Huis), Plesmanlaan 121, Amsterdam, The Netherlands

R.L. Ceriani, John Muir Cancer and Aging Research Institute, 2055 North Broadway, Walnut Creek, CA 94596

G.M. Clark, Department of Medicine/Oncology, University of Texas Health Science Center, San Antonio, TX 78284

A.M. DeMarzo, University of Colorado Health Sciences Center, Department of Pathology (B216), 4200 E. 9th Avenue, Denver, CO 80262

S. DeNardo, University of California Davis, Radiotherapy, Sacramento, CA 95816

D.G. Dienhart, Division of Medical Oncology, Department of Medicine, University of Colorado Cancer Center, Denver, CO

A. Dion, Center for Molecular Medicine and Immunology, Newark, New Jersey, U.S.A.

T. Duhig, Imperial Cancer Research Fund, P O Box 123, Lincoln's Inn Fields, London WC2A 3PX, U.K.

D.P. Dylewski, Department of Botany and Microbiology, Auburn University, Auburn, AL

A. Eaton, Department of Immunology, Cetus Corporation, 1400 53rd Street, Emeryville, CA 94068

D.P. Edwards, University of Colorado Health Sciences Center, Department of Pathology (B216), 4200 E. 9th Avenue, Denver, CO 80262

P.A. Estes, University of Colorado Health Sciences Center, Department of Pathology (B216), 4200 E. 9th Avenue, Denver, CO 80262

J. Flatgaard, Department of Purification Process Development, Cetus Corporation, 1400 53rd Street, Emeryville, CA 94068

P. Furmanski, Department of Biology, New York University, New York, New York 10003

S.J. Gendler, Imperial Cancer Research Fund, P O Box 123, Lincoln's Inn Fields, London WC2A 3PX, U.K.

A.M.C. Gennissen, Department of Tumor Biology, The Netherlands Cancer Institute (Antoni van Leeuwenhoek Huis), Plesmanlaan 121, Amsterdam, The Netherlands

S. Glenn, Coulter Immunology, Hialeah, FL

R. Gonzalez, Division of Medical Oncology, Department of Medicine, University of Colorado Cancer Center, Denver, CO

Ph. Hageman, Department of Tumor Biology, The Netherlands Cancer Institute (Antoni van Leeuwenhoek Huis), Plesmanlaan 121, Amsterdam, The Netherlands

S. Hallam, Genetic Systems, Seattle, Washington

M. Hareuveni, Department Microbiology/Cell Biology, Tel Aviv University, Ramat Aviv, Israel 69978

C.A. Hart, Medical Microbiology, University of Liverpool, Prescott St., Liverpool, U.K.

C. Hartmann, Division of Medical Oncology, Department of Pharmacy, University of Colorado Cancer Center, Denver, CO

J. Hilkens, Department of Tumor Biology, The Netherlands Cancer Institute (Antoni van Leeuwenhoek Huis), Plesmanlaan 121, Amsterdam, The Netherlands

J. Horev, Department Microbiology/Cell Biology, Tel Aviv University, Ramat Aviv, Israel 69978

S.T. Hsieh-Ma, Department of Immunology, Cetus Corporation, 1400 53rd Street, Emeryville, CA 94068

T. Johnson, Division of Medical Oncology, Department of Radiology, University of Colorado Cancer Center, Denver, CO

C.S. Johnston, Bone Marrow Transplant Program, University of Colorado Health Sciences Center, Denver, Colorado 80262

R.B. Jones, Bone Marrow Transplant Program, Division of Oncology, University of Colorado Health Sciences Center, Denver, Colorado 80262

R. Kasliwal, Division of Medical Oncology, Department of Radiology, University of Colorado Cancer Center, Denver, CO

T.W. Keenan, Department of Biochemistry, Virginia Polytech Institute, Blacksburg, VA, and Department of Botany and Microbiology, Auburn University, Auburn, AL

I. Keydar, Department Microbiology/Cell Biology, Tel Aviv University, Ramat Aviv, Israel 69978

J.G. Konrath, Abbott Laboratories, Diagnostic Division, North Chicago, Illinois 60064

P. Kotkes, Department Microbiology/Cell Biology, Tel Aviv University, Ramat Aviv, Israel 69978

J. Kuniyoshi, John Muir Cancer and Aging Research Institute, 2055 North Broadway, Walnut Creek, CA 94596

C.A. Lancaster, Imperial Cancer Research Fund, P O Box 123, Lincoln's Inn Fields, London WC2A 3PX, U.K.

D. Larocca, John Muir Cancer and Aging Research Institute, 2055 North Broadway, Walnut Creek, CA 94596

R. Lathe, AFRC, University of Edinburgh, King's Bldgs, Edinburgh, U.K.

J. Lear, Division of Medical Oncology, Department of Radiology, University of Colorado Cancer Center, Denver, CO

C. Lekutis, Cell and Molecular Biology Division, Lawrence Berkeley Laboratory, University of California, Berkeley, CA 94720

M.J.L. Ligtenberg, Department of Tumor Biology, The Netherlands Cancer Institute (Antoni van Leeuwenhoek Huis), Plesmanlaan 121, Amsterdam, The Netherlands

S. Litvinov, Department of Tumor Biology, The Netherlands Cancer Institute (Antoni van Leeuwenhoek Huis), Plesmanlaan 121, Amsterdam, The Netherlands

C. Longley, AMC Cancer Research Center and Hospital, Lakewood, CO

D.M. Lynch, Abbott Laboratories, Diagnostic Division, North Chicago, Illinois 60064

G.L. Manderino, Abbott Laboratories, Diagnostic Division, North Chicago, Illinois 60064

E. Martin Jr., Ohio State University, College of Medicine, Department of Surgery, Columbus, Ohio

K. McCarthy, Medical Microbiology, University of Liverpool, Prescott St., Liverpool, U.K.

W.L. McGuire, Department of Medicine/Oncology, University of Texas Health Science Center, San Antonio, TX 78284

C.C. McInerney, Abbott Laboratories, Diagnostic Division, North Chicago, Illinois 60064

I.F.C. McKenzie, Research Centre for Cancer and Transplantation Department of Pathology, University of Melbourne, Parkville, Victoria, 3052, Australia

G.J. Miller, Division of Medical Oncology, Department of Pathology, University of Colorado Cancer Center, Denver, CO

I.A. Mizrahi, Hybritech Incorporated, San Diego, California and Liege, Belgium

R. Molina, Laboratory of Clinical Biochemistry, Medical School, Hospital Clinico, Barcelona, Spain

C.A. Nieroda, National Cancer Institute, Laboratory of Tumor Immunology and Biology, Bethesda, Maryland

S.K. Nordeen, University of Colorado Health Sciences Center, Department of Pathology (B216), 4200 E. 9th Avenue, Denver, CO 80262

S. Onate, University of Colorado Health Sciences Center, Department of Pathology (B216), 4200 E. 9th Avenue, Denver, CO 80262

A.J. Parodi, Instituto de Investigaciones Bioquimicas "Fundacion Campomar" Antonio Machado 151, 1405 Buenos Aires, Argentina

G. Parry, Cell and Molecular Biology Division, Lawrence Berkeley Laboratory, University of California, Berkeley, CA 94720

N. Peat, Imperial Cancer Research Fund, P O Box 123, Lincoln's Inn Fields, London WC2A 3PX, U.K.

L. Pemberton, Imperial Cancer Research Fund, P O Box 123, Lincoln's Inn Fields, London WC2A 3PX, U.K.

W.P. Peters, Division of Hematology/Oncology, Duke University Medical Center, Durham, North Carolina 27710

J.A. Peterson, John Muir Cancer and Aging Research Institute, 2055 North Broadway, Walnut Creek, CA 94596

L.W. Przywara, Abbott Laboratories, Diagnostic Division, North Chicago, Illinois 60064

J. Reeder, Department of Immunology, Cetus Corporation, 1400 53rd Street, Emeryville, CA 94068

D.B. Ring, Department of Immunology, Cetus Corporation, 1400 53rd Street, Emeryville, CA 94068

T. Shi, Department of Immunology, Cetus Corporation, 1400 53rd Street, Emeryville, CA 94068

E.J. Shpall, Bone Marrow Transplant Program, Division of Oncology, University of Colorado Health Sciences Center, Denver, Colorado 80262

K. Singer, Cell and Molecular Biology Division, Lawrence Berkeley Laboratory, University of California, Berkeley, CA 94720

N.I. Smorodinsky, Department Microbiology/Cell Biology, Tel Aviv University, Ramat Aviv, Israel 69978

A.L. Sorrell, Abbott Laboratories, Diagnostic Division, North Chicago, Illinois 60064

A.P. Spicer, Imperial Cancer Research Fund, P O Box 123, Lincoln's Inn Fields, London WC2A 3PX, U.K.

T.A.W. Splinter, University Hospital, Rotterdam

P. Stewart, Genetic Systems, Seattle, Washington

K. Stob, Genetic Systems, Seattle, Washington

J. Stubbs, San Francisco State University, San Francisco, CA 94132

A.K. Tandon, Department of Medicine/Oncology, University of Texas Health Science Center, San Antonio, TX 78284

J. Taylor-Papadimitriou, Imperial Cancer Research Fund, P O Box 123, Lincoln's Inn Fields, London WC2A 3PX, U.K.

R.L. Thillen, Abbott Laboratories, Diagnostic Division, North Chicago, Illinois 60064

I. Tsarfaty, Department Microbiology/Cell Biology, Tel Aviv University, Ramat Aviv, Israel 69978

R. Urrea, John Muir Cancer and Aging Research Institute, 2055 North Broadway, Walnut Creek, CA   94596

H.L. Vos, Department of Tumor Biology, The Netherlands Cancer Institute (Antoni van Leeuwenhoek Huis), Plesmanlaan 121, Amsterdam, The Netherlands

G. Walkup, John Muir Cancer and Aging Research Institute, 2055 North Broadway, Walnut Creek, CA   94596

M. Weiss, Department of Medicine, Tel Hashomer, Israel

C. Williams, Center for Molecular Medicine and Immunology, Newark, New Jersey, U.S.A.

D.H. Wreschner, Department Microbiology/Cell Biology, Tel Aviv University, Ramat Aviv, Israel 69978

P. Xing, Research Centre for Cancer and Transplantation Department of Pathology, University of Melbourne, Parkville, Victoria, 3052, Australia

M.J. Yerna, Hybritech Incorporated, San Diego, California and Liege, Belgium

D. Yuzuki, San Francisco State University, San Francisco, CA   94132

J. Zaretsky, Department Microbiology/Cell Biology, Tel Aviv University, Ramat Aviv, Israel 69978

S. Zrihan, Department Microbiology/Cell Biology, Tel Aviv University, Ramat Aviv, Israel 69978

INDEX